Yuriy I. Posudin | Nadiya P. Massjuk | Galyna G. Lilitskaya

Photomovement of *Dunaliella* Teod.

Yuriy I. Posudin | Nadiya P. Massjuk
Galyna G. Lilitskaya

# Photomovement of *Dunaliella* Teod.

VIEWEG+TEUBNER RESEARCH

Bibliographic information published by the Deutsche Nationalbibliothek
The Deutsche Nationalbibliothek lists this publication in the Deutsche Nationalbibliografie;
detailed bibliographic data are available in the Internet at http://dnb.d-nb.de.

1st Edition 2010

Cover design: KünkelLopka Medienentwicklung, Heidelberg
Printed on acid-free paper
Printed in Germany

ISBN 978-3-8348-0974-2

# *In Memory of Professor Nadiya Massjuk*
# *1931-2009*

One of the authors of this monograph, Professor Nadiya Massjuk, Dr. Sci. Biol. and Leading Researcher of M. Kholodny Institute of Botany of the National Academy of Sciences of Ukraine, passed away on 13 March, 2009.

Her scientific interests were related to algology, particularly biodiversity, flora, systematics, ecology, geography, origin, evolution, phylogeny, the role of algae in the world of living organisms, and applied algology.

She was interested in the biology of algal photomovement from the point of view of diversity, phylogeny of phytoflagellates, classification, biotechnology of cultivation of carotene-containing algae, and carotenoid production.

Dr. Massjuk was an author of the classic monograph "Morphology, Systematic, Ecology, Geographical Distribution of Genus *Dunaliella* Teod and Perspectives of its Practical Applications" (Kiev, Naukova dumka, 1973) in which the results provide essential basic information on the genus *Dunaliella*, the main principles of systematics of the genus and elucidation of its species, subspecies, versions and forms.

Dr. Massjuk's published work (over 260 books and articles) and her impact on her friends and collegues has left an invaluable and lasting legacy to the scientific community.

May she rest in peace.                                                    Colleagues

# Preface

This monograph represents 30 years of scientific cooperation on the study of the basic biology of photomovement in algae between the National University of Life and Environmental Sciences of Ukraine (Prof. Yuriy Posudin) and the M.G. Kholodny Institute of Botany of National Academy of Sciences of Ukraine (Prof. Nadiya Massjuk and Dr. Galyna Lilitskaya). It reviews the historical development and current state of the art in the biology of photomovement in algae. Problems in terminology and a logical basis for classification of photomovement in microorganisms are discussed. The research has focused on two species of *Dunaliella* Teod., *D. salina* Teod. and *D. viridis* Teod., as the principal organisms investigated.

The results of experimental investigations on the critical factors controlling and modulating photomovement are described and include the effects of various abiotic factors, critical aspects of photomovement such as photoreception (i.e., location and structure of photoreceptor systems, composition of photoreceptor pigments, mechanisms of photoreception and photoorientation), sensory transduction of absorbed light into signals that govern the activity of the motor apparatus, and flagellar activity.

Various aspects involved in the utilization of these species as models for studying photomovement, such as testing aquatic media and the effects of surface-active substances, salts of heavy metals, and pesticides on algal photomovement parameters are described. Vector methods for testing are proposed for assessing the action of various chemicals. Likewise, the potential of using the two species as organisms for transgenic alteration, such as enhanced production of β-carotene, ascorbic and dehydroascorbic acids, glycerin and other valuable organic compounds are described.

The results of photomovement investigations are assessed relative to the evolutionary biology of algae and their phylogenetics, systematics, taxonomy, ecology and geography. Critical aspects of photomovement biology that remain to be investigationed in flagellates are discussed.

The monograph is intended for algologists, protistologists, hydrobiologists, biophysicists, physiologists, ecologists and biotechnologists, teachers, post-graduate students and students of related biological specialities.

The authors express their deep and sincere gratitude to Professor Francesco Lenci and Doctor Giuliano Colombetti (Institute of Biophysics CNR, Pisa, Italy) for stimulating our interest in the photobiology of microorganisms and introducing the authors to the fascinating world of algal photomovement.

The authors are grateful to Professor Felix Litvin (Moscow State University, Russia) and Professor Boris Gromov (St.-Pertersbourg State University, Russia) for their continued interest in the investigation of photomovement in *Dunaliella* and fruitful discussions of the results.

The authors are much indebted to Prof. D.P. Häder (Friedrich-Alexander University, Erlangen, Germany), Prof. A. Flores-Moya (University of Malaga, Malaga, Spain), Prof. H. Kawai (Kobe University, Kobe, Japan), Prof. C. Wiencke (Alfred Wegener Institute for Polar and Marine Research, Bremerhaven, Germany) and Prof. D. Hanelt (Hamburg University, Hamburg, Germany) for providing the opportunity to conduct research on the photobiology and photomovement of algae in their laboratories.

The authors would like to express their grateful thanks to Prof. Ami Ben-Amotz (National Institute of Oceanography, Israel) for illustrative materials and Prof. Shogo Nakamura (Toyama University, Japan) for an electron micrograph of *Dunaliella*.

Special gratitude to Dr. Igor Zaloilo for developing the computer versions of figures in the book.

The authors would like to express their very grateful thanks to Professor Stanley J. Kays and Betty Schroeder (The University of Georgia, USA) for technical assistance.

Yuriy Posudin<br>
Nadiya Massjuk<br>
Galyna Lilitskaya

# Contents

# List of Figures and Photographes

# List of Tables

*"The only generalization that can be made for photomovement is its diversity"*

W. Haupt, 1983

# Introduction

In a broad context the term *photomovement* encompasses any movement or its alteration induced by light. Photomovement is the result of the *photoregulation of movement* – which includes an entire complex of elementary processes caused by a light stimulus such as photoreception, primary reactions of the photoreceptor pigments, and the sensory transduction of the light stimulus into a physiological signal that governs the activity of the motor apparatus and results in the photoorientation of the organism.

The study of photomovement and the photoregulation of movement in microorganisms is of considerable interest due to the importance of these phenomena and that they are closely tied to fundamental biological processes such as photosynthesis, photoreception, energy transformation, membrane-coupled and membrane-mediated phenomena. The investigation of photomovement and its photoregulation are also closely tied to the elucidation of the basic principles of intracellular developmental processes, as well as ontogenesis, embryogenesis, and morphogenesis. A better understanding of light mediated responses impacts our understanding of light's role in the ecology and biocenology of these organisms since light is an important factor in their spatial and temporal distribution. While photomovement has an independent function, it also conveys information on the complexity of related environmental factors (e.g., temperature, pH, biogenesis of compounds, oxygen content, the presence of other microorganisms [Kritsky, 1982; Sineschekov and Litvin, 1982]).

The investigation of photomovement mechanisms is also of interest from the standpoint of bionics, evolutionary biology, morphology, phylogeny, and systematics. It is known, for example, that the structure of the motor apparatus and photoreceptor is an important systematic character at higher taxonomic levels (divisions and classes) in phycology [Sedova, 1977; Topachevsky and Massjuk, 1984; van den Hoek et al., 1995; Graham and Wilcox, 2000; Massiuk and Kostikov, 2002]. Thus, it is possible to assume the specificity of the mechanisms of photoperception and photoregulation of photomovement among members of different divisions or classes of algae. Finally, the study of photomovement has the potential for stimulating the practical application of this technology in areas such as biomonitoring of the environment, biotechnology, and the use of these organisms for the synthesis of useful natural products.

There have been a number reviews on light induced movement of microorganisms [Halldal, 1958, 1961; Haupt, 1959, 1983; Bendix, 1960; Hand and Davenport, 1970; Nultsch, 1975; Wolken, 1977; Lenci and Colombetti, 1978; Miyoshi, 1979; Nultsch and Häder, 1979, 1988; Diehn, 1979, 1980; Feinleib, 1980; Colombetti and Lenci, 1982; Lenci, 1982; Poff and Hong, 1982; Sineshchekov and Litvin, 1982, 1988; Häder, 1987*a*, 1987*b*, 1987*c*; 1994; 1996*a*; Lenci et al., 1984; Colombetti and Petracchi, 1989; Doughty, 1991; Nultsch and Rueffer, 1994; Donk and Hessen, 1996; Häder and Lebert, 2000; Lebert and Häder, 2000; Sineshchekov and Govorunova, 2001*a*; Hegemann and Deininger, 2001; Hegemann et al., 2001; Williams and Braslavsky, 2001; Sgarbossa et al., 2002; Checcucci et al., 2004]. In addition, there have been a number of scientific conferences and schools that communicate recent advances in our fundamental understanding of the subject (e.g., "Biophysics of Photoreceptors and Photobehaviour of Microorganisms" (Pisa, 1975), "Photoreception and Sensory Transduction in Aeural Organisms" (N.Y., 1980), "Sensory Perception and Transduction in Aeural Organisms" (N.Y., 1985), "Biophysics of Photoreceptors and Photomovement in Microorganisms" (Tirrenia, 1990), "Light as Energy Source and Information Carrier in Plant Photophysiology" (Volterra, 1994); International Conferences "Actual Problems of Algology" (Chercassy, 1987; Kiev, 1999); "Photosensory Receptors & Signal Transduction" (Ventura, 2004), just as periodical Congresses of the European Society for Photobiology that are organized each two years since 1986, and annual meetings of American Society for Photobiology and of the Japanese Society for Photomedicine and Photobiology). Likewise, several conferences were dedicated to algal biotechnology (Third Asia-Pacific Conference on Algal Biotechnology, 1997, Phuket, Thailand and "Algae and Their Biotechnological Potential", 2000, Hong Kong).

New strains of motile microorganisms continue to be identified. Experimental analysis of photomovement includes methods such as videomicrography, phototaxigraphy, Doppler laser spectroscopy, high-speed cinematography, and electrophysiological measurements. An automated system of registration of different photomovement characteristics and the collection and analysis of information utilized to assess differences in photomovement of organisms are now widely used. Meanwhile, the development and application of new experimental approaches and instrumentation to assess photomovement have stimulated considerable interest.

The study of the photomovement of microorganism is confronted with number of problems due in part to the great diversity in types of photoreactions and photoreceptor systems within and among various microorganisms, variation in the absorption spectra of photoreceptor pigments, and the difficulty in isolation of these pigments.

The study of sensory transduction of a quantum of light absorbed by a pigment molecule and its conversion into a signal that controls the movement of the cell is extremely complicated. As a consequence, the mechanism of photoregulatory control of movement in microorganisms is sometimes referred to as a "black box" due to the mysteries that remain to be elucidated.

While many well-known photobiological processes, such as photosynthesis or the biophysics of vision, are sufficiently uniform that they allow making generalizations about many of the details across a diverse range of organisms, the situation is quite different for photomovement of organisms. The elucidation of the basic photoregulatory biology of one type of microorganism is not necessarily applicable to another. This situation was most aptly described by the prominent photobiologist W. Haupt: "The only generalization that can be made for photomovement is its diversity" [Haupt, 1983].

Due to the tremendous diversity among organisms in their biology of photomovement, we have focused on theoretical, experimental and applied problems that are related to the photomovement of unicellular green alga of *Dunaliella salina* Teod. and *Dunaliella viridis* Teod.

Intense investigation in any field usually results in the enrichment, revision, and alteration of old terminology since new information often requires new terms to be properly understood. At the present, alterations in terminology are occurring in the biology of microorganism photomovement. As a consequence, we have paid special attention to both the terminology and classification of photomovement.

Our primary focus with regard to experimental and methodological approaches has been the investigation of the location and structure of the photoreceptor system, the composition of photoreceptor pigments, the mechanisms of photoreception and photoorientation, the processes of sensory transduction, and the activity of the motor apparatus in the two species. Comparison of photomovement parameters between two species of the same genus is likewise of taxonomic interest. The authors assessed the experimental and methodological techniques needed to facilitate understanding the key processes of photomovement in these species since they had not been previously studied. It was also imperative to understand the effect of environmental factors such as ultraviolet and visible radiation, temperature, pH, and electrical fields on the photomovement parameters in these species.

The potential of algal biotechnology is likewise addressed. Both species represent possible organisms for the commercial production of β-carotene (provitamin A), ascorbic and dehydroascorbic acids, glycerol, feed for fish production, and other products. Assessment of changes in photomovement by these organisms can also potentially be used as biosensors for assessing the composition of aquatic media.

A comparative analysis of both general and specific differences in photomovement among these flagellated algae species and representatives of different orders (classes) of algae is also reported.

The *main objective* of this monograph is to critique the current understanding of photomovement in the unicellular green algae species *D. salina* and *D. viridis*.

The *specific aims* of this work are:

1. Review the historical development and current state of the art of investigations on algal photomovement;

2. Describe theoretical problems in terminology and the logic of the existing method for the classification of photomovement in these microorganisms;

3. Elucidate the primary characteristics of *D. salina* and *D. viridis*;

4. Critique the experimental methods utilized for the measurement of photomovement of these species and the effects of abiotic factors on photomovement;

5. Describe the processes of photoreception – location and structure of phootoreceptor systems, composition of photoreceptor pigments, mechanisms of photoreception, and photo-orientation of the two species;

6. Describe the processes of sensory transduction of absorbed light into signals that govern the activity of the motor apparatus of the two species;

7. Assess the possible application of *D. salina* and *D. viridis* as models for testing the quality of aquatic media and estimating the effects of surface-active substances, salts of heavy metals, and pesticides on photomovement in algae;

8. Assess the potential of the two species of *Dunaliella* for transgenic alteration to enhance the synthesis of β-carotene, ascorbic and dehydroascorbic acids, glycerol and other valuable organic compounds;

9. Assess the implications of photomovement on evolutionary biology, phylogenetics, systematics and taxonomy, ecology and geography of algae;

10. Critique critical areas for future research on the biology of photomovement in flagellates.

# Chapter 1

# Photomovement of Algae – Historical Overview of Research and Current State of the Art

Interest in understanding the mystery surrounding the movement of living organisms dates from ancient times. The first published work in the field [De Motu Animalium ("On the Motion of Animals")] was by Aristotle (384–322 B.C.) who was interested in similarities in motion among animals. Leonardo da Vinci (1452–1519), a distinguished painter, architect and engineer, also studied the mechanics of movement in organisms (biomechanics). His Codex on the Flight of Birds was a precise study of the mechanics of flight and air movement. The same problems captured the interest of Giovanni Alfonso Borelli (1608–1679), a famous Italian mathematician, astronomer and compatriot of Galileo Galilei. He authored the first book on biomechanics [De Motu Animalium I and De Motu Animalium II ("On the Motion of Animals"), 1679] that was dedicated to muscular movement and body dynamics. He also studied bird flight and the swimming of fish [Thurston, 1999].

The nature and mechanisms of movement of living organisms preoccupied the attention of many famous scientists – I.M. Sechenov (1829-1905), I.P. Pavlov (1849–1936), P.F. Lestgaft (1837–1930), A.A. Ukhtomsky (1875–1942), N.A. Bernstein (1896–1966) and others.

There has been a progressive increase in interest in motile behaviour of microorganisms since 1674 when Antonie van Leeuwenhoek [Mosolov and Belkin, 1980] first observed, using a microscope he developed, the movement of *Euglena* and *Volvox* [cited by: Wolken, 1975]. An article by Ludolph Christian Treviranus (1779–1864), a German botanist, was the first work dedicated to the investigation of algae. Zoospores of *Draparnaldia glomerata* (Vaucher) CA Agardh and *Ulothrix subtilis* Kutzing accumulated near the illuminated edge of the vessel or at the opposite side [Treviranus, 1917].

Christian Gottfried Ehrenberg (1795–1876), a German scientist, studied over a 30 years period thousands of new species, including flagellates such as *Euglena*, ciliates such as *Paramecium aurelia* Müller and *Paramecium caudatum* Ehr., a group of unicellular protists called diatoms, and many species of radiolaria. Of particular interest was his manuscript published in 1838 describing the red eye (eyespot) or stigma of *Euglena*, an organelle that plays an important role in the photomovement of the algae.

Charles Darwin wrote in 1872 *"How a nerve comes to be sensitive to light, hardly concerns us more than how life itself originated; but I may remark that, as some of the lowest organisms, in which nerves cannot be detected, are capable of perceiving light, it does not seem impossible that certain sensitive elements in their sarcode should become aggregated and develop into nerves, endowed with this special sensibility"*.

Experiments by F. Cohn (1865*a*) demonstrated that zoospores of some algae, just as the cells of *Euglena*, exhibited phototaxis in response to blue-green but not red light. This was the first indication of spectral sensitivity in microorganism photomovement.

A.S. Famintzin (1843–1918) published "Action of light on algae and some other organisms close to them" (St.-Petersbourg, 1866) and was conferred the title of Doctor of Botany. The author distinguished two types of locomotion in protozoa; those that have cilia (zoospores) and pseudopodia (amoeboid organisms). Cilia are present in flagellates such as: *Volvox, Gonium, Stephanosphaera, Euglena,* and *Chlamydomonas*.

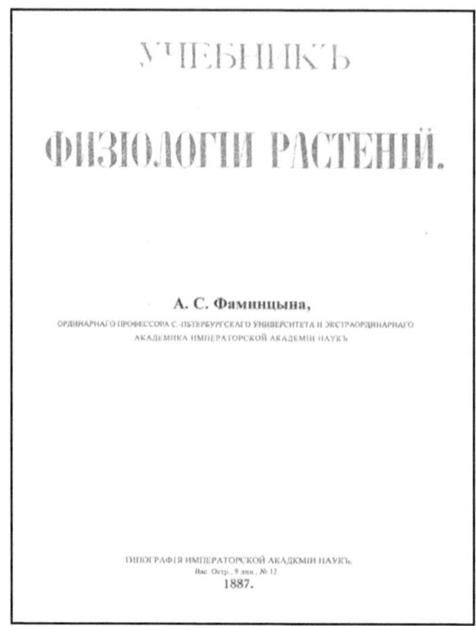

**Fig.1.1.** A. Famintzin "Text-Book of Plant Physiology" (St.-Petersbourg, 1887)

Famintzin (1887*a*, c. 19) further characterized this phenomenon: "*It was observed long ago that lateral illumination of the vessel with liquid, where zoospores are swimming, provoked the accumulation of them along the edge of the vessel forming a green strip*" (translation of Yu. Posudin). Famintsin concluded that light induced the movement of zoospores and that the light-induced movement of algae depended upon the light intensity, temperature, and the composition of aquatic medium [Famintzin 1867*a*, *b*, 1887*a*]. A number of articles describing photomovement in desmids and blue-green algae were published in the 1880s which represented the general understanding of the biology of photoorientation mechanisms during that time period.

Eduard Adolf Strasburger (1844–1912), a famous Polish-German Professor of Botany, confirmed that various microorganisms use different mechanisms of photoorientation. He believed that zoospores of *Haematococcus* respond to light gradients, while the motile reproductive cells of *Botrydium* respond to the direction of the light. Strasburger was the first to use the term "phototaxis" to distinguish between the light-induced transfer of mobile (phototaxes) and fixed (phototropisms) organisms and to distinguish between positive and negative phototaxes. He also was the first to use coloured glass filters to study spectral peculiarities of zoospore photomovement and likewise discovered the ability of colourless microorganisms to respond to light [Strasburger, 1878].

Theodor Wilhelm Engelmann (1843–1909), a German botanist, physiologist, and microbiologist, published in 1882 the effects of different wavelengths (or colors of light) on photosynthetic activity and showed that the conversion of light energy to chemical energy took place in the chloroplast [Drews, 2005; Engelmann, 1882*a,b*]. Engelmann also made a number of valuable contributions to the investigation of photomovement in algae (1882*a,b*). Using the technique "projected microspectrum", he demonstrated the dependence of the photoreaction in microorganisms on the wavelength of the light stimulus. Despite the limited qualitative precision of these early experiments, it was possible to estimate the action spectra of photoreactions in microorganisms. Engelmann determined the spectral sensitivity of *Euglena viridis* (O.F. Mueller) Ehrenberg to be around 470 to 490 nm. He described the shock reactions of *Euglena* in response to the application of a narrow beam of light and found that the front of the cell (where the stigma is located) was more sensitivit than other areas [Engelmann, 1882*a,b*].

A series of articles were published toward the end of 1890s that were dedicated to the effect of external physical factors on photomovement in algae, for example, electric fields [Vervorn, 1889], temperature [Wildeman, 1893] and X-rays [Shaudinn, 1899] (cited by: Jahn, Bovee, 1968). These were followed by a number of studies at the beginning of 20th century that focused on photomovement terminology [Rothert, 1901; Nagel, 1901; Pfeffer, 1904]. Loeb and Maxwell (1910) discovered the ability of cells to aggregate when exposed to the blue portion of the electromagnetic spectrum. During this period, *Euglena gracilis* G.A. Klebs became the classic model for photomovement investigations. A number of investigators focused on this organism in their study of the mechanisms of photoreception and photoorientation. Stigma morphology was addressed as was the colour of the cup-shaped plastids [Francé, 1893] present in hexagonally-packed layers [Wolken, 1956] or aggregates of pigmented granules (globules) [Hall, Jahn, 1929; Gojdics, 1934] (all authors are cited by: Jahn, Bovee, 1968).

Research dedicated to the study of the mechanisms of photoorientation in *Euglena* was also published during this period [Jennings, 1906; Mast, 1911]. The ability of *Euglena* to move toward the direction of the light and photophobic reactions of algae ("Schreckbewegung" [Engelmann, 1882*a,b*] or "avoiding reaction" [Jennings, 1906]) were described. The sensitivity of *Euglena* to the light in the blue portion of the spectrum was also established. Mast supposed that the photosensitive pigments in *Euglena* were located near the stigma [Mast, 1911]. He stressed the necessity of using "spectrally pure" (monochromatic) light and its precise calibration for estimating the stimulating efficiency of light at various wavelengths (a procedure now referred to as "registration of the action spectrum"). He also measured the optical transmission of the stigma of *Euglena* and his results ($\tau = 0,28$) are similar to those obtained by investigators half a century later ($\tau = 0,32$) [Wolken, 1967]. The action spectrum for phototaxis in *Euglena* was measured by Mast in 1927. The spectrum is characterised by a wavelength band between 410–540 nm with a maximum at 485 nm. Carotenoids were thought to be the photoreceptor pigments [Mast, 1927].

With regard to the function of the stigma in *Euglena*, Arnoldi concluded *"Its (the eye) function has been mysterious up to now, but it has certain similarity to the eyespot of some protozoa, which plays the role of a rudimentary organ of sense"* [Arnoldi, 1908] (translation of Yu. Posudin).

Beginning with pioneer work of A. Fisher [1894, cited by: Jahn, Bovee, 1968], where the morphology of flagellar system of *Euglena* was described, an increasing number of investigations on the structure and function of the locomotor apparatus in algae were published, the results of which have only recently been expanded through the application of modern techniques.

An attempt to establish a cause and effect relation between the light stimulus and response in microorganisms was addressed in a number of investigations [Buder, 1917; Mainx, Wolf, 1929]. A valuable contribution to the development of ideas concerning the mechanisms of photoorientation in *Euglena* was introduced by T. Buder (1917) who was the first to propose that the location of the photoreceptor was in the paraflagellar body (a hypothesis that remains correct to this day). Buder also utilized light passing through the condenser of a microscope to measure photomovement parameters. He stressed the importance of the spectral composition of the light and introduced the term *"schwaches, tief rubinrotes licht"* (weak, deep ruby-red light), used in the microscopy study of microorganism mobility. Buder identified the problems associated with the quantitative analysis of photomotile reactions, investigated flagellar activity during phototaxis and photophobic reactions, proposed the probable location of the photosensitive system in the cell, and described the relationship between flagellar beating and the position of stigma relative to the source of light. Many of the questions proposed by Buder have yet to be answered.

In 1936, Sergey Chakhotin (1883–1973) studied the functions of the photoorientation apparatus in *Euglena* using ultraviolet microirradiation which he developed [cited by Posudin, 1995]. A *Euglena* cell was placed in the illuminated portion of a quartz capillary tube. As the cell moved along the capillary, it reversed its direction of movement when reaching the "light-dark" boundary. Ultraviolet irradiation on the stigma stimulated its excitation and contraction of the stigma. After exposure to UV irradiation the cell moved past the "light-dark" boundary without reversing its movement, indicating a deactivation of the photoorientation mechanism in the cell. The use of monochromatic light demonstrated the sensitivity of stigma to blue-green light. Chakhotin concluded that the stigma of *Euglena* was a primitive sensory organ that controlled the movement of the cell. Ultraviolet irradiation appeared to "blind" the cell with the frontal part of the cell being more sensitive to the irradiation than the back.

B.V. Perfiliev (1915) studied photomovement in the blue-green algae (*Oscillatoria geminata* Menegh. and *Synechococcus aeruginosus* Naeg.) and diatoms (*Pinnularia streptoraphe* Cl. and *Anomoeoneis sculpta* Pfitzer), particularly in relation to spectral sensitivity. He supposed that "*...movement is probably the common property of blue-green algae...*" (translation of Yu. Posudin) based on the stimulation of photomovement in blue-green algae by the red region and in diatoms by the blue region of the electromagnetic spectrum. The role of light on the behaviour of green algae, diatoms and desmids is mentioned by A.A. Elenkin (1925). He also noted the ability of chloroplasts of *Mougeotia genuflexa* (Roth) C.A. Agardh to move inside the cell in response to light.

In the 1930s, A. Luntz determined the action spectra of phototaxis of *Eudorina, Volvox, Chilomonas*, and *Chlamydomonas reinhardtii* P.A. Dang. [Luntz, 1931 a,b]. An attempt to precisely measure the of absolute light energy values was a distinctive feature of his investigations. The action spectra for phototaxis in green flagellates (*Eudorina, Volvox* and *Chlamydomonas*) had a maximum at 492 nm while the colorless cryptophyte algae *Chilomonas* was sensitive at 366 nm.

The fundamental work of Per Halldal at the beginning of the 1960s [Bjorn L. et al., 2007] determined the action spectra for representatives of *Volvocales, Dinophyceae*, and *Ulva*-gametes [Halldal, 1958]. The action spectra for positive phototaxis were determined for *Dunaliella viridis* Teod. and *Dunaliella f. euchlora* Lerche, while the action spectrum for a negative phototaxis was characterized in *Dunaliella salina* Teod. The projected spectrum method was used in these investigations. The phototactically active spectral region in the three *Dunaliella* species was between 400 to 540 nm with a maximum at 493±3 nm and a small shoulder at around 435 nm. Halldal indicated that "*the stigma is a photoreceptive organ associated with the orientation of topo-phototactic algae...*" [Halldal, 1958].

The 1960s and 1970s were marked by the transition from naturalistic descriptions of photomovement to basic photochemical explanations of the phenomenon [Clayton, 1964; Haupt, 1966; Tollin, 1969; Nultsch, 1970; Feinleib and Carry, 1967; Hand and Davenport, 1970]. A number of investigators [Wolken and Shin, 1958; Wolken, 1967, 1971, 1977;] published detailed studies on the photomovement in *Euglena gracilis*, in particular the effect of external factors on the velocity and direction of movement, measurement of the action spectra for the photoreactions, elucidation of the nature of photoreceptor pigments, and analysis of the function of the flagellar apparatus. Due in part to the progressive accumulation of data, *E. gracilis* became the model system of choice for studying photomovement in algae. This was followed by the first detailed review of the literature on photomovement in which problems of terminology, experimental and methodological approaches for studying photoreception, and sensory transduction were critiqued [Diehn, 1979].

The second half of the 20th century was marked by the study of photomovement in both individual cells and populations [Feinleib, 1977; Ascoli et al., 1978; Barghigiani et al., 1979; Colombetti and Lenci, 1982]. The study of individual cells is more labour intensive and complex but it makes it possible to determine all motile reactions and photomovement parameters, as well as response characteristics to specific light stimuli. The study of populations, however, allows obtaining data on the movement of millions cells with a high level of precision, though due to the absence of photomovement data on individual cells, interpretation of the results can be complex. The subsequent years have seen a progressively increasing interest in alga photomovement [Williams, 2001; Iseki and Watanabe, 2004].

We have seen the formation of several highly productive schools of scientific inquiry into the biology of photomovement in various countries of the world under the aegis of prominent photobiologists. A German school led by the photobiologists W. Nultsch (Marburg, Germany) studied photomovement using a green alga (*Chlamydomonas reinhardtii* P.A. Dang.) as the model [Nultsch, 1962, 1970, 1983]. Several extensive reviews on the analysis of experimental and methodological approaches to the investigation of algal photomovement were published [Nultsch, 1975, 1980; Nultsch and Häder, 1979, 1988]. Nultsch described the considerable confusion in the scientific literature due to a plethora of terminology utilized in the studies of photomovement in microorganisms [Nultsch, 1973, 1975].

Also in Germany, W. Haupt (Erlangen, Germany) studied motile responses to light in organisms and cellular organelles, and the phototropic and photonastic movement of plants. He authored the first detailed reviews on the diversity of photobiological reactions among alga species and also focused on the problem of terminology [Haupt and Feinleib, 1979; Haupt, 1983, 1986; Haupt and Seitz, 1984].

D.-P. Häder (Erlangen, Germany) and a group of scientists under his leadership studied photomovement and gravitaxis in various algae. He authored a number of reviews on the subject [Häder, 1979, 1987*a*, *b*, 1996*a*; Häder and Lebert, 2000; Lebert and Häder, 2000]. His research group used a computerized system of videomicrography for the quantitative analysis of photomovement parameters [Häder, 1994*b*]. Häder and M. Lebert also published a monograph "Photomovement" (2001) on the subject.

A fourth group of German scientists was lead by P. Hegemann (Regensburg). They investigated light-sensitive processes in green algae using *C. reinhardtii* as the model. They utilized optical, spectroscopic and electrophysiological methods to study the nature of the photoreceptor. Hegemann (together with co-authors) published several reviews [Hegemann, 1997; Hegemann and Harz, 1998; Hegemann and Fisher, 2001; Hegemann and Deininger, 1999, 2001; Hegemann et al., 2001].

In Italy a group of investigators at the Institute of Biophysics CNR (Pisa, Italy) that included Francesco Lenci, Giuliano Colombetti, Francesco Ghetti, Paolo Gualtieri and others have studied photomovement for a number of years [Lenci, 1982; Lenci and Colombretti, 1978; Gualtieri, 2001]. The principal research focus of the scientists has been on photocontrolled biological phenomena such as the ability of microorganisms to use light to transmit information that affects their motile behaviour under environmental light conditions. The molecular processes of photoreception and sensory transduction of light into a signal that controls movement of the organism have been studied using *E. gracilis* as a model. Microspectroscopic fluorometric analysis was utilized for investigating the photoreceptor pigments which resulted in the discovery of fluorescence by an organelle located near the base of the flagellum [Benedetti, Checcucci, 1975].

It is necessary to mention a brilliant methodological approach that was used to identify the photoreceptor pigments in alga. The paraflagellar body in *E. gracilis* was illuminated using a tunable dye-laser focused on it using a microscope. The technique made it possible to determine the *in vivo* fluorescence excitation spectrum of the paraflagellar body and compare it with absorption spectra of flavins. They concluded that these pigments participate in the photomovement of alga [Colombetti et al., 1980,1981; Ghetti et al., 1985].

An additional technological breakthrough has been the use of laser Doppler spectroscopy developed by scientists at the Institute of Biophysics CNR. The method allows the linear and rotational velocity of movement and the frequency of flagella beatings to be measured [Ascoli, 1975; Ascoli et al., 1978, 1978]. Current research at the Institute is directed toward photomovement in algae and protozoa.

A Polish school, led by E. Mikolajczyk, has studied photophobic reactions in *E. gracilis*, in particular, the dependence of these reactions on the wavelength of light and conditions under which the alga are grown. They have also studied photomovement in the colourless euglenophyte algae *Astasia longa* and *Peranema trichophorum* [Mikolajczyk, 1984*a, b*, 1986].

In the United States, the photobiologist M.E. Feinleib (Tufts University, USA) studied photomovement in *Chlamydomonas reinhardtii* [Feinleib, 1977, 1978, 1980, 1985]. Likewise, Pill-Soon-Song (University of Nebraska, USA) studied the molecular mechanisms of sensory transduction in protozoans (e.g., *Stentor coeruleus* Ehr.) with special emphasis on the photocontrol functions of stentorins and blepharismins [Song, 1983, 1985].

The Japanese school of photobiologists used a unique measuring complex – an Okazaki Large Spectrograph [Watanabe et al., 1982]. The spectrograph was developed in 1980 for the measurement of the action spectra of various photobiological reactions in the 250–1200 nm range at a high resolution (i.e., 0.8 nm/cm). Algae such as the cryptophytes *Cryptomonas* sp., *Cryptomonas rostratiformis* Skuja ex T. Willén, *Chroomonas nordstedtii* Hansgirg, *Chroomonas coerulea* (Geitler) Skuja [Watanabe and Furuya, 1982] and the green algae *C. reinhardtii* [Kondo et al., 1988] and *D. salina* [Wayne et al., 1991] were the primary species studied.

A second group of photobiologists under leadership of H. Kawai (Kobe University, Japan) studied the nature of the photoreceptor pigments in the gametes and zoospores of golden  and brown algae using microspectrofluorometry of the flagellar apparatus [Kawai, 1988, 1989, 1992; Kawai et al., 1991, 1996; Yamano et al., 1993].

A third Japanese research team at the University of Toyama has been led by the algologist S. Nakamura. They proposed using flagellar regeneration of *Dunaliella* sp. to assess seawater pollution [Horike et al., 2002]. Together with T. Takahashi they likewise studied the photoreceptor system (eyespot, structure and localization) in *C. reinhardtii* [Nakamura et al., 2001; Suzuki et al., 2003]. At the University of Tokyo, the laboratory of R. Kamiya studied photoreception and the mechanisms of flagellar beating during the phototactic activity in *Chlamydomonas*. They were interested in the structure and function of eukaryotic flagella and used mutants lacking flagella to address certain questions at the universities of Tokyo and Tsukuba [Isogai et al., 2000; Yoshimura and Kamiya, 2001; Kamiya, 2002; Okita et al., 2004; Fujiu et al., 2009].

A Russian school on photobiologists (e.g., F. Litvin, O. Sineshchekov, E. Govorunova et al.) applied electrophysiological methods to study the light-induced excitation of the photoreceptor in alga that instigates a cascade of fast electrical events in the cell membrane. The measurement of photoreceptor currents allowed identifying the photoreceptor pigment in two species of green algae (*Haematococcus pluvialis* Flotow, *C. reinhardtii*) that appeared to be rhodopsin [Sineshchekov and Litvin, 1982, 1988; Sineshchekov, 1991*a,b*; Sineshchekov and Govorunova, 1999, 2001*a,b,c*].

Research on photomovement in the Ukraine, led by Yu. Posudin, N. Massjuk, and G. Lilitskaya began in 1980. The main research focus of the group was on problems in terminology and a logical basis of classification of photomovement in microorganisms. Their experimental investigations on photomovement used two green algae species from genus *Dunaliella* (*Dunaliella salina* Teod. and *Dunaliella viridis* Teod.) and the effect of abiotic factors on them. They also studied the processes involved in photoreception (i.e., location and structure of the photoreceptor system, nature of the photoreceptor pigments, mechanisms of photoreception and photoorientation), sensory transduction of light received by the photoreceptor and its conversion into a signal controlling the movement of alga, the potential for using photomovement as a means of biotesting of aquatic media, and the potential for the use of certain species for transgenic alteration to address biotechnological problems [Massjuk et al., 2007].

The systematic investigation of photomovement in algae is currently in progress in Europe (Germany, Italy, Poland, Russia, Ukraine), Asia (Japan) and the United States. The following representatives of various divisions of cyanoprocariotic algae – *Cyanophyta* (*Oscillatoria geminata, Synechococcus aeruginosus*), eucariotic algae – *Euglenophyta* (*Euglena gracilis, Astasia longa, Peranema trichophorum*), *Dinophyta* (*Peridinium gatunense*), *Bacillariophyta* (*Pinnularia streptoraphe, Anomoeoneis sculpta*), *Cryptophyta* (*Cryptomonas* spp., *Chroomonas* spp.), *Chlorophyta* (*Dunaliella salina, D. viridis, Chlamydomonas reinhardtii, Chloromonas* sp., *Haematococcus pluvialis, Stephanosphaera* sp., *Gonium* sp., *Eudorina* sp., *Volvox* sp.), and spores and gametes of green (*Chlorophyta*), golden (*Chrysophyta*), yellow-green (*Xanthophyta*), red (*Rhodophyta*) and brown (*Phaeophyta)* algae are currently or have been used as models in these investigations. Special attention is being directed toward further elucidating the mechanisms of photoreception and photocontrol of movement and to the nature of the photoreceptor and locomotor apparatus. In addition, continuing problems with terminology and the classification of various motile reactions in microorganisms are also subjects of interest.

# Chapter 2

# Terminology and the Fundamentals of Classification of Light-Induced Behaviour in Freely Motile Microorganisms

The problems concerning terminology and classification of the phenomena associated with the motile properties of microorganisms have been studied by a number of researchers [Halldal,1958; Jahn and Bovee, 1968; Nultsch, 1973, 1975, 1980; Diehn et al., 1977; Feinleib, 1977, 1978, 1980; Lenci and Colombetti, 1978; Diehn, 1979; Häder, 1979, 1987*a*, 1987*b*, 1987*c*, 1996*a*, 1996*b*; Nultsch and Häder, 1979; Posudin, 1982, 1985; Sineschekov and Litvin, 1982; Colombetti and Lenci, 1982; Colombetti et al., 1982; Lenci, 1982; Haupt, 1983; Burr, 1984; Lenci et al., 1984; Burr, 1984; Posudin et al., 1988, 1990, 1991, 1992*a*, 1992*b*, 1993, 1995, 1996*a,b,c*; Massjuk et al.,1988; Massjuk and Posudin,1991*a*; Kawai and Kreimer, 1992; Witman, 1994; Kreimer, 1994; Martynenko et al., 1996; Posudin and Massjuk, 1996; Lenci et al., 1997; Hegemann and Harz, 1998; Sineschekov and Govorunova, 1999; Lebert and Häder, 2000; Gualtieri, 2001; Häder and Lebert, 2001; Hegemann and Fisher, 2001; Hegemann et al., 2001; Massjuk and Posudin, 2002]. The definition of terms and concepts, however, remains unacceptably ambiguous. On occasion a new meaning is attached to an old term, e.g., "phototaxis" or "photokinesis" which is interpreted differently by various authors (e.g., compare Diehn et al., 1977; Nultsch and Häder, 1979). Likewise, terms with different meanings are used as synonyms (e.g. *"behavioural light response"*, *"behavioural response"*, *"light controlled cell motility"*, *"light controlled movement"*, *"light response"*, *"light-induced behavioural response"*, *"light-induced responses of freely moving organisms"*, *"locomotive and motile response"*, *"motile behaviour"*, *"motile response to light"*, *"movement behaviour"*, *"photobehaviour"*, *"photobehaviour response"*, *"photoinduced behaviour"*, *"photomotile response"*, *"photomotion"*, *"photomovement response"*, *"photomovement"*, *"photoreaction"*, *"photoregulation of movement"*, *"photoresponse"*) – see Posudin, 1982, 1985; Sineschekov and Litvin, 1982; Wayne, 1991; Kreimer, 1994; Hegemann, 1997; Holland et al., 1997; Matsuda, 1998; Horigushi et al., 1999; Sineschekov, Govorunova, 1999; Lebert and Häder, 2000; Govorunova et al., 2000, 2001; Haupt, 2001; Häder and Lebert, 2001; Tahedl and Häder, 2001.

Currently there are several terminological systems that describe the motile behaviour of organisms (e.g. [Fraenkel and Gunn, 1961; Burr, 1984]). The wide diversity in the systems studied has been a source of terminological confusion. In addition, existing classification systems and definitions of the terms have a number of errors hindering precise communication among scientists in photobiology.

The objective of this Chapter is to describe a logical basis for the classification of photomovement phenomena in freely motile organisms (independent of their structure and systematic position), and to define the appropriate terminology.

## 2.1. State of the Art

Problems of classification and terminology first received attention in the early part of the 19th century when the photomotile reactions of organisms were initially noted by Treviranus [Treviranus, 1817]. Strasburger (1878) was the first author to suggest the distinction between photoreactions of freely motile (phototaxis) and fixed (phototropism) organisms; so, according to his interpretation *phototaxis is any movement of motile organisms in space caused by light*. In this original sense, the term fit into a larger system of terms denoting the displacement of freely motile microorganisms in space due to different environmental factors [e.g., chemical (chemotaxis), thermal (thermotaxis), gravitational (gravitaxis), mechanical (mechanotaxis)]. Phototaxis found its place among various terms for photobiological reactions in a family of terms describing general functional-physiological reactions [Konev and Volovsky, 1979].

The term *"phototaxis"* was derived from the Greek words – $\varphi\omega\xi$, $\varphi\omega\tau\acute{o}\xi$ (light) and $\tau\acute{a}\xi\iota\zeta$ (arrangement, order, orientation). Early in the 19th century the term was used in a stricter sense, i.e. the description of movement that is oriented in relation to the direction of light [Nagel, 1901] which unfortunately has led to "considerable confusion" in the terminology [Nultsch, 1975]. As a consequence, the term *"strophic phototaxis"* was used to indicate oriented movement [Rothert, 1901]. Subsequently the term "photo-topotaxis" was proposed to denote photoreactions that are related to the orientation of movement to the direction of the incident light [Pfeffer, 1904]. This terminology was accepted by investigators and has been widely used during the following decades (see Halldal, 1958).

Fraenkel and Gunn (1961) proposed distinguishing *primary photoorientation* of organisms in space and time when a stimulus is absent and the organism is inactive, from *secondary photoorientation* that is concerned with the change in position of an organism in which active movement is instigated by and is oriented toward a light stimulus. They defined *taxes* as the movement of an organism directly toward or away from the source of stimulation. If the movement of the organism is bilateral and oriented simultaneously toward and away from a source of stimulation, the response was called *tropo-taxis* ($\tau\rho\sigma\pi\eta$ – turn). *Telo-taxis* ($\tau\varepsilon\lambda\sigma\varsigma$ – end, consummation, result) indicates orientation in the direction of the source of stimulation which occurs without a bilateral distribution of the organism; a response characterized by regular deviations as part of the orientation mechanism is termed *klino-taxis* ($\kappa\lambda\iota\nu\omega$ – bend, incline).

Diehn defined the term "phototopotaxis" in 1970 and proposed a more strict interpretation of the term "phototaxis" which would now mean the photooriented movement of organisms. In his early work, Nultsch (1975) also indicated that the *" Greek word "taxis" in its original sense denotes only a distinct spatial array, and so the term "phototaxis" should mean any array of organisms in space caused by light and is not restricted to their direct movement"*. However, he was inconsistent in subsequent publications accepting the narrow interpretation of the term "phototaxis", that he had criticized earlier [Nultsch and Häder, 1979; Nultsch, 1980]. Following Diehn's proposal [Diehn, 1970], the narrow interpretation of the term "phototaxis" as movement oriented relative to the stimulus direction was widely accepted [e.g., Fraenkel and Gunn, 1961; Diehn et al., 1977; Feinleib, 1977, 1978, 1980; Kawai and Kreimer, 1992; Lenci and Colombetti, 1978; Mikolajczyk and Diehn, 1976, 1979; Häder, 1979; Sineschekov and Litvin, 1982; Posudin, 1982, 1985; Burr, 1984; Flores-Moya et al., 2002].

It should be noted that the narrow interpretation of a term, which has a broader meaning, is in violation of the requirements of formal logic and leads to disagreements in meaning. The new interpretation of the term *"phototaxis"* resulted in the appearance of many different equivalents that were used to fill a gap in terminology: *"behavioural responses", "light induced response of freely moving microorganisms", "motile behaviour", "motility behaviour patterns", "orientation reactions", "photobehaviour", "photomotion", "photomovement", "photoorientation", "photoresponses", "phototactic movement", "phototactic orientation", "phototactic reactivity", "phototactic response", "phototopotaxis", "response strategies", "response type"*, etc. [Halldal, 1958; Nultsch, 1973, 1983; Feinleib, 1977, 1980; Lenci and Colombetti, 1978; Nultsch and Häder, 1979; Kuznitski and Mikolajczyk, 1982; Sineschekov and Litvin, 1982; Posudin, 1982, 1985; Morel-Laurens, 1987; Feinleib, 1980;

Pfau et al., 1983; Häder and Lebert, 1998; Horiguchi et al., 1999; Lebert and Häder, 2000; Häder and Lebert, 2001].

The plurality and diversity of these terms are undesirable. New designations are either bulky or not exact and concrete, and they may be interpreted both in a broader meaning (e.g. including phototropism, photoregulation) and in a more narrow one. Some authors use these expressions as synonyms, while others distinguish photocontrol and its consequences (photomovement, photomotion) as well as photoresponses and behavioural consequences, increasing the frequency of disagreements in terminology.

Taking into account the above-mentioned considerations we propose keeping the primary interpretation of the term *"phototaxis"* as any light-induced movement of freely motile organisms. Photooriented movement of organisms is termed *"phototopotaxis"* and subdivided into two basic patterns: *positive* (toward the light source) and *negative* (away from it) [Halldal, 1958; Nultsch, 1975; Feinleib, 1980]. Occasionally a third pattern, *transverse phototopotaxis* (perpendicular orientation relative to the direction of the light) is used [Diehn et al., 1977; Diehn, 1979; Häder, 1979]. This includes any linear movement of organisms that is oriented to the direction of the light except in the case of diffused light or simultaneous illumination from all sides.

The term *"photomovement"* should have a broad meaning, i.e., *any motility response or its alteration induced by light*. Thus, this term involves either phototaxis (in the broader meaning) or phototropism. Our interpretation of this term does not contradict the definition given by W. Nultsch and D.-P. Häder (1979), *"The term photomovement denotes any movement or change of movement elicited by light..."*. As a consequence, the terms "photomovement" and "phototaxis" are compatible, but not identical. The former is subordinating, while the latter is subordinate.

It is quite clear that the control of movement and photomovement in organisms are absolutely different and are distinguished by a cause and effect relationship. Photomovement in organisms occurs by way of controlled movement. Control of movement is realized through the interaction of external factors (e.g., light stimulus – its intensity, spectral characteristics, direction of propagation) and internal regulatory mechanisms of the organism (e.g., photoreception, sensory transduction, function of the motor apparatus). Spatial displacement due to photomovement takes place as a result of such an interaction. Thus, the meanings "photomovement" and "control of movement" are not identical and not compatible from the point of view of formal logic.

In the latter part of the 19th century, Strasburger was the first to define the reaction of microorganisms to a sudden change in light intensity, a response called *"Schreckbewengung"* (*"motion of fright"*). Engelmann [1882a,b] observed a sharp change in the direction of movement of some microorganisms in response to a sudden increase in light intensity. This response was termed *"phobic"* (from Greek word $\varphi\delta\gamma\beta\delta\xi$ – fear). Later, several terms were proposed for indicating the shock reaction of microorganisms to changes in the intensity of light: *"apobatic phototaxis"* [Rothert, 1901], *"phobism"* (from the Greek word "fear") [Massart, 1902, cyt. by Nultsch, 1975], *"discrimination sensitivity"* [Nagel, 1901] and *"photophobotaxis"* [Pfeffer, 1904].

Within the scientific literature there are synonyms for the term *"photophobic reaction"* such as *"photophobic response"*, *"light induced stop response"*, *"photoshock"*, *"photoshock response"*, *"photoshock cell response"*, *"photophobic, stop or ecclitic response"*, *"photophobotaxis"*, and *"stop response"* [Beckmann and Hegemann, 1991; Hegemann et al., 1991; Govorunova et al., 1997; Holland et al., 1997; Sineschekov and Govorunova, 1999; Lebert and Häder, 2000; Häder and Lebert, 2001].

Engelmann (1882) termed the restoration of movement in microorganisms, that were immobile in the dark, after switching on the light stimulus *"photokinesis"* (from Greek word $\kappa\iota\nu\eta\sigma\iota\zeta$ – motion). The term *orthokinesis* indicates that the linear velocity of the organism is altered by the stimulus while *klinokinesis* indicates the frequency or rate of turning of the organism induced by the stimulus. These terms continue to be used although their interpretation has undergone significant alterations. Thus, there are at least four different interpretation for the term "photokinesis" which indicates an essential discrepancy in the interpretation of "phototaxis" and "photophobic response" [Burr, 1984].

A number of new terms (see, f.e., [Fraenkel and Gunn, 1961]) have subsequently emerged the appearance and interpretation of which are discussed in a review by Nultsch (1975). The author noted *"these disagreements in terminology have, in the past, led to errors and misinterpretation, so that the creation of new terms would further confuse the situation rather than clarify it"*.

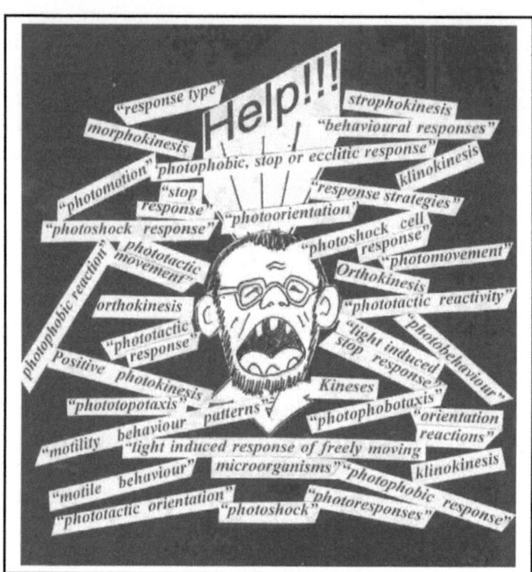


**Fig.2.1.** Chaos in Terminology

This fact resulted in the creation of the Committee on Behaviour Terminology that was convened in 1976 at the conference on Sensory Transduction in Microorganisms in Santa Barbara, California.

The recommendations of the Committee were published and signed by prominent photobiologists from various countries (e.g., B. Diehn, M. Feinleib, W. Haupt, E. Hildebrand, F. Lenci, W. Nultsch). The committee's recommendations follow and are presented in detail.

There are three main nomenclature sections: 1. Stimuli; 2. Responses; and 3. Behavioural consequences. The first section defines a *stimulus* as any quantity of energy or matter which, when interacting with the organism, can elicit a response. The nature of the stimulus is indicated using a prefix as in photo- (stimulus is a radiant energy affecting specific receptor molecules), thermo- (thermal energy), galvano- (ionic electric current), electro- (electrical field), geo- or gravi- (gravitational force), mechano- (mechanical force), magneto- (magnetic field), chemo- (molecular species acting upon specific receptor molecules). The recommendations were in agreement with the requirements of formal logic (since nothing can be said against them) and were considered to be generally accepted [Burr, 1984; Massjuk and Posudin, 1991*a*]. The authors were cognizant of the nature of the stimulus since some physical factors (e.g. radiation or electrical energy) are capable of generating chemical substances that may act as a direct stimuli. They also proposed distinguishing the dependence of a response upon an increase ("step-up") or decrease ("step-down") in the stimulus intensity.

This section is devoted to the definition of *15*as any stimulus-induced alteration in the activity of the organism's motor apparatus which may (but not always) result in an alteration in the movement or orientation of the organisms [Diehn et al., 1977].

The responses can be classified as:

*"Kineses"* – responses in which the steady-state rate of activity of the organism is controlled by the absolute magnitude of the stimulus intensity. In *positive kinesis* the activity rate is greater in the presence of the stimulus than in its absence; in *negative kinesis* the activity rate is lower in the presence of stimulus than in its absence.

*"Phobic responses"* denote the transient alteration in the activity of the organism that is caused by a change in stimulus intensity. The authors delineate the types of response levels in the organism: active but not due to the activity of the motor apparatus of the organism [Diehn et al., 1977]. Therefore, the definition that was proposed by the Committee for the term "response" does not correspond exactly to this meaning. They proposed distinguishing two main types of the responses: *steady-state* (*kineses*) and *transient* (*phobic responses*) [Diehn et al., 1977; Diehn, 1979]. Both types of responses are responsible for a change in the linear velocity and the direction of movement. The former is controlled by the absolute magnitude of the stimulus intensity, while the latter is controlled by a temporal stimulus gradient. Interpretation of these terms changed when compared to the primary variant [Engelmann, 1882]. In addition, according to Diehn et al. (1977), phobic responses that have a long adaptation time are nearly indistinguishable from kinesis. In such cases the choice of one or another term depends on the  investigator. Therefore the response classification suggested by the Committee (kinesis and phobic responses) lacks a common efficient basis for division from the standpoint of formal logic.

The classification of responses according to the presence or absence of adaptation did not suit the zoologists [Burr, 1984] whom found that it accounts for only one of the light stimulus parameters – its intensity (absolute magnitude or temporal gradient) but ignores other parameters such as spectral composition, polarization, etc. The fact is that the same authors use terms such as "action spectrum of photokinesis" in their other publications [Nultsch, 1973; Nultsch and Häder, 1979; Colombetti et al., 1982], testifies to the dependence of movement upon the wavelength of the light stimulus.

Unfortunately, the term *"behavioral consequences"* is not defined [Diehn et al., 1977]. Therefore, it is not clear which criterion is used as the basis for the separation of "responses" and "behavioral consequences". The authors considered there to be two types of "consequences". The first concerns single cells when their movement is oriented relative to the stimulus direction; the second applies to populations of cells that accumulate in the region with the higher intensity of the stimulus or their dispersal from it. In other words, the response of an organism affected by the stimulus must be considered at individual (microeffect) or aggregation (macroeffect) levels.

The following prefixes were proposed for determining the dependence of movement parameters (of all kineses and phobic responses) on the action of the stimulus: *ortho-, klino-, stropho-* and *morpho-* [Diehn, 1979]. Therefore *orthokinesis* is when the linear velocity of movement is increased (positive orthokinesis) or decreases (negative orthokinesis); *klinokinesis* – the rate of random spatial alteration in the direction of movement; *strophokinesis* – the frequency or amount of turning; and *morphokinesis* – the rate of alteration in shape or contour of the microorganism.  Each of these kineses is induced by a light stimulus.

Thus, according to Diehn et al. (1977) and Diehn (1979), taxis are the result of a response or of a series of responses by the motor apparatus rather than just a single response. We have discussed earlier that it is undesirable to use a broad meaning for the term «taxis» in the narrow interpretation of organism photoorientation.

In spite of the fact that the publication [Diehn et al., 1977] was the result of the collective thinking of the entire Committee, only one of its member [Diehn, 1979] actually followed the terminology recommendations. The other authors [Feinleib, 1978, 1980; Lenci, andColombetti, 1978; Nultsch and Häder, 1979; Nultsch, 1980; Lenci, 1982; Colombetti and Lenci, 1982; Haupt, 1983; Lenci et al., 1984] accepted the classification previously suggested by Nultsch (1975) as did other photobiologists (see, e.g. [Sineschekov and Litvin, 1982] ). This classification was presented in detail in the review by Nultsch and Häder (1979) in which the division of "responses" and "behavioural consequences" proposed by the Committee was not accepted. They adopted three types of responses for freely motile organisms [Nultsch and Häder, 1979]:

1. *"Photokinesis"* – the effect of light intensity on the speed of movement;

2. *"Phototaxis"* – movement oriented with respect to light direction;

3. *"Photophobic response"* – a reaction caused by a temporal change in light intensity ($dI/dt$), often seen as a stop response followed by a reversal of movement.

It should be noted that this classification has nothing in common with the recommendations of the Committee even though the authors were members of it. The narrow meaning of phototaxis is not considered the result of a response (or a series of responses) by an individual microorganism (see, e.g. [Diehn et al., 1977]), but as a response that has its own mechanism. Contrary to recommendations of the Committee [Diehn et al., 1977], the interpretation of the term "photokinesis" and the sense of the meaning of the term "photophobic response" changed.

The classification of Nultsch and Häder suffers from a number of errors in logic [Nultsch, 1975; Nultsch and Häder, 1979]. The requirements of formal logic are not met in the definitions of "photokinesis" and "phototaxis" and their relationship to the meanings of generic and specific distinction. Thus, photokinesis is interpretated not as a type of a response, but as "the effect"; phototaxis – not as a type of a response – but as a "movement". Such definitions deviate from the meaning of "photoresponse" due to their interpretation through the set of other meanings ("effect", "movement").

This classification is characterized by the following deficiencies: 1. a common basis of division is absent; 2. members of the division do not exclude each other (e.g., the definition of photophobic reaction coincides with the definition of phototaxis); 3. it represents a major alteration in the division (e.g., phototaxis and photophobic describe reactions at of different levels).

## 2.2. Parametrical Classification Principles for Photomovement in Organisms

It should be noted that classifications developed by microbiologists were often severely criticized by zoologists. For example, Burr (1984) commented *"In reviewing the literature on movement behavior I was impressed by the amount of disagreement or confusion as to the meaning of moving terms"*. He was the first to note the necessity of taking into account the influence of various light stimulus parameters on each type of movement behaviour in microorganisms. It is a pity that he did not complete his proposed classification system. Having indicated the impossibility of strictle delimitating adapting and non-adapting reactions, he suggested using the terms irrespective of the adapting characteristics of movement behaviour. Recognizing the necessity of the propositions of Burr (1984), we can use the term *"light-dependent moving reactions of freely motile organisms"* (photoreactions, photoresponses) to mean the sudden movement responses by an organism to an alteration in the light stimulus (e.g., appearance or disappearance of the light, changes in its spectral composition, intensity or direction).

From our perspective, the term *light-dependent behaviour* of freely motile organisms has a broader meaning. It includes various light-induced displacements of freely motile microorganisms in space, independent of changes in the light stimulus over time (a meaning that is equivalent to the term "phototaxis" in its original sense). Thus, according to our interpretation the terms light-dependent movement behaviour of freely motile organisms (phototaxis) and photoreactions have compatible but not identical meanings, the former is subordinating and the latter subordinate.

*Light-induced motile behaviour* of biological objects is a broad physical phenomenon – movement. Some types of movement can be described by traditional physical parameters such as velocity ($\upsilon$), direction ($\vec{r}$) and trajectory ($l$). These can be subdivided into other, less general parameters for a more detailed characterization of the various types of movement. For example, velocity may be linear or angular. Both linear and angular velocities may be constant or may vary as a function of time. Certain classes of linear and angular velocity may also be divided according to specific limits in velocity inherent to a particular organism or group of organisms. Direction of propagation can be oriented or not oriented relative to the source of light and the orientation can be positive (toward the source of light) or negative (away from the source of light). Likewise, trajectory of movement can be rectilinear, curvilinear, parabolic, zigzag, helical, or circular (around its own longitudinal or transverse axis or around an external axis).

According to the previous assessments, the light stimulus therefore is characterized by parameters such as intensity ($I$), direction ($\vec{s}$), spectral composition ($\lambda$), polarization ($P$), and duration and frequency of light pulses. Similar to movement parameters, light parameters may be divided into the intensity of light that is characterized by absolute magnitude ($I$) and its gradient in space ($dI/dx$) and time ($dI/dt$).

According to the parametrical character of either the movement of organisms or a light stimulus, we propose the classification of phototaxes (the dependence of the organisms movement on a light stimulus) on the basis of the *parametrical principle* for both individual freely motile organisms and populations of organisms as is presented schematically in Tables 2.1 and 2.2. Table 2.1 demonstrates the phototaxis of individual freely motile organisms (i.e., individual or microeffect), Table 2.2 the phototaxis of populations of organisms (i.e., group or macroeffect).

The well-known dependences are shown in each case, e.g. $\upsilon$ $(I)$ is the dependence of linear velocity of a single cell on the absolute magnitude of intensity of the light stimulus (or photokinesis according to [Nultsch and Häder, 1979] or ortokinesis according to [Fraenkel and Gunn,1961]); $n(I)$ is the dependence of angular velocity of movement (frequency of random rotations) on the absolute intensity of the light stimulus (klinokinesis) [Fraenkel and Gunn,1961]; $\vec{r}(\vec{s})$ is the dependence of movement direction on the direction of the light stimulus (or phototaxis according to [Diehn et al., 1977]); $l(dI/dt)$ is the dependence of movement on temporal changes in the light stimulus intensity (or photophobic response according to [Diehn et al.,1977]).

**Table 2.1.** Photoresponses of individual organisms

| Parameters of light stimulus / Parameters of movement | Intensity $I$ | Gradients of intensity $dI/dt$ | $dI/dx$ | Direction $\vec{s}$ | Wavelength $\lambda$ | Polarization $P$ |
|---|---|---|---|---|---|---|
| Velocity of individuals: <br> • linear $\upsilon$ <br> • angular $n$ (frequency of spatial changes of trajectory of individuals: oscillations, rotations, trembling) | $\upsilon\,(I)$ <br> $n(I)$ | $\upsilon\,(dI/dt)$ <br> $n(dI/dt)$ | $\upsilon\,(dI/dx)$ <br> $n(dI/dx)$ | $\upsilon\,(\vec{s})$ <br> $n(\vec{s})$ | $\upsilon\,(\lambda)$ <br> $n(\lambda)$ | $\upsilon\,(P)$ <br> $n(P)$ |
| Direction of movement of individual $\vec{r}$ | $\vec{r}\,(I)$ | $\vec{r}\,(dI/dt)$ | $\vec{r}\,(dI/dx)$ | $\vec{r}\,(\vec{s})$ | $\vec{r}\,(\lambda)$ | $\vec{r}\,(P)$ |
| Trajectory of movement of individual $l$ | $l(I)$ | $l(dI/dt)$ | $l(dI/dx)$ | $l(\vec{s})$ | $l\,(\lambda)$ | $l(P)$ |
| Frequency of spatial changes in trajectory of individual (oscillations, trembling, rotations) $n$ | $n(I)$ | $n(dI/dt)$ | $n(dI/dx)$ | $n(\vec{s})$ | $n(\lambda)$ | $n(P)$ |

We propose based upon the preceding analysis the following terminology [Massjuk et al., 1988; Massjuk and Posudin, 1991*a*; Massjuk et al., 1991; Massjuk and Posudin, 2002]:

*Photomovement* – movement or change in movement of organisms induced by light.

*Phototaxis* – any movement or change in character of the movement of freely motile (not fixed) organisms, not obligatorily oriented relative to light source.

**Table 2.2.** Photoresponses of populations and colonies of organisms

| Parameters of light stimulus / Parameters of movement | Intensity | Gradients of intensity | | Direction | Wavelength | Polarization |
|---|---|---|---|---|---|---|
| | $I$ | $dI/dt$ | $dI/dx$ | $\vec{s}$ | $\lambda$ | $P$ |
| Concentration $N$ of individuals (optical density) in population or colony | $N\,(I)$ | $N\,(dI/dt)$ | $N\,(dI/dx)$ | $N(\vec{s}\,)$ | $N(\lambda)$ | $N(P)$ |
| Shape (spatial distribution) of individuals in population or colony $S$ | $S\,(I)$ | $S\,(dI/dt)$ | $S\,(dI/dx)$ | $S(\vec{s}\,)$ | $S\,(\lambda)$ | $S(P)$ |
| Trajectory of movement of population or colony $L$ | $L(I)$ | $L(dI/dt)$ | $L(dI/dx)$ | $L(\vec{s}\,)$ | $L\,(\lambda)$ | $L(P)$ |
| Relative number of individuals performing a response $N/N_0$ | $N/N_0(I)$ | $N/N_0(dI/dt)$ | $N/N_0(dI/dx)$ | $N/N_0(\vec{s}\,)$ | $N/N_0\,(\lambda)$ | $N/N_0(P)$ |

*Photocontrol of movement* – the entire complex of elementary processes that are induced by a light stimulus, particularly photoreception, primary reactions of photoreceptor pigments, sensory transduction of the light stimulus into a physiological signal(s) that governs through the function of the motor apparatus, photoorientation, and velocity of movement of organisms.

*Photokinesis* – dependence of the velocity of movement of an individual organism and their groups on any parameter of the light stimulus.

*Phototopotaxis* – dependence of the direction of the movement of an individual organism and their groups on any parameter of a light stimulus.

The dependences that are included in Tables 2.1 and 2.2 may be direct or indirect, positive or negative, though the forms of the dependence may not always be evident. Likewise, it is possible to further divide these terms into sub-classes. In particular, *photoortokinesis* (the dependence of the linear velocity of movement on the parameters of the light stimulus) and *photoklinokinesis* (dependence of the quantity of rotations per unit of the time on parameters of the light stimulus) can be distinguished. Some dependences have not yet been given a specific term, e.g., the trajectory of movement, the concentration of organisms, the shape of "ensemble", the relative number of organisms, that undergo the photoreactions mediated by a light stimulus (see Tables 2.1, 2.2). A number of the light stimulus parameters may also be further categorized (e.g., rhythm of the light stimulus, its constant or variable character) or the possible interaction(s) of various parameters (e.g. wavelength and intensity, change of velocity and direction of propagation).

Thus, the parametrical principles of classification of light-induced behaviour for organisms we propose not only places in the appropriate order the available terminology and the data of interaction of the light stimulus with the specific features of movement of living organisms, but also forecasts and allows for characterizing peculiarities and facilitates the development of a program of additional research needed for the completion of the classification utilizing strict basic logic. This principle may also be applied to different types of taxes induced by other physical factors. With regard to photoreactions, they may be expressed through: 1. a change in the velocity (*kinetic reaction*); 2. a change in the direction and trajectory of movement (*vector reaction*) and 3. a simultaneous change in the velocity and direction of movement (*photophobic reaction*). Thus we distinguish such meanings as photomovement and control of movement, and within the limits of photomovement – phototaxes (in the original broad sense) and phototropisms. The parametrical principle is proposed for the classification of the light-induced behaviour of single cells as well as for populations of organisms. The broader meaning of "phototaxis" involves the term subordinate to it – "photoreaction", that consists of three phenomelogical types: kinetic, vector and phobic.

## 2.3. Summary

A critical assessment of the terminology and classification of different types of light-induced behavior in freely motile organisms has shown considerable diversity and ambiguity in the existing classification systems that has led to confusion. We present a new parametrical classi-fication of light-dependent behavior for either individual motile cells (individual effect or micro-effect, Table 2.1) or their aggregations (group effect or macroeffect, Table 2.2). The original meaning of the term *phototaxis* as any light-induced movement of freely motile organisms in space is retained. Light-dependent *reactions* of motile organisms (*photoresponse, photoreac-tion*) indicate any immediate motion responses of the organisms to any change in the light sti-mulus.

*Motility* of biological objects represents a special case in the general physical phenome-non of movement (mobility). Therefore, motility of organisms can be described by traditional parameters such as velocity ($\upsilon$), direction ($\vec{r}$) and trajectory ($l$) of movement. The light sti-mulus, in turn, is characterized by such parameters as intensity ($I$), direction ($s$), spectral com-position ($\lambda$), and polarization ($P$) and the duration, frequency, and shape of light pulses. Consi-dering the parametrical characteristics of both factors (light and movement), we believe that any classification of the dependence of microorganism movement (phototaxis) on light should be based on parametrical principles (Tables 2.1, 2.2). We propose *photokinesis* to be any depen-dence of velocity of individual organisms or their groups on any parameters of the light stimu-lus, and *phototopotaxis* as any dependence of the direction of movement of individual organ-isms or their groups on any parameters of the light [Massjuk et al., 1988; Massjuk and Posudin, 1991*a*; 2002].

The proposed classification can be further developed by accounting for additional para-meters of movement and light (e.g., the rhythm of a light flux), their possible interactions (e.g., wavelength and intensity of the light, velocity and direction of movement), or specific features of some parameters (e.g., the velocity of movement can be linear or angular, the light intensity can be characterized by its absolute value ($I$) or its gradient in space ($dI/dx$) and time ($dI/dt$)). The suggested principles not only promote further improvement of the existing terminology but also identify the direction of future research.

# Chapter 3

# Investigations with Species of *Dunaliella* Teod.

## 3.1. History of the Discovery and Description of the Genus *Dunaliella*

A mysterious phenomenon, the "red flowering" of salt watersheds located in low latitudes worldwide, was known long before the genus *Dunaliella* was described. The Academy of Sciences in Paris requested A. Payen ascertain the origin of the phenomenon. He believed the red coloration of the salt water was caused by a tremendous increase in the population of brine shrimp (*Artemia salina* L.) and their subsequent destruction due to the high concentration of salts. The die-off was accompanied by a distinctive violet odour. Subsequently, Michel Felix Dunal (1789-1856), a professor of botany in Montpellier, France, questioned this conclusion (1838). He found microscopic algae that he named *Protococcus salinus* Dunal and *Haematococcus salinus* Dunal in the salt bogs of Montpellier on the South coast of France. According to Dunal the algae were responsible for the coloration of the salt water. A special committee of the Academy of Science confirmed his conclusion although discussion about the cause of the coloration continued for nearly another century.

The systematic position of the algae was debated for years due in part to the absence of adequate illustrations and a detailed description. Joly (1840) combined all the species described by Dunal under the name *Monas dunalii* Joly. Cohn (1865*b*) subsequently put this species into the genus *Chlamydomonas*, under the name *C. dunalii* Cohn. Dujardin (1841) then assigned this alga to the genus *Diselmis*, as *D. dunalii* Duj., while Hansgirg (1866) considered it a representative of *Sphaerella*, i.e., *S. lacustris* var. *dunalii* Hansg.

Emmanuel Constantin Teodoresco (1866–1949), a Romanian botanist, published the results of his investigation in 1905 in which he concluded that the coloration of the salt watersheds was due to a unicellular algae that lack a rigid wall and reproduce and copulate in a motile state. According to Teodoresco these algae should be transferred into a family of polyblepharids under the new genus *Dunaliella* Teod. He proposed that the species *Dunaliella salina* Teod.was an example of a species that was well known at that time; *Dunaliella viridis* Teod. was identified one year later [Teodoresco, 1906].

A publication by Clara Hamburger with a description of *Dunaliella* was presented the same year [Hamburger, 1905].

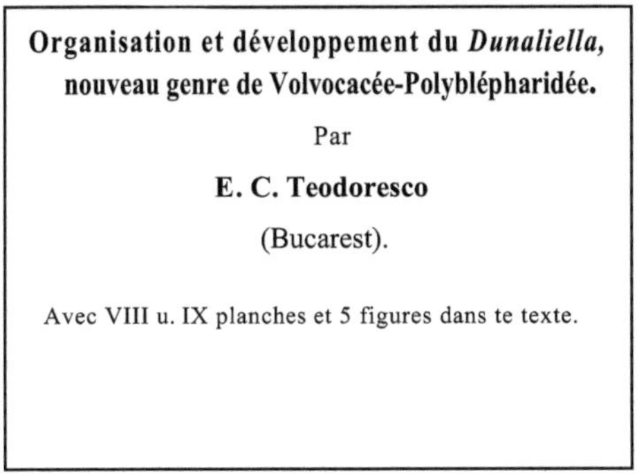

**Organisation et développement du *Dunaliella*, nouveau genre de Volvocacée-Polyblépharidée.**

Par

**E. C. Teodoresco**

(Bucarest).

Avec VIII u. IX planches et 5 figures dans te texte.

**Fig.3.1.** Title-page of the article of E.C. Teodoresco (1905) where he described *Dunaliella*

La diagnose du nouveau genre peut être donnée de la façon suivante:

**_Dunaliella[1]_) n. g.**

Zoospores vivant însolées, de forme généralement allongée-ellipsoïde ou cylindrique, souvent plus ou moins étranglée vers le milieu; **corps po ssédant des p ropriétés f aiblement métaboliques** et devenant sphérique dans l'eau salée très diluée; **enveloppe** mince, lisse, **dépourvue de cellulose,** entourant directement le protoplasma, **extensible et s uivant les changements de forme d u corps; deux longs flagellums** dépassant en longueur le corps tout entier; chromatophore on forme de cloche, coloré en vert; hématochrome inprégnant non seulement le chromato-phore mais encore le corps tout entier des individus âgés; pyrénoïde gros, entouré par une amylosphère; noyau vers le milieu du corps; un point oculaire rouge, allongé, situé au niveau du noyau, **multiplication pendant la m arche par division longitudinale e n de ux ind ividus;** re-production sexuelle par l'u n i o n, pendant la marche, de deux gamètes égaux ou à peu prés égaux; zygote ne passant pas (?) a l'etat de repos dans les conditions favorables de v ie; aplanosplores?

Le genre ne comprends qu' une seule espèce:

_Dunalidla salina._   Caractères du genre.   Longueur des zoo-spores âgées à hématochrome 16 μ à 24 μ (— 28μ), épaisseur 9,5 μ à 13,3 μ;  longueur niinima des zoospores vertes  13,3 μ, épaisseur mimma des mêmes 6,3 μ.

**Fig.3.2.** Description of _Dunaliella salina_ by E. Teodoresco (1905)

Based on comprehensive studies, the genus _Dunaliella_ was placed in the Familia _Dunaliella-ceae_, Order _Dunaliallales_, Class _Chlorophyceae_, Division _Chlorophyta_, Regnum _Viridiplantae_. The genus _Dunaliella_ has seventy or more species and 30 specific and intraspecific names, some appearing _nomen nudum_ and are definitely synonyms. _Dunaliella cordata_ Pascher and Jagoda, deemed distinctly different from the other members of the _Dunaliella_, was determined to be a separate genus, _Papenfussiomonas_ Desikachari. Currently the genus _Dunaliella_ includes 28 species with 33 intraspecific taxons (Table 3.1)[1].

**Table 3.1.** Intraspecific taxons of _Dunaliella_ Teod.

| Taxon | Size of cells (length×width, μm) | Basionyms, synonyms |
|---|---|---|
| 1 | 2 | 3 |
| Subgenus _Pascheria_ Massjuk | | |
| _Dunaliella acidophila_ (Kalina) Massjuk | 7.6-12,7 × 1.7-2.3 | _Spermatozopsis acidophila_ Kalina |
| _D. flagellata_ Skvortzov | 15.0 × 19.0 | |
| _D. lateralis_ Pascher et Jahoda | 7.0-11.0 × 4.0-6.,0 | |
| _D. obliqua_ (Pascher) Massjuk | 9.0-14.0 × 6.0-9.0 | _Apiochloris obliqua_ Pascher |
| _D. paupera_ Pascher | 9.0-12.0 × 7.0-9.0 | |

---

[1] This table is composed on the basis of monograph of Massjuk, 1973.

[2] _Dunaliella bardawil_ (Ben-Amotz et al., 1982_a_) is not taken into account since we were unable to find references detailing either its description or a diagnosis of the species. H.R. Preisig (Avron and Ben-Amotz, 1992) considers this name a synonym of _D. salina_ Teod.

**Table 3.1.** Intraspecific taxons of *Dunaliella* Teod. (continued)

| Taxon | Size of cells (length×width, µm) | Basionyms, synonyms |
|---|---|---|
| | Subgenus *Dunaliella* | |
| *Dunaliella maritima* Massjuk | Sectio *Tertiolectae* Massjuk | *Dunaliella parva* sensu Butcher, non Lerche; *D. parva* f. *eugameta* Lerche |
| *D. polymorpha* Butcher | 5.0-19.0×2.5-15.0 | |
| *D. quartolecta* Butcher | 5.0-29.0×2.5-21.0 | |
| *D. tertiolecta* Butcher | 7.0-9.0×4.0-6.0 | |
| | 5.0-18.0×4.5-14.0 | |
| *D. parva* Lerche | Sectio *Dunaliella* | |
| *D. pseudosalina* Massjuk et Radchenko | 9.9-16.0×4.0-10.0 | |
| | 11.0-23.0×6.0-16.0 | |
| *Dunaliella salina* Teodorescu | | *Haematococcus salinus* Dunal, *Protococcus salinus* Dunal, *Monas dunalii* Joly, *Diselmis dunalii* Dujardin, *Chlamydomonas dunalii* Cohn, *Sphaerella lacustris* var. *dunalii* Hansgirg, *Dunaliella kermesiana* sensu Labbe, *D. Bardawil*[2] |
| | 5.0-29.0×2.5-21.0 | |
| - ssp. *salina* | 5.0-29.0×4.0-21.0 | |
| -- f. *salina* | 5.0-23.0×4.0-19.0 | |
| -- f. *magna* Lerche | 7.5-29.0×7.5-21.0 | |
| -- f. *oblonga* Lerche | 7.0-28.0×5.0-13.0 | |
| - ssp. *sibirica* Massjuk et Radchenko | 6.0-24.0×2.5-20.,0 | |
| | Sectio *Virides* Massjuk | |
| *D. baas-beckingii* Massjuk | 16,0-20,0×4,0-6,0 | *Dunaliella* sp. 5 (Ruinen, 1938) |
| *D. bioculata* Butcher | 6,5-13,0 × 3,5-8,0 | *D. marina* nomen nudum (Kombrink, Wöber, 1980) |
| *D. carpatica* Massjuk | 10,0-19,0×7,0-13,0 | |
| *D. gracilis* Massjuk | 32.0-40.0×4.0-5.0 | *Dunaliella* sp. 3 (Ruinen, 1938) |
| *D. granulata* Massjuk | 5.0-18.0×3.0-15.0 | |
| *D. media* Lerche | 10.0-20.0×3.0-5.0 | |
| *D. minuta* Lerche | 3.0-13.0×1.5-10.0 | |
| *D. minutissima* Massjuk | 2.8-6.0 | *Dunaliella* sp. 2 (Ruinen, 1938) |
| *D. ruineniana* Massjuk | 24.0-28.0 × 12.0 | *Dunaliella* sp. 4 (Ruinen, 1938) |
| *D. terricola* Massjuk | 3.5-13.0 × 2.0-6.,5 | |
| *D. viridis* Teodorescu | 3.0-18.0×2.0-15.2 | *D. salina* f. *viridis* (Teod.) Butcher |
| - var. *viridis* | | |
| - - f. *viridis* | 5.1-17.0 ×3.0-15.2 | |
| - - f. *euchlora* (Lerche) Massjuk | 3.0-17.8×2.0-11.4 | *D. euchlora* Lerche 1937 |
| - var. *palmelloides* Massjuk | 9.0-12.0 × 5.0-7.0 | |
| | Sectio *Peirceinae* | |
| *D. asymmetrica* Massjuk | 6.0-13.0 × 3.5-9.0 | |
| *D. jacobae* Massjuk | 6.0-12.0 × 4.0-5.0 | *Dunaliella* sp. 1 (Ruinen, 1938) |
| *D. peircei* Nicolai et Baas-Becking | 7.0-25.0×3.0-12.0 | |
| *D. turcomanica* Massjuk | 4.0-12.7 × 3.0-9.0 | |

The original and subsequent scientific literature on the genus *Dunaliella* are summarized in a monograph by Nadia Massiuk ["Morphology, Systematic, Ecology, Geographical Distribution of the Genus *Dunaliella* Teod. and Perspectives of its Practical Utilization", Kiev, Naukova Dumka, 1973 (in Russian)].

The total extent, phylogenetic relations and position of the genus in the taxonomy of green algae was discussed, the main tendencies of intra-specific evolution indicated, and an original system for the classification of the species within the genus proposed. Problems dealing with the extent of species and principles of intra-specific systematics for algae are considered as well as determinant tables and criteria for the identification of *Dunaliella* species were proposed. The potential for the commercial utilization of representatives of the genus *Dunaliella* as fish forage and as a source of raw material containing carotene was described. Classical comparative-morphological, biochemical, physiological, ecological, geographic and biological criteria were widely used for the first time. The investigations were conducted at the population level and used variational statistics.

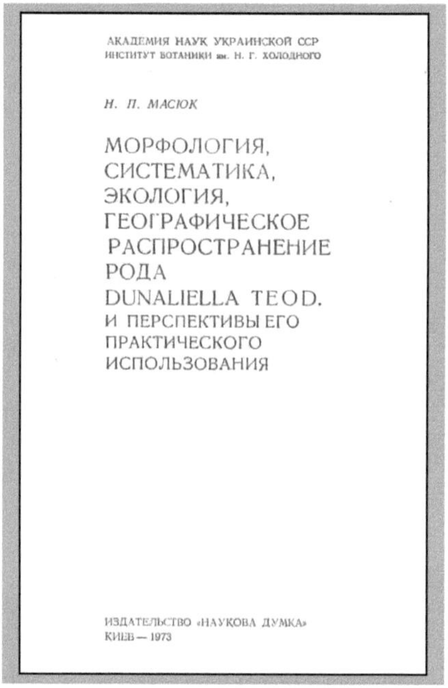

АКАДЕМИЯ НАУК УКРАИНСКОЙ ССР
ИНСТИТУТ БОТАНИКИ им. Н. Г. ХОЛОДНОГО

Н. П. МАСЮК

МОРФОЛОГИЯ,
СИСТЕМАТИКА,
ЭКОЛОГИЯ,
ГЕОГРАФИЧЕСКОЕ
РАСПРОСТРАНЕНИЕ
РОДА
DUNALIELLA TEOD.
И ПЕРСПЕКТИВЫ ЕГО
ПРАКТИЧЕСКОГО
ИСПОЛЬЗОВАНИЯ

ИЗДАТЕЛЬСТВО «НАУКОВА ДУМКА»
КИЕВ — 1973

**Fig. 3.3.** Title-page of the monograph by Nadia Massiuk entitled "*Morphology, Systematic, Ecology, Geographical Distribution of the Genus Dunaliella Teod. and Perspectives of its Practical Utilization*" (Kiev, Naukova Dumka, 1973).

A comprehensive review of the central topics relevant to *Dunaliella* were presented in a multi-author book entitled "*Dunaliella*: Physiology, Biochemistry, and Biotechnology" edited by Mordhay Avron and Ami Ben-Amotz (CRC Press, 1992). Morphological and taxonomic problems within *Dunaliella*, the function of flagellar apparatus and cell motility, photosynthesis, ATPases and ion transport, mechanisms of osmoregulation, β-carotene biosynthesis, acidophilism, and biotechnology of Dunaliella are discussed. A new book, "The alga *Du-*

*naliella*: Biodiversity, Physiology, Genomics and Biotechnology" (A.Ben-Amotz, J. Polle, S. Rao, eds.), was also recently published (2009).

Additional information on the alga can be found in reviews by M. Ginzburg ["*Dunaliella*: a green alga adapted to salt" (1987)], M. Borowitzka ["The mass culture of *Dunaliella salina*" (1990)], A. Borovkov ["Green microalga *Dunaliella salina* Teod. (Review)" (2005)], and A. Oren ["A hundred years of *Dunaliella* research: 1905-2005" (2005)].

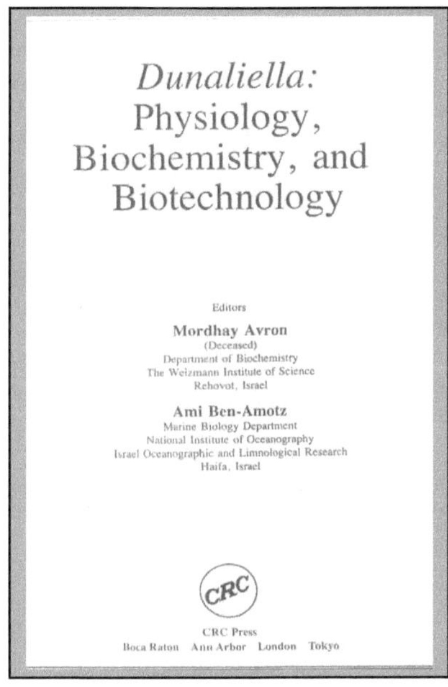


**Fig. 3.4.** Title-page of multi-author review *"Dunaliella: Physiology, Biochemistry, and Biotechnology"*(Mordhay Avron and Ami Ben-Amotz, eds.) CRC Press, 1992.

This monograph is dedicated to the study of photomovement in *Dunaliella* Teod., in particular, two species – *D. salina* Teod. and *D. viridis* Teod.

## 3.2. Characteristics of the Test Species

Unialgal cultures of two species of *Dunaliella*, *D. salina* Teod. strain №10 and *D. viridis* Teod. strain №42, from the collection of the N.G. Kholodny Institute of Botany, Ukrainian Academy of Sciences [Massjuk and Tereshchuk, 1983], were used in this study. Information about the genus *Dunaliella* may be found in a monograph by Massjuk (1973) that contains extensive references detailing the major advances in research on the genus.

The principal peculiarity of *Dunaliella* spp. in comparison with other green flagellates that demonstrate photomovement is the absence of a rigid cell wall. In addition, the cells are covered with a thin colourless cytoplasmic membrane (plasmalemma) that from the exterior has an irregular surface due to the presence of glycoproteins [Melkonian and Preisig, 1984]. The shape of the cells can change during movement due to the absence of a dense cell coat.

The front of the cell has papilla where two isocontic, isomorphic and isodynamic flagella are found. The flagella display smooth ciliary the beating of which results in both translational and rotational (clockwise and counter clockwise) movement of the cell.

The basal body of *Dunaliella* spp. are interconnected by distal striated fiber. It is thought that these connecting fibers may be involved in coordinating the flagellar beating [Melkonian and Preisig, 1984]. A system of the flagellar roots (of the 4-2-4-2 type) is attached to each basal body. The roots act as a cytoskeleton providing a regular framework around the cell and conferring uniform distribution of the mechanical strain caused by flagellar beating [Melkonian and Preisig, 1984].

According to Melkonian (1978) the flagellar roots determine the precise location of the cellular organelles, e.g. the eyespot's (stigma) position is relative to the flagellar apparatus. The stigma is located in the anterior peripherical portion of the chloroplast. The stigma in *D. viridis* is large, distinct, and bright-red coloured, while the stigma of *D. salina* is diffuse, pale-red coloured and barely visible.

The species of *Dunaliella* described herein are distinguished by differences in the cell shape and size. The cells of *D. salina* are 5 to 29 µm in length and 4 to 20 µm wide, while *D. viridis* is 3 to 18 µm in length and 2 to 15 µm wide [Massjuk, 1973]. The length of the flagella of *D. salina* is approximately equal to the length of the cell, while in *D. viridis* flagella are 1,3 times longer than the cell (Fig. 3.5).

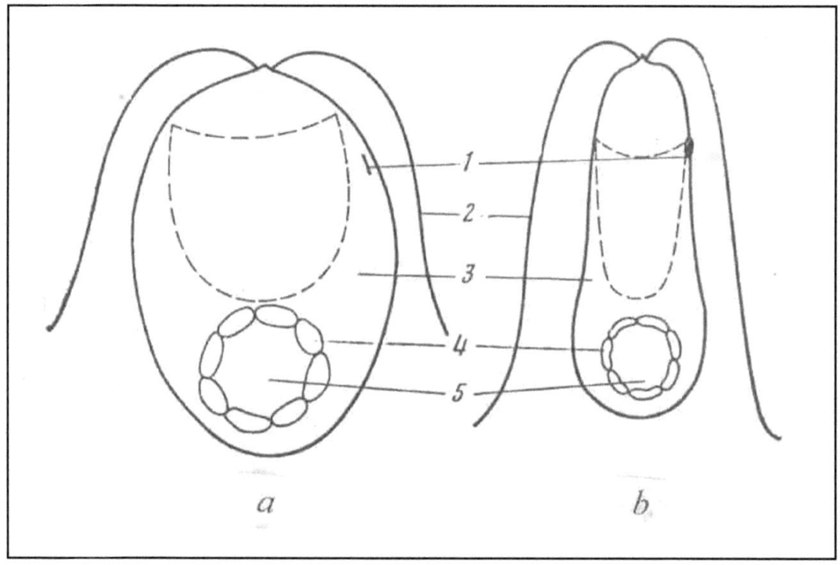


**Fig. 3.5.** General schematic of two species of *Dunaliella*: *a* – *D. salina*; *b* – *D. viridis* where: *1* – stigma; *2* – flagella; *3* – chloroplast; *4* – starch; *5* – pyrenoid (after Posudin et al., 1988).

An image of *Dunaliella* sp. from an electron microscope is given in Photograph 3.1.

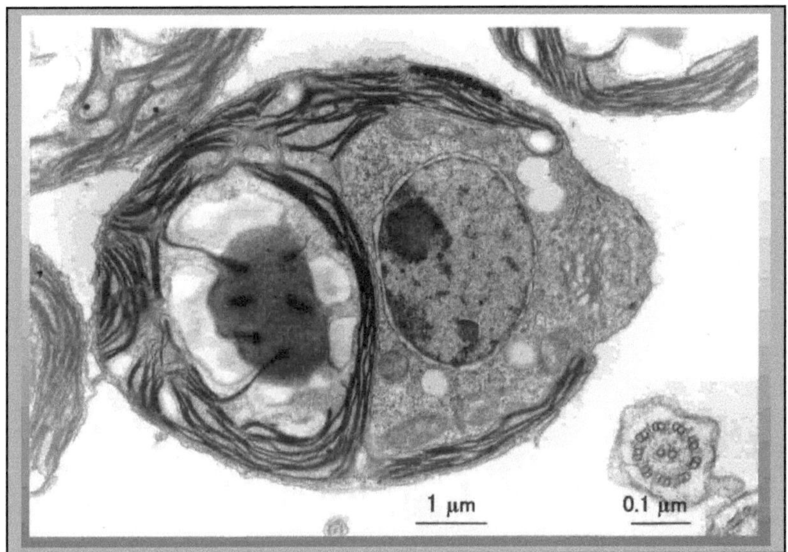

**Photograph 3.1.** An image of *Dunaliella* sp. from an electron microscope. Courtesy of Prof. Shogo Nakamura (Toyama University, Japan)

Species such as *D. salina* can accumulate certain oils in the stroma of the chloroplasts in which carotenoids, primarily β-carotene, are soluble giving a red coloration. Red forms of *D. salina* are prevalent in the salt brines on the Crimea in southern Ukraine. However, if the salt concentration in the brine decreases, under the low illuminance, the red cells quickly lose the carotene and turn from red to green in color.

## 3.3. Cultivation of the Species

The algae can be readily grown in a nutrient medium containing the following salts: 116 g/l NaCl; 50 g/l $MgSO_4 \cdot 7H_2O$; 2,5 g/l $KNO_3$ ; and 0,2 g/l $K_2HPO_4$. The medium has a specific density of $1 g/cm^3$ and pH 6–7 (the pH is adjusted using an alkali [Massjuk, 1973]). The cultivation temperature was $21 \pm 1^0C$, the illumination 3000 lx, and a L:D photoperiod of 12:12 h. Three to six day old cultures were used in the following experiments with fresh drops of algal suspension being transferred to a microscopic slide for measurement of the photomovement experiments.

# Chapter 4

# Investigation of Photomovement in *Dunaliella*

## 4.1. Methods of Investigating the Photomovement Parameters in *Dunaliella*

### 4.1.1. Experimental Installation

To study photomovement in *Dunaliella*, a special experimental videomicrography system was developed. It allows the observation and measurement of the velocity and direction of movement of individual cells as modulated by light stimulus parameters. The system utilizes a microscope connected to a light source, monochromator and videosystem (Fig. 4.1).

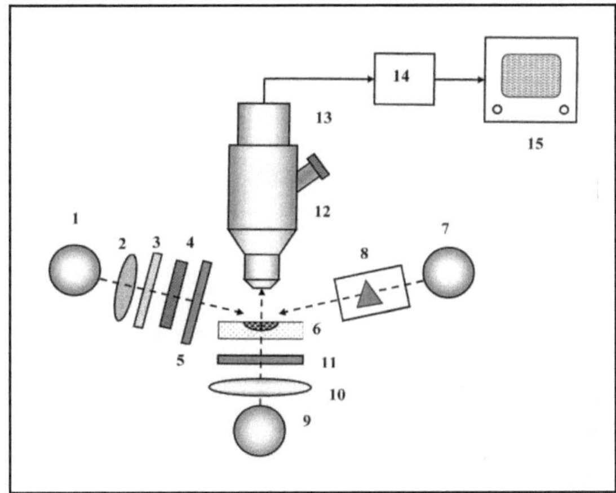

**Fig. 4.1.** A schematic of experimental videomicrography for studying photomovement in algae: *1* – source of white light; *2* – collimator; *3* – glass infrared filter; *4* – liquid infrared filter; *5* – interference filter; *6* – sample in a concave slide; *7* – halogen lamp; *8* – monochromator; *9* – source of white light; *10* – condenser; *11* – polarizer; *12* – microscope; *13* – videocamera; *14* – coupling unit; *15* – monitor [Posudin et al. 1992, 1996,*a*].

Radiation from a white light (300 W filament lamp) source *1* passes through a collimator *2*, glass infrared filter *3*, liquid infrared filter *4* (a 1 cm deep layer of distilled water), and to record the light curves, through an interference filter *5*. The light is directed at an angle of 30° to the slide plane, thus illuminating the algal suspension on a concave slide *6*. Special attention was paid to avoid air bubbles under the glass cover slip in that they may cause chemotactic reactions by the cells to oxygen. The spectral dependence of phototopotaxis was studied using the halogen lamp *7*, with the radiation passed through a monochromator *8*. To study the effect of the polarized light, radiation from the light source *9* was directed through a condenser *10* and a polarizer *11* before striking the sample. A xenon lamp (2 kW) was employed as the source for high light intensities. Photomovement parameters (e.g., linear and rotational velocity, number of cells moving toward or away from the light source, motility of the cells) modulated by light stimulus parameters (e.g., intensity, polarization, spectral composition) were estimated using a recording system that consisted of a microscope *12*, video camera *13*, coupling unit *14* and monitor *15* [Posudin et al., 1996*a*].

The light intensity was determined using a light meter and the illuminance with a lux-meter. The experiments videomicrography system was developed by the Biophysics Department of National University of Life and Environmental Sciences of Ukraine (Fig. 4.2).

**Photograph 4.1.** An experimental videomicrograph developed by the Biophysics Department at National University of Life and Environmental Sciences of Ukraine for investigating the photomovemt of *Dunaliella* [Posudin et al., 1992, 1996*a*].

### 4.1.2. Measuring the movement velocity of the cells

The movement velocity of selected cells on the calibrated monitor screen was estimated using an object-micrometer and the cell passage time for a straight line segment of the trajectory (142 μm). This gave a measure of the dependence of cell movement velocity on any light stimulus parameter. Data were based on an average of 10 or more velocity measurements using different cells in each suspension sample. Thus, the mean values for the velocity of cell linear movement ($n > 10$) is indicated by dots together with their standard deviations.

The photokinetic response of cells to a change in the light intensity was estimated in relative units using the formula $R_I = (\upsilon_I - \upsilon_0)/\upsilon_0$, where $\upsilon_I$ and $\upsilon_0$ are cell movement velocities at the given and minimal (under the experimental conditions) light intensity, respectively ($< 1$ W/m$^2$) [Posudin et al., 1988].

### 4.1.3. Measuring the phototopotaxis

Trajectories for moving cells were marked on the polyethylene film covering the monitor screen. The cell trajectory represented a set of vectors, each value of which was equal to the section of the route that the cell passed by in 1 second, and the direction together with the make-up of the light distribution (the angle of the light was from 180° to 0°) at the angle $\alpha_i$ of the slide plane. All vectors fixed after switching on the light source for 5 minutes were assigned to 12 sectors on a polar diagram, whose coordinates were angle $\alpha_i$ and the quantity of cells $N_i$ occupying the given sector (Fig. 4.2) [Posudin et al., 1991].

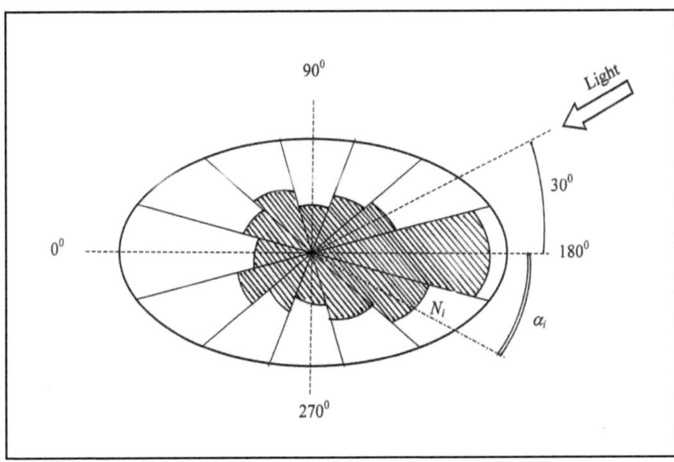


**Fig. 4.2.** Geometry of the interaction of light directed at an angle of $30^0$ to a slide plane containing algal cells that results in a change in the angular distribution of the moving cells. $N_i$ is the number of cells in $i$-sector; $\alpha_i$ is the angle that the sector forms with the projected light in the direction of the slide plane.

Several statistical methods were used for assessing the distribution of the moving cells (e.g., Rayleigh test, $V$-test method, $\chi^2$-criterion, method of moments) that permitted determinating the primary direction of movement and estimating phototopotaxis quantitatively [Mardia, 1972; Batschelet, 1981; Häder et al., 1981]. Thus, the angular distribution of cells moving due to the effect of light could be estimated quantitatively using vector $\vec{R}$, whose components were the mean values for the sum of sine and cosine of the measured angles $\alpha_i$, i.e $\vec{R} =$ ( $\frac{1}{N}\sum \sin \alpha_i ; \frac{1}{N}\sum \cos \alpha_i$ ), where $N$ is the number of the cells measured. The vector is characterized by differing values for length $r$ (in this case $0 < r < 1$) and angle $\theta$ that obey the following ratios [Häder et al., 1981; Mardia, 1972]:

$$r = \sqrt{\left(\frac{\sum \sin \alpha_i}{N}\right)^2 + \left(\frac{\sum \cos \alpha_i}{N}\right)^2} \; ; \; \theta = \frac{\arccos \dfrac{\sum \cos \alpha_i}{N}}{R}. \quad (4.1)$$

If the stimulating light is absent, the angular distribution of the moving cells should be isotropic in character; in this case $r = 0$, and angle $\theta$ is not indefinable.

The statistical method utilized was based on the estimation of an estimate of vector $r$ that is called the *Rayleigh test* which provides a comparison of the value for $z = Nr^2$ with a certain tabular value of $z_r$. The Rayleigh test permits estimating the significance of the difference in the observed angular distribution of the cells with illumination versus isotropic. The data obtained with the Rayleigh test were later used for a *V-test* that is preferable when there is a dominant cell movement direction as indicated by the angle $\theta_0$. Using the $V$-test method,

it is possible to estimate the value of $V = \sqrt{\dfrac{2}{N}}\, Nr\cos(\theta - \theta_0)$. At $V > 0$, a positive phototopotaxis is observed, while at $V < 0$, a negative. The ability to determine the sign of phototopo-
taxis is the primary advantage of the $V$-test in comparison with the Rayleigh test.

We also used a method for analyzing the angular distribution, the $\chi^2$- *criterion* [Rokit-
sky, 1973; Batschelet, 1981]. In this case the entire polar diagram is divided into three ($K = 3$)
groups. The first group included cells moving within a sector of 60° toward the direction of
the stimulating light; the second group, also within 60°, indicates cells moving in the opposite
direction; and the third group includes the remainder of the cells. The $\chi^2$- criterion is estimated as follows:

$$\chi^2 = \sum_{i=1}^{K} \frac{(o_i - e_i)^2}{e_i}, \qquad (4.2)$$

where $o_i$ are the observable frequencies, $e_i$ are frequencies of the expected series, and $K$ is the
number of the test groups. Under an isotropic distribution, the number of cells moving in the
above-mentioned directions would be 25, 25 and 50 %, respectively.

All the methods have the same disadvantage: it is impossible to estimate the bimodality of the angular distribution due to the effect of stimulating light. Objects, such as *Dunaliella*, are characterised by the ability of different cells to move simultaneously toward and
away from the light source. The *method of moments* [Mardia et al., 1972], in contrast, permits
determining the relative number $\bar{\mu}$ of cells moving toward and away from the light source
according to the equation:

$$( 2\bar{\mu} - 1 )A(K) = \bar{C} \cos \lambda_0 \bar{S} \sin \lambda_0, \qquad (4.3)$$

where $A(K)$ is the tabular value (Appendix 2.2 in Mardia [1972]); $\bar{C} = \dfrac{1}{N} \sum \cos \alpha_i$;

$\bar{S} = \dfrac{1}{N} \sum \sin \alpha_i$; $\lambda_0 = \dfrac{1}{2} \operatorname{arctg}(\bar{S} / \bar{C})$.

The phototopotaxis action spectrum is estimated quantitatively by the parameter $F(\lambda)$,
that characterizes the relative quantity of the cells moving toward the light source or away
from it. This parameter is defined as $F(\lambda) = R(\lambda) / N(\lambda)$, where $R(\lambda) = (n_\downarrow - n_\uparrow) / (n_\downarrow + n_\uparrow)$; $n_\downarrow$
and $n_\uparrow$ are the number of cells moving within 60° toward and away from the light source,
respectively, during the initial 5 min after the light is switched on. $N(\lambda)$ is the quantity of photons striking the sample, in this case $N(\lambda) \sim I \cdot \lambda$, where $I$ is the intensity of the stimulating light,
and $\lambda$ is the light wavelength. We have used the linear sections of the dependence of parameters $R$ and $F$ on the stimulating light intensity $I$ for different wavelengths [Posudin et al.,
1991].


### 4.1.4. Fourier Transform of Angular Distribution of the Cells

An angular histogram of moving *Dunaliella* cells in response to lateral illumination has been
used to determination the direction of the movement. However, in addition to the cells that are
demonstrating the proper photoorientation response, there are cells that are moving randomly
(e.g., due to scattering, collisions) and thereby impacting the overall angular distribution of
the cells. These cells create a "noise" that can be reduced by using a mathematical technique
based on the Fourier analysis of the histogram [Emerson, 1980; Zimmermann, 1981; Häder,
1986a; Häder and Lipson, 1986; Häder and Grienbow, 1987]. The technique involves the
construction of histogram, a fast Fourier transform, decomposition of the complex signal into
a set of harmonics that form a discrete frequency spectrum, and the elimination of high fre-

quency harmonics with low amplitudes. Reversing the technique (Fourier synthesis) results in a smoothed histogram. Fourier transform makes it possible to construct histograms devoid of noise and thereby estimate real tendencies in the direction of movement of the microorganism [Posudin et al., 1991].

The microscope field of vision was divided into 48 angular ($7.5^0$) sectors on a polar diagram and the trajectories of the moving cells were plotted into the appropriate sector on the polyethylene film covering the monitor screen. The angle $\alpha_i$ between the trajectory of movement of the cell and direction of the light stimulus was determined.

When restricted to a number of $N = 7$ for the Fourier transform of the angular distribution of *Dunaliella* it resulted in a coarse histogram. A greater number increases the noise, significantly distorting the shape of histogram [Posudin et al., 1991*b*].

## 4.2. Results of Measurement of Photomovement Parameters in Dunaliella

### 4.2.1. Photokinesis and Photokinetic Reactions

The primary goal of this stage of the investigation was to study photokinesis and photokinetic reactions in the two species of *Dunaliella*, in particular, the dependence of the absolute values and the relative change in cell linear and rotational velocity on changes in the characteristics of the light stimulus. The dependence of mean values of the velocity of linear movement in both *Dunaliella* species on the intensity of white light is given in Fig. 4.4. The mean values for the velocity of linear movement of the cells *(n* >10) are indicated by dots and their standard deviations. The maximum velocity values were 48 ± 2 μm/s for *D. salina* and 36 ± 2 μm/s for *D. viridis* and were obtained at 20°C within an illumination range of 150-550 lx which corresponds to a light intensity of 0.22-0.81 W/m$^2$ [Posudin et al., 1988]. Further increases in intensity resulted in a decrease in the linear velocity for both species. The kinetic responses of the *Dunaliella* species to the change in intensity essentially did not differ (Fig. 4.3).

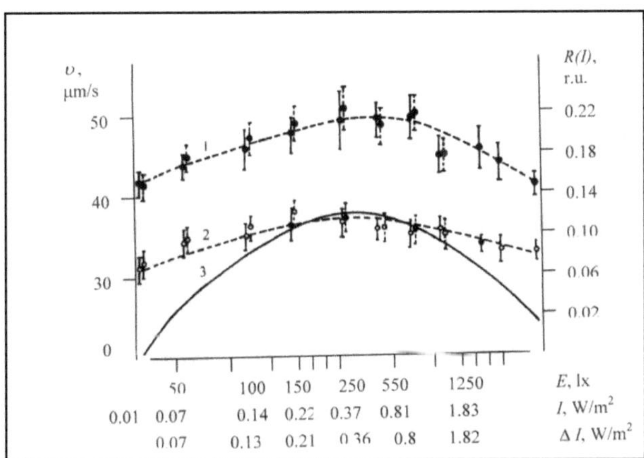

Fig. 4.3. Dependence of the linear velocity (μm/s) mean of *Dunaliella salina* ( 1 ) and *Dunaliella viridis* ( 2 ) on the intensity $I$ (W/m$^2$) or illuminance $E$ (lx) of white non-polarised light (dotted deviations) and polarized white light (solid deviations) and photokinetic reactions $R(I)$ (relative units) of both species ( 3 ) on the change in the intensity $\Delta I$ (W/m$^2$) of the light. Axis of ordinate indicates: the linear velocity of cell movement (left); photokinetic reaction (right). Axis of abscissa indicates the intensity $I$ of light, illuminance $E$ of the sample, and change of the intensity $\Delta I$. Vertical bars reprersent one standard error [Posudin et al., 1988].

The illumination intensity that resulted in a maximum velocity for *Dunaliella* movement was similar to the value reported for *Euglena gracilis* G.A. Klebs (illuminance 300 lx) [Wolken and Shin, 1958] but was lower that than at the optimum illumination for *Anabaena variabilis* Kütx. (1000 lx) [Nultsch, 1975].

Motile cells of cyanobacterium *Synechocystis* sp. also demonstrate photomovement in response to a light stimulus.Under vertical irradiation, *Synechocystis* decreased the frequency of the mean vectorial gliding speed depending upon the applied fluence rate, whereas the deviation distribution width of the speed increased. This strongly suggests the involvement of photokinesis.The maximum photokinetic activity at 420 nm and 680 nm supports the continuation hypothesis that the cyanobacteria's photokinesis is governed by the energy-generating chlorophyll pigments since the action spectrum for photokinesis resembles the absorption spectrum of chlorophyll [Chung Young-Ho et al., 2004].

In general the cell movement velocity of the both *Dunaliella* species is 1 to 2 orders of magnitude higher than that of blue-green and red algae that do not possess a flagellate apparatus [Nultsch, 1980]. It is to a certain extent lower than in some species of *Chlamydomonas* [Racey et al., 1981] and *E. gracilis* [Jahn and Bovee, 1968; Haupt, 1959]. It is important to note, however, that the flagellated algae display a velocity in the same order as flagellated bacteria that have different flagella structure [Gromov, 1985] and use a different source of energy for movement [Evtodienko, 1985] (Table 4.1).

**Table 4.1.** Velocity of linear movement of the cells in selected microorganisms

| Taxon | Velocity of cell movement, μm/s | References |
|---|---|---|
| *Porphyridium cruentum* | 0.05 | Nultsch, 1980 |
| *Anabaena variabilis* | 0.5 | " |
| *Dictyostelium discoideum* | 0.1 | " |
| *Micrasterias denticulata* | 1.0 | " |
| *Pinnularia nobilis* | 2.8 | " |
| *Nitzschia palea* | 6.0 | " |
| *Navicula peregrina* | 18.0 | " |
| *Dunaliella salina* | 48±2 | Posudin et al., 1988 |
| *Dunaliella viridis* | 36±2 | Posudin et al., 1988 |
| *Chlamydomonas* sp. | 200.0 | Racey et al., 1981 |
| *Euglena gracilis* | 160 | Wolken, Shin, 1958 |
| *Euglena gracilis* | 84 | Bovee, 1968 |
| *Euglena rubra* | 20 | " |
| *Thiospirillum jenense* | 87 | " |

| Taxon | Velocity of cell movement, μm/s | References |
|---|---|---|
| *Chromatium okenii* | 46 | " |
| *Pseudomonas aeruginosa* | 56 | " |
| *Escherichia coli* | 16 | " |
| *Bacillus licheniformis* | 21 | " |
| *Sporosarcina urea* | 28 | " |

We found that the cell movement velocity essentially does not change during measurement, which lasts for several minutes per sample. At the same time, the cells respond rapidly to any change in the intensity of the stimulating light's spectral composition.

A statistically significant variation in the linear movement velocity was found in both species in response to irradiation with polarized versus non-polarised white light at the same intensity (Fig. 4.3). This indicates a non-crystallite, non-dichroic photoreceptor in these species [Posudin et al., 1988].

The dependence of the velocity of movement of the two *Dunaliella* species on the wavelength of the stimulating light was not established in our experiments. Likewise, there has been contradictory data concerning the wavelength dependence of photokinesis in other algae species. Some authors [Ascoli, 1975; Häder and Häder, 1989] explain such a dependence in *E. gracilis* as being due to the participation of the photosynthetic pigments (chlorophyll *b* and/or β-carotene), however, there is experimental evidence that contradicts this hypothesis. For example, there is a positive photokinetic reaction of *Astasia longa* Pringsheim which is devoid of a photosynthetic apparatus [Mast, 1911]. Diehn (1973) reported a complete immobilization of *E. gracilis* in the blue region of the electromagnetic spectrum. The photokinetic reactions of *Chlamydomonas* have not been differentially established. Feinleib and Carry [1967] confirmed the absence of these reactions, while [Nultsch and Throm [1975] observed a positive photokinetic reaction in *Chlamydomonas* after prolonged dark adaptation. It is not altogether clear what photoreceptors are responsible for the photokinetic reactions in *E. gracilis* and *Chlamydomonas*, photosynthetic (as in prokaryotic organisms [Häder, 1979]) or specialized blue-light receptors. Quantitative estimation of the velocity of rotational movement of the two *Dunaliella* species indicates that maximum velocities are $0.52 \pm 0.04$ s$^{-1}$ for *D. salina* and $0.54 \pm 0.04$ s$^{-1}$ for *D. viridis* at 20°C (Fig. 4.4).

The cells can rotate around either the longitudinal axis of the ellipsoid or along the spiral trajectory. The maximum rotational velocities were obtained within the illumination range of 150-550 lx. This is in sharp contrast to the frequency of cell body rotation of *Chlamydomonas reinhardtii* P.A. Dang.which is in the 1-5 Hz range [Yoshimura and Kamiya, 2001].

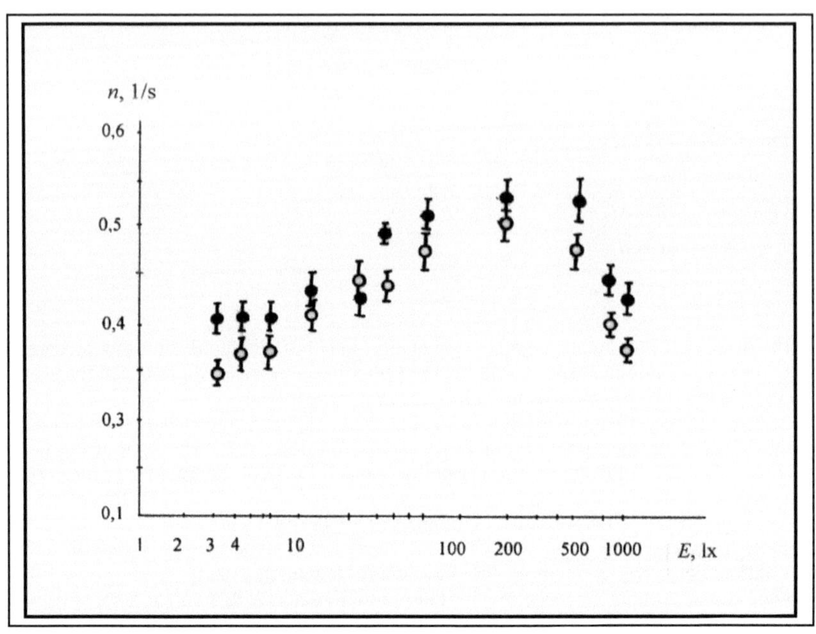


**Fig. 4.4.** Dependence of the velocity *n* of rotational movement of *Dunaliella viridis* ( - • - ) and *D. salina* ( - o - ) on the illuminance *E* of the sample by white light.

## 4.2.2. Phototopotaxis

The primary objective of this portion of the investigation was to establish the quantitative characteristics of phototopotaxis in the two *Dunaliella* species in response to the intensity and spectral composition of the light stimulus using the statistical methods described previously. Diagrams of the angular distribution for the *Dunaliella* species moving within the 12 sectors in response to different levels of illumination are shown in Fig. 4.5*a,b*.

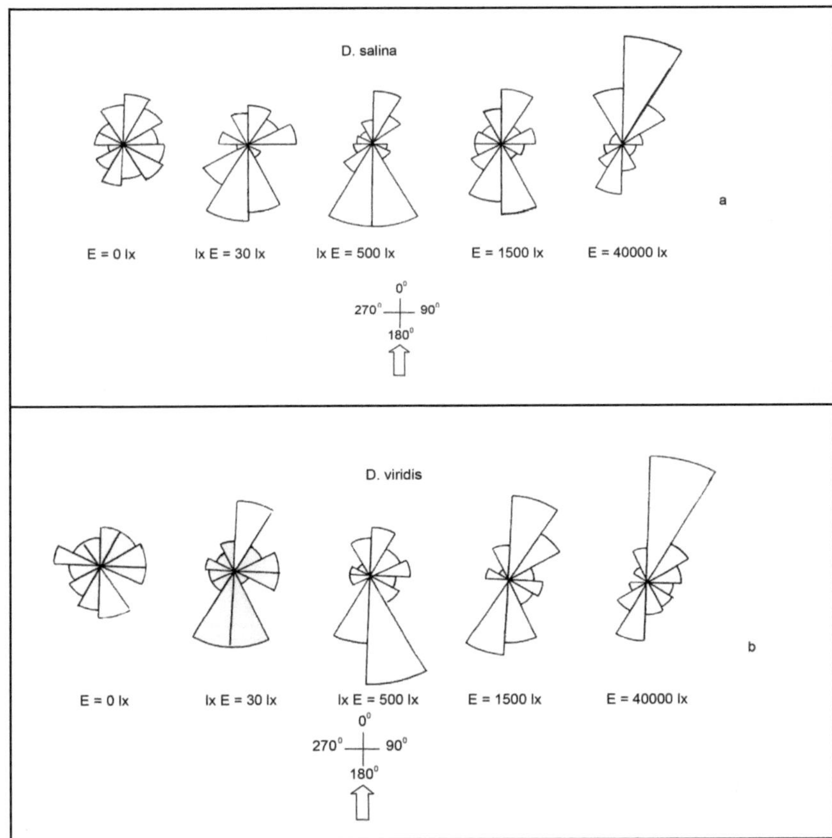


**Fig. 4.5.** Diagrams of the angular distribution of two species of *Dunaliella* Teod. under different levels of illumination: *a – D. salina*; *b – D. viridis*. The direction of the stimulating light ranges $180^0$ to $0^0$.

There is an absence of a light effect in an isotropic distribution of the moving cells while anisotropy of the angular distribution is observed due to lateral light. Statistical analysis of the data using various methods (Rayleigh test, $V$-test, $\chi^2$-criterion, method of moments) and the angular distribution of moving cells are given in Table 4.2 for *D. salina* and in Table 4.3 for *D. viridis*.

**Table 4.2.** Analysis using different statistical methods on the dependence of the angular distribution of moving cells of *Dunaliella salina* on illumination intensity (*NM* – non-meaningful, *M* — meaningful differences in the angular distribution from the anisotropic for the given level of significance *P*).

| Method of analysis | $E$, lx | | | | |
|---|---|---|---|---|---|
| | 0 | 30 | 500 | 1500 | 40000 |
| | $N$ | | | | |
| | 105 | 112 | 144 | 104 | 591 |
| Rayleigh test | | | | | |
| Parameters of vector $\vec{R}$ : | | | | | |
| Angle $\theta$ (degree), | – | 171 | 190 | 174 | 251 |
| $r$-value | 0.09 | 0.12 | 0.27 | 0.08 | 0.27 |
| Parameter $z=Nr^2$ | | | | | |
| Real value | 0.85 | 1.59 | 10.73 | 0.75 | 44.56 |
| Tabular value | | 2.99 | | 4.57 | |
| | | ( $P=0.05$ ) | | ( $P=0.01$ ) | |
| Probability level: | | | | | |
| for $P=0.05$ | NM | NM | M | NM | M |
| for $P=0.01$ | NM | NM | M | NM | M |
| $V$-test | | | | | |
| Values of parameter $V$: | | | | | |
| Real value | −1.30 | −1.78 | +4.63 | +1.22 | −8.51 |
| Tabular value | | 1.64 | | 2.32 | |
| | | ( $P=0.05$ ) | | ( $P=0.01$ ) | |
| Probability level: | | | | | |
| for $P=0.05$ | NM | M | M | NM | M |
| for $P=0.01$ | NM | NM | M | NM | M |
| $\chi^2$-criterion | | | | | |
| Values of parameter $\chi^2$: | | | | | |
| Real value | 3.73 | 4.59 | 18.75 | 1.83 | 16.5 |
| Tabular value | | 5.99 | | 9.21 | |
| | | ( $P=0.05$ ) | | ( $P=0.01$ ) | |
| Probability level: | | | | | |
| for $P=0.05$ | NM | NM | M | NM | M |
| for $P=0.01$ | NM | NM | M | NM | M |
| Method of moments | | | | | |
| Value of parameter $\bar{\mu}$ | 0.50 | 0.63 | 0.76 | 0.60 | 0.71 |

**Table 4.3.** Analysis using different statistical methods on the dependence of the angular distribution of moving-cells of *Dunaliella viridis* on illumination intensity (*NM* – non-meaningful, *M* — meaningful differences of the angular distribution from the anisotropic for the given level of significance *P*).

| Method of analysis | $E$, lx | | | | |
| --- | --- | --- | --- | --- | --- |
| | 0 | 30 | 500 | 1500 | 40000 |
| | $N$ | | | | |
| | 119 | 134 | 152 | 158 | 604 |
| Parameters of vector $\vec{R}$ : | Rayleigh test | | | | |
| Angle $\theta$ (degree), | – | 140 | 147 | 82 | 44 |
| $r$-value | 0.10 | 0.15 | 0.24 | 0.15 | 0.19 |
| Parameter $z=Nr^2$ | | | | | |
| Real value | 1.11 | 3.01 | 9.12 | 2.88 | 2.18 |
| Tabular value | | 2.99 | | 4.57 | |
| | | ( $P$ =0.05 ) | | ( $P$ =0.01 ) | |
| Probability level: | | | | | |
| for $P$ =0.05 | NM | M | M | NM | M |
| for $P$ =0.01 | NM | NM | M | NM | M |
| Values of parameter $V$: | $V$-test | | | | |
| Real value | +1.49 | +2.46 | +4.27 | –0.74 | –4.82 |
| Tabular value | | 1.64 | | 2.32 | |
| | | ( $P$ = 0.05 ) | | ( $P$ = 0.01 ) | |
| Probability level: | | | | | |
| for $P$ =0.05 | NM | M | M | NM | M |
| for $P$ =0.01 | NM | M | M | NM | M |
| Values of parameter $\chi^2$: | $\chi^2$-criterion | | | | |
| Real value | 8.38 | 4.82 | 12.25 | 3.26 | 13.95 |
| Tabular value | | 5.99 | | 9.21 | |
| | | ( $P$ =0.05 ) | | ( $P$ = 0.01 ) | |
| Probability level: | | | | | |
| for $P$ =0.05 | M | NM | M | NM | M |
| for $P$ =0.01 | NM | NM | M | NM | M |
| | Method of moments | | | | |
| Value of parameter $\overline{\mu}$ | 0.50 | 0.63 | 0.68 | 0.38 | 0.67 |

Our experiments allowed elucidating the dependence of angular distribution in the two species on the intensity of the lateral light in the range 0–40,000 lx (pH 6–7, 21 °C). Using the Rayleigh test, it was possible to confirm that the angular distribution of the cells differed from isotropic in response to lateral light at an illumination of $E$=500 lx and 40,000 lx for *D. viridis*. At $E$=30 lx the level of significance was $P$=0.01, Table 4.3. Similar results also are obtained using the *V*-test. The latter statistic proved to be more sensitive since it allowed analysis of the anisotropic distribution of *D. salina* ($P$=0.05 at $E$=30 lx, Table 4.2).

Comparison of the results between the two statistical tests (Rayleigh test and *V*-test) confirmed that under laboratory conditions *D. viridis* begins to demonstrate its photoorientation ability at lower levels of illuminance (30 lx) than *D. salina*, a response that is in accordance with observations in nature [Massjuk, 1973]. The *V*-test's advantage is its ability to distinguish positive ($V > 0$) from negative ($V < 0$) phototopotaxis [Mardia, 1972; Häder, et al., 1981]. Both species demonstrated a positive *V*-test for phototopotaxis at an illuminance of 30 and 500 lx and negative phototopotaxis at 40,000 lx (Fig. 4.6, Tables 4.2, 4.3). Isotropicity of the angular distribution of the cells at higher (i.e., 1500 lx) illumination may be explained by the transition of certain cells to a negative phototopotaxis. Therefore the number of cells moving toward and away from the light source was approximately equal.

It is necessary to note that the two species of *Dunaliella* demonstrate different behaviours in the concentrated saline basins found in southern Ukraine where the illuminance of the surface salt-water reaches 100000 lx or more. *D. viridis* cells concentrate in the shadowy parts near the bottom, while *D. salina* cells gathered near the surface where the illumination was highest [Масюк, 1973]. Differences in response between laboratory and natural populations is no doubt due in part to the fact that the cells differ in color, i.e., red vs. green. The accumulation of carotene in vegetative cells of *D. salina* under natural conditions increases considerably their tolerance to high illuminance. The transition from positive to negative phototopotaxis was not observed in natural populations of *D. salina in vivo* even under an illuminance intensity that exceeded by 100 times the threshold for the transition established under laboratory conditions [Massjuk,1973]. The unique behaviour of the red form of *D. salina* can be explained not only by increased tolerance caused by the elevated synthesis of carotene but also by the cells ability to accumulate a considerable amount of glycerol [Ben-Amotz et al.,1982; Enhuber and Glimmer,1980; Wegmann,1979; Posudin, Didyk, 2007]. Glycerol decreases the density of the cells, increasing their buoyancy. It also performs a protective function like carotene. The mechanisms that determine the difference in photomovement of red cells of *D. salina* in natural settings needs further elucidation.

The application of $\chi^2$-criterion (Tables 4.2 and 4.3) confirmed the conclusions derived from the Rayleigh test and *V*-test. The deviation of the real distribution of mobile cells in both species from isotropic was observed under a lateral illumination of 500 lx and 40,000 lx. The analysis made it possible to estimate with a high level of probability ($P$=0,05 for *D. salina* and $P$=0,01 for *D. viridis*) the relative quantity of cells moving in a given direction. This additional information can be obtained via application of the method of moments that permits estimation of the bimodality of angular distribution of the species due to lateral illumination, i.e. relative number $\bar{\mu}$ of cells moving toward and away from the light source corresponds to the level of illuminance at 500 lx and 10,000 lx (Tables 4.2 and 4.3).

Application of these statistical methods made it possible to determine the dominant directions during phototopotaxis, the level of the anisotropy of angular distribution of the moving cells under lateral illuminance (Rayleigh test), the sign of phototopotaxis (*V*-test), the relative number of cells moving at given direction ($\chi^2$-criterion), and the bimodality of angular distribution of the cells (method of moments). It is necessary, therefore, to use all the statistical methods to adequately characterize the movement. The presence of differences in photomovement parameters between the two species and their dependence upon the external conditions and the composition of aquatic medium indicates that such parameters can be used to identify ecological peculiarities among species and as taxonomic indices [Posudin, 2007].

It is useful to compare photopotaxis of *Dunaliella* with other algae. The phototopotaxis maximum occurs at an illumination of 125 lx in *Ochromonas danica* Pringsheim [Häder et al., 1981], 50 lx in *E. gracilis* [Häder et al., 1981], 50-200 lx in *Phormidium ambiguum* Gomont [Nultsch, 1962], 200 lx in *Nitzschia communis* [Nultsch, 1971], and 500-1000 lx in *C. reinhardtii* [Feinleib, 1974; Nultsch et al., 1971]. Illumination at which the transition from positive phototopotaxis to the negative occurs is over 125 lx in *Ochromonas* [Hader et al., 1981], 250 lx in *Euglena* [Hader et al., 1981], $10^3–10^4$ lx in *Phormidium* [Nultsch, 1962], 4000 lx in *Micrasterias denticulata* Brébisson ex Ralfs [Neuscheler, 1967] and about $10^5$ lx in *Chlamydomonas* [Nultsch et al., 1971]. The red form of *D. salina* that is high in carotene and glycerol, responds like *Chlamydomonas* displaying a high level of photoresistivity [Massjuk, 1973]. The diversity in photomovement characteristics among various algae species could be due to differences in experimental conditions, faulty experimental methodology, or actual biological differences among the taxa and strains studied. Based on the data obtained, *D. viridis* is more sensitive to the influence of weak light than *D. salina,* a result that concurs with our observations in nature [Massjuk, 1973].

At very high illumination (ca. 40,000 lx) the cells of both *Dunaliella* species demonstrate a negative phototopotaxis. The action spectrum of phototopotaxis $F(\lambda)$ of *D. salina* and *D. viridis* (Fig. 4.6) falls within 400-520 nm and is characterized by two maxima – at 410-415 nm and 465-475 nm.

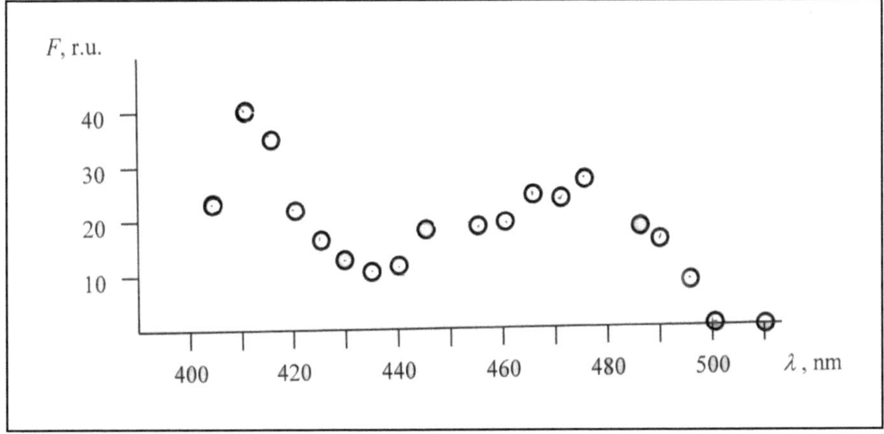

**Fig. 4.6.** Phototopotaxis action spectrum for two species of *Dunaliella* Teod. [Posudin et al. 1991].

A similar spectral interval for the phototopotaxis action spectrum in *D. salina* (400-520 nm) was observed by other authors [Wayne et al., 1991]. They determined that the action spectrum for phototaxis had a maximum at 450-460 nm and proposed that carotenoproteins or rhodopsins acted as the photoreceptor pigments. The different spectral positions of two maxima (i.e., near 460 nm and 520 nm) can be explained by differences in the alga strain tested and/or or differences in salt condition within the medium [Wayne et al., 1991].

In the 400-550 nm range, species of *Peridinium, Gonyaulax, Platymonas, Stephanoptera* [Halldal, 1958], *Cryptomonas* [Watanabe and Furuya, 1974], and *Chlamydomonas* [Nultsch, 1971] have been shown to orient themselves relative to the direction of the light. In the green alga *Ochromonas danica*, both green and bleached cells accumulate under blue light, though only green cells do so under red light [Di Pasquale et al., 1980].

### 4.2.3. Results of Fourier Transform of Angular Distribution of the Cells

The procedure of Fourier transform of the angular distribution for different levels of illuminance makes it possible to segment the angular histogram into discrete sets of harmonics with different amplitudes and phases. Each amplitude corresponds to the quantity of cells moving in a certain direction that is determined by corresponding phase. An inverse Fourier transform, in contrast, avoids the high frequency harmonics.

If the light stimulus is absent ($E = 0$ lx), the *Dunaliella* histogram for the angular distribution is characterized by maxima in virtually each direction (Fig. 4.7$a$) though the amplitudes of the first 7 harmonics did not demonstrate sufficient differences (Fig. 4.7$b$). The phases of these harmonics (Fig. 4.7$c$) indicate movement of the cells in the direction of $135^0$, $235^0$, $180^0$ such that it is difficult to distinguish a dominant direction (Fig. 4.7$d$).

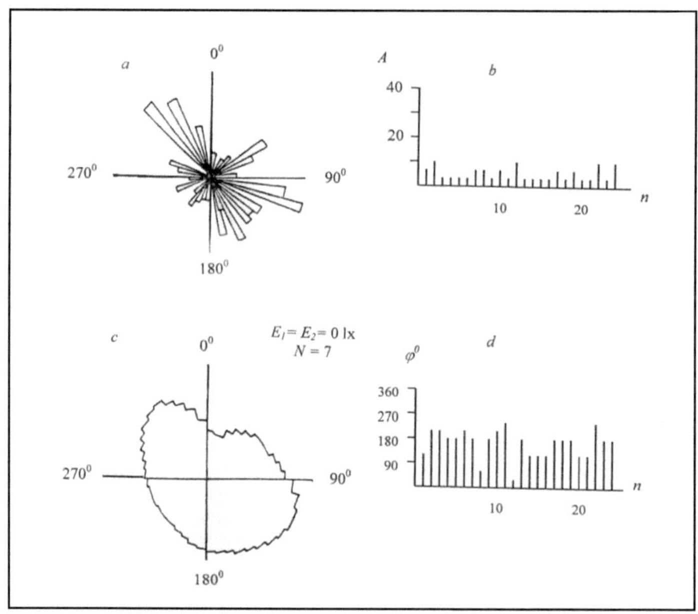

**Fig. 4.7.** Fourier-transform of the angular distribution of motile cells of *Dunaliella* Teod. in the absence of a light stimulus ($E = 0$), where: $a$ – real histograms of angular distribution; $b$ – amplitudes of harmonics ($A$); $c$ – histograms of an inverse Fourier-transform for the first 7 harmonics; $d$ – the phases of harmonics; $n$ – number of sectors in the angular distribution [Posudin et al., 1991].

If the light stimulus ($E = 500$ lx) is increased, the histogram of angular distribution indicates a prominent maxima in the real distribution of moving cells (Fig. 4.8*a*). It is therefore possible to distinguish the first and second harmonics (Fig. 4.8*b*). The phases of the harmonics indicate the cell movement in a direction of $270^0$ with the population more oriented toward the source of light (Fig. 4.8*c*). The inverse Fourier transform makes it possible to observe the primary movement of the cells in a smoothed histogram indicating a positive phototopotaxis for *Dunaliella* (Fig. 4.8*d*).

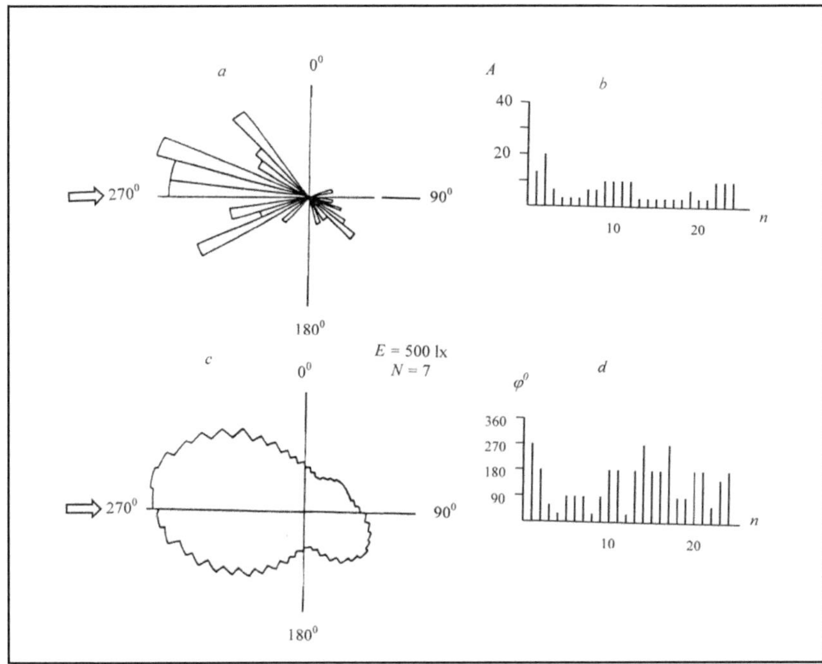



**Fig.4.8.** Fourier-transform of the angular distribution of motile cells of *Dunaliella* Teod. at an illuminance $E$ of 500 lx [Posudin et al., 1991].

Increasing the illuminance up to 40,000 lx resulted in the appearance of at least five intense maxima in the distribution (Fig. 4.9*a*). It is possible to distinguish the first four harmonics (Fig. 4.9*b*) which have phases close to zero (Fig. 4.9*c*). The smoothed histogram for these harmonics indicates a negative phototopotaxis for *Dunaliella* (Fig. 4.9*d*).

Fourier transform establishes the dependence of angular distribution of the cells on the level of lateral illuminance by white light. The advantage of this method is the ability to construct histograms for the angular distribution of the cells that are deprived the random factors and to estimate the dominant tendencies in the direction of movement of the organisms in response to the light stimulus at varying intensities.

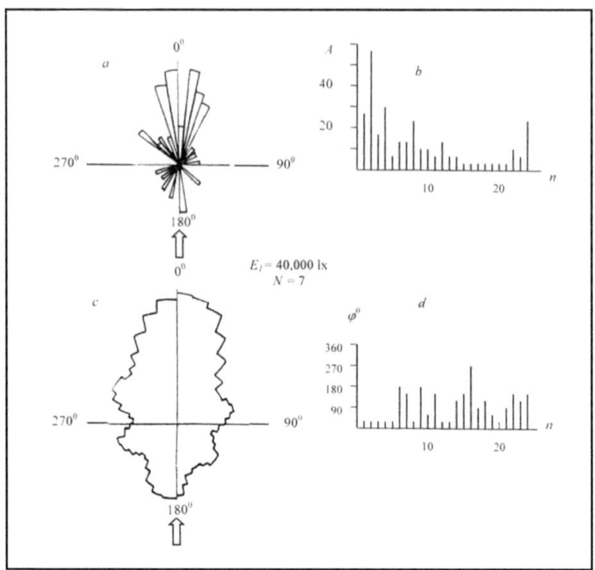


**Fig. 4.9.** Fourier-transform of the angular distribution of motile cells of *Dunaliella* Teod. at an illuminance *E* of 40,000 lx.

## 4.3. Summary

Species of *Dunaliella*, as well as other flagellate algae move freely in an aquatic environment in response to light, i.e. *photomovement* (*phototaxis*). The species are capable of photokinetic and photovector reactions, even though photokinetic reactions in *Chlamydomonas* are challenged in the literature. We did not observe photophobic reactions in *Dunaliella* (in contrast to *Chlamydomonas* and *Euglena*), though previous reports suggest the possibility of such reactions [Wayne et al., 1991].

An average velocity of cell translational movement in *Dunaliella*, both of which are hyperhalobic species, was 36 ± 2 μm/s (*D. viridis*) and 48 ± 2 μm/s (*D. salina*). The average velocity of movement of the two species varied over a wide range, thus eliminating the possibility of making distinctions between the species using the parameter. The modal value for the average velocity of movement of *Dunaliella bioculata* Butcher was 105 ± 5 μm/s. These values exceed by 1-3 orders of magnitude those in microorganisms not possessing a flagellar apparatus and are within the limits known for others flagellates, both prokaryotic and eukaryotic.

However, the average velocity of movement of both species of *Dunaliella* was lower (sometimes by an order of magnitude) than those in the marine species, *D. bioculata* and the freshwater species *C. reinhardtii*, and *E. gracilis*. These differences are most likely caused by differences in viscosity of the media.

The average velocity values for rotary movement of cells in the *Dunaliella* species were nearly identical i.e., 0.52±0.04 rotations (revolutions) per second in *D. salina* and 0.54±0.04 rotations per second in *D. viridis*. However, these values are lower than in the freshwater species *C. reinhardtii*.

The maximum values for the average velocity of linear and rotary photomovement in *Dunaliella* cells were observed within the limits of the white-light intensity (0.22–0.81 W/m$^2$), illumination (150–550 lx), temperature (20–30°C), and pH (8). No statistically significant variation in linear movement velocity was found in either species due to irradiation with polarized light versus non-polarized white light of the same intensity (Fig. 4.3). This suggests a non-crystallite non-dichroic photoreceptor in the species [Posudin et al., 1988].

Phototopotaxis in the two hyperhalobic species of *Dunaliella* under laboratory conditions was observed at an illuminance of 500 lx (positive) and 40,000 lx (negative). The transition from positive to negative phototopotaxis occurred at 1,500 lx. These parameters are within the limits known for other algae species. However, sensitivity thresholds to weak and strong illuminance and transitions from positive to negative phototopotaxis differed substantially among alga species. This allows assessing shade-tolerance, sun-tolerance, and resistance to high light exposure among species.

*Dunaliella viridis* is more sensitive to weak light (30 lx) than *D. salina*, a finding that corresponds to behavioral peculiarities between the two species in nature. In laboratory cultures both species were more sensitive to high illuminance than *C. reinhardtii* which has a transition to negative phototopotaxis at 100,000 lx.

The transition from positive to negative phototopotaxis in the *Dunaliella* species differs from chlamydomonads. Unlike chlamydomonads, the change in flagellar beating from ciliary to undulate mode was not observed in either species. The beating of only one flagellum was observed which caused a turning of the cell and movement in the direction opposite to the direction of the light source.

In Crimean hyperhaline watersheds under high illuminance (>100000 lx), natural populations of the red form displayed a complete absence of negative phototopotaxis. Thus, hyperhalobic species (*D. salina* and *D. viridis*) differ in their sensitivity to both high- and low-light intensities explaining in part differences in the ecological niches occupied by the species.

The action spectrum for phototopotaxis was identical for the two *Dunaliella* species. It is between 400–520 nm and has two maxima: at 410–415 nm and 465–475 nm. The phototopotaxis spectrum for *Dunaliella* differs somewhat from those of *C. reinhardtii* and *Haematococcus pluvialis* Flotow that display a wide band in the 400–600 nm range and a maximum at 500 nm. This indicates differences in their photoreceptor systems and the composition of their photoreceptor pigments.

# Chapter 5

# Effect of Abiotic Factors on Photomovement Parameters of *Dunaliella*

Motile microorganisms are exposed to the influence of a number of abiotic factors such as mechanical (mechanical shocks, hydrostatic pressure), gravitational, thermal, electromagnetic (ultraviolet, visible, infrared, and microwave radiation), electrical and magnetic fields, and ionizing radiation. They are also influenced by the chemical, gas and ion composition and the pH of the aquatic media, biogenous elements and other organisms, each of which can affect their photomovement responses [Jahn and Bovee, 1968; Marbach and Mayer, 1970; Kritsky, 1982; Sineschekov and Litvin, 1982; Colombetti et al., 1982]. The effects of light on the photomovement parameters of two species of *Dunaliella* were described in the Chapter 4.

Motile microorganisms respond to various abiotic factors in their environment gravitating toward conditions that enhance their survival and population growth. Thermal [Gimmler et al., 1978; Lynch, 1984; Lynch et al., 1984; Poff, 1985; Norman and Thompson, 1985; Yang, 1988; Ramazanov et al., 1988; Ben-Amotz, 1996; Krol et al., 1997] and chemical [Berg, 1985] gradients, gravitational [Häder,1987*b*], electrical [Mast, 1911; Häder, 1977] and magnetic [Esquivel and de Barros, 1986; Yamaoka et al., 1992] fields, solar radiation [Nultsch and Häder,1988; Richter et al., 2007], and ionizing radiation [Saraiva, 1972] have all been shown to modulate their behavior. Several articles have elucidated the effect of multiple abiotic factors on the physiology and behaviour of algae [Mil'ko, 1963; Jimenez and Niell, 1991; Thakur and Kumar, 1998*b*; Gomez and Gonzalez, 2005; Zhang et al., 2006].

The effects of abiotic factors such as temperature, electrical fields, medium pH, and ultraviolet and ionizing radiation, as well as the influence of physical factors (i.e., optical radiation, temperature, electrical fields) on the photomovement parameters of two species of *Dunaliella* are discussed in this Chapter.

## 5.1. Effect of Temperature

The effect of temperature on cell movement velocity was assessed over a temperature gradient from 16 to 35 °C using a controlled temperature bath and microscope. Precise measurement of the temperature of the algal suspension has shown that changes in temperature caused by switching on the lights did not exceed one hundredth of a degree. This observation provides evidence for the absence of any significant sample heating due to the light treatments [Posudin et al., 1988]. The response to the change in the temperature is calculated as $R_t = (\upsilon_t - \upsilon_0)/\upsilon_0$, where $\upsilon_t$ and $\upsilon_0$ are cell movement velocities at the given and minimum (16 °C) temperature (under the conditions of our experiments), respectively [Posudin et al., 1988].

Maximum values for cell velocity were reached around 25 °C. Kinetic reactions for both *Dunaliella* species due to the temperature change (16 °C initial temperature) did not differ significantly (Fig. 5.1).

The value of photokinetic reaction to the increasing temperature between 16-25 °C increased until $R(t) = 0.19$. The sharpest increase occurred between 16 and 20 °C. Further increases in temperature up to 35 °C lead to a diminishing value for the photokinetic reaction $R(t) = 0.12$ (see Fig. 5.1).

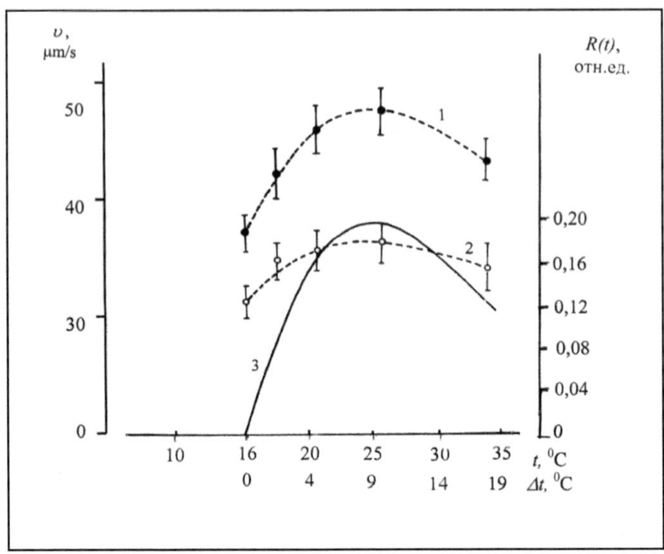


**Fig. 5.1.** Dependence of mean linear velocity υ of *D. salina* ( 1 ) and *D. viridis* ( 2 ) on the temperature *t* and kinetic reactions *R (t)* in both species ( 3 ) on the change of temperature Δ*t* [Posudin et al., 1988].

The dependence of linear cell movement velocity on temperature can be in part explained by a decrease in the viscosity of medium with increasing temperature between 16 to 25 °C and a progressive inhibition of the flagellar apparatus between 25 and 35 °C. It is important to note that the temperature within this range did not affect phototopotaxis.

## 5.2. Effect of Electrical Fields

A special rectangular cuvette was constructed for investigating the effect of electrical fields on photomovement. The cuvette consisted of an observation chamber (40×10×4 mm) that contained the algal suspension, two electrode chambers that were separated from the observation chamber by gelatine and a 0.3 M solution of *KCl* to prevent electrolysis. Gold electrodes, positioned at a 90° angle to the lateral light source, were attached to an electrical source. The distance between the parallel electrodes was 30 mm [Posudin et al., 1991].

Fourier-transform (see Section 4.1.4) was used to determine the level of phototopotaxis inhibition by the electrical field. The method allowed analyzing changes in the amplitude of the principal harmonics to elucidate the possible participation of membrane electrical potentials in algal photomovement.

Application of an electrical field (10-20 V/cm) inhibited phototopotaxis in *D. salina* during lateral illumination with white light (500 lx illuminance). Fig. 5.2 displays histograms of angular distribution in the absence and presence of an electrical field (20 V/m). Fourier analysis allowed estimating the reduction in amplitude of the principal harmonics (the first harmonic 3 times, the second 3.7 times, etc.). The histogram of angular distribution demonstrates the inhibition of phototopotaxis in *D. salina* when the electrical field was switched on and its recovery 2 minutes after the electrical field was switched off.

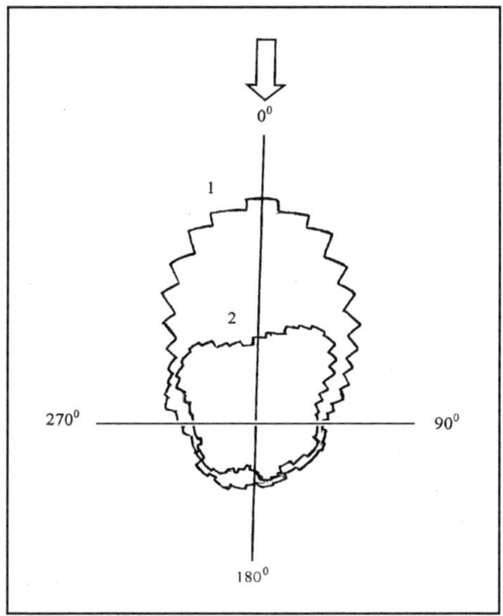


**Fig. 5.2.** Effect of an external electric field of 20 V/cm applied to the algal suspension on the angular distribution of the cells and intensity of phototopotaxis of *D. salina*: 1 – field is switched on; and 2 – field is switched off. Arrows indicate the direction of propagation of the light stimulus (illuminance 500 lx) [Posudin et al., 1991].

External electrical fields influence light-induced movement of microorganisms. The effect of an electrical field can be explained via the participation of bioelectrical processes in photo-movement, a phenomenon that has been confirmed by a number of investigators [Marbach and Mayer, 1971; Häder, 1977; Litvin et al., 1978; Nultsch and Häder, 1979; Sineshchekov and Litvin, 1982; Dolle and Nultsch, 1988]. Marbach and Mayer (1971), for example, demonstrated that under normal conditions, 85 % of the cells aligned to the light; this number was reduced to 20 % when the electrical field was switched on. In addition, the effect was reversible; when the electrical field was switched off, *Chlamydomonas* recovered to 73 % aligned cells. The authors suggested that the phototactic response, activated through the transmission of an electrical stimulus from the receptor site to the flagella, is disturbed by an external electrical field.

Participation of bioelectrical processes in the photomovement of various kinds of microorganisms is supported by the results of a number of investigations. Extracellular measurement of biopotentials supports the connection between bioelectrical responses and photomovement in the green alga *Haematococcus pluvialis* Flotow [Litvin et al., 1978].

The effect of external electrical fields on photo-accumulations of *Phormidium uncinatum* (Ag.) Gom. in light traps has also been studied [Häder, 1977; Nultsch and Häder, 1979]. The amplitude (3-7 V) of the electrical field and the shape of electrical waves affect the motility of the trichomes. They proposed that the sensory transduction of photophobic reactions in blue-green algae is mediated by changes in endogenous membrane potential.

Phototopotaxis in *Chlamydomonas reinhardtii* P.A. Dang. is reversibly inhibited by an applied electrical field [Nultsch and Häder, 1979]. Nultsch (1983) proposed that absorption of light by the photoreceptor molecules is accompanied by their excitation and conformational alterations in the photoreceptor proteins. This leads to opening of calcium channels in the plasma membrane of the stigma and corresponding increase in the flow of calcium ions into the cell. The plasma membrane is depolarised locally resulting in a corresponding opening of calcium channels in the flagellar membrane, increasing the flow of calcium ions into the flagellar axoneme increasing flagella beating.

Our experimental results and those described in the literature support the hypothesis that the inhibition of phototopotaxis in green algae by an external electrical field is due to one of two phenomena – phototopotaxis and galvanotaxis, as was observed in *Chlamydomonas* [Marbach and Mayer, 1971; Nultsch and Häder, 1979; Dolle and Nultsch, 1988] or the participation of light induced changes in membrane potential during photomovement was demonstrated in *Haematococcus* [Litvin et al., 1978; Sineshchekov and Litvin, 1982]. It is evident that photomovement in microorganisms is affected by external electrical fields and the same relationship between photoresponse and changes of photoinduced potentials have been found in bacteria and protozoa.

Our data on the inhibition of phototopotaxis in *Dunaliella* by external electrical fields and those described for *Chlamydomonas* [Marbach and Mayer, 1971; Dolle and Nultsch, 1988] and *Haematococcus* [Litvin et al., 1978; Sineshchekov and Litvin, 1982] support the idea of the participation of an electrical potential (in particular, an action potential that appears as a response to the light stimulus) occurring during the photoregulation of algal movement. The application of an external electrical field disturbs the propagation of the potential from the receptor to the flagellar apparatus causing an inhibition of phototopotaxis.


## 5.3. Effect of pH

The effect of the concentration of hydrogen ($H^+$) ions in an aquatic medium, measured in pH units, is a critical environmental factor influencing the viability of organisms. There is a wealth of literature on the effect of medium pH on growth, development, reproduction, biomass production of algae [Massjuk and Yurchenko, 1962; Weggmann, 1968; Weggmann and Metzner, 1971; Malis-Arad et al., 1980; Rao, et al., 1982; Goldman et al., 1982 *a,b*; Ghasi, et al., 1983; De Busk and Ryther, 1984; Lukas et al., 1986; Gimmler and Weis, 1992; Lustigman et al., 1995; Thakur and Kumar, 1998; Thakur et al., 2000] and their distribution in nature [Massjuk, 1973; Lopez-Archilla and Amils, 1999; Lopez-Archilla et al., 2001; Topics…, http//www.bio.uni-potsdam.de/oeksys/fsvte.htm]. This is especially so for algae living in highly acidic environments [Gross, 2000]). Changes in ultrastructure [Ma et al., 1999], biomass and  pigment synthesis [Ghazi et al.,1983; Celekli and Doenmez, 2000], photosynthetic rate [Wegmann, 1968; Wegmann and Metzner, 1971; Gimmler and Weis, 1992], fermentation activity [Myronuk et al., 1980], ion transport [Balnokin et al., 1983; Lucas et al., 1986; Pick, 1992], $Ca^{2+}$ [Quarmby, 1996] and $Na^+$ [Weiss and Pick, 1990] influx, phosphate uptake [Hirsch et al., 1993], velocity of cytoplasmic movement [Masashi and Teruo, 1982], accumulation of chemicals [Yamaoka et al., 1992]  and heavy metals [Riisgaard, 1980; Takimura et al., 1989; Turker and Balcioglu, 2001; Sacan et al., 2001], changes in zeta potential [Gimmler et al., 1991], and cell harvesting [Horiguchi et al., 2003] have been documented.

The relationship between the medium pH and the concentration of hydrogen ions has been studied inside algal cells [Beardall and Entwisle, 1984; Gimmler et al., 1988; Goyal and Gimmler, 1989; Katz et al., 1991, 1992; Weiss and Pick, 1996; Braun and Hegemann, 1999] within the cytosol and vacuole [Kuchitsu et al., 1989]. Effect of trans-membrane electrical potential [Remis et al., 1992, 1994] and membrane proton pumps in acidophile species [Sanders et al., 1981] on the regulation of intracellular pH is also under investigation. Photoinduced pH changes in suspensions of an acid-resistant green algae were discussed by Remis et al. (1994). Special attention has been paid to mechanisms, including those at the molecular level, that provide $H^+$ homeostasis within the cytosol of acidophile and hyperhalobic species of algae [Gimmler et al., 1988; Sekler et al., 1991, 1994; Gimmler et al., 1991; Gimmler and Weis, 1992; Pick, 1992, 1999; Weiss and Pick, 1996; Ohta et al., 1997; Messerli et al., 2005; Topics, http://www.bio.uni-potsdam.de/oeksys/fsvte.htm; Pick et al., http://www.weizmann.ac.il/ Biological_Chemistry/ scientist /Pick/uri_pick.html; Pick et al., http://bioinformatics. weizmann.ac.il/_Is/uri_pick/uri_pick.html].

While phototrophs of representatives of the genus *Dunaliella* can grow over much of the wide range of pH values found in nature [Raven, 1990], little is known about the influence of pH on photomovement parameters [Nultsch, 1977]. The pH range in which *D. salina* [Massjuk and Yurchenko, 1962] and *D. viridis* [Baas-Becking, 1930] cells maintain their motility has been determined, though the dependence of motility, velocity of movement and phototopotaxis on medium pH has not been studied. As a consequence, the dependence of various photomovement parameters in *D. salina* on the pH of the medium are presented in this section [Massiuk and Posudin, 2007].

The dependence of photomovement parameters in *D. salina* (velocity of linear movement $\upsilon$ in μm/s, phototopaxis $F$ and relative quantity of mobile $N_m/N_0$ or immobile $N_{im}/N_0$ cells) on pH was investigated between pH 2.95 and 9.50 created by varying the concentration KOH or HCl. Medium pH was measured using a pH-meter at the end of the first, seventh and twentieth day after the onset of the experiment.

The cells displayed motility between a pH of 2.95 and 9.50 at the end of the first day. The cells were immobile, deformed and completely destroyed at pH 2.95 ($N_{im}/N_0$ =100 %). The algae varied in their photomovement, exhibiting different pH-optima for the various parameters (Fig. 5.3). Optimal pH for mobility ($N_m/N_0$ = 100 %; $N_{im}/N_0$ = 0 %) was 6.8; for phototopotaxis ($F$ = 0.7) 7.35; and for linear velocity of movement ($\upsilon$ = 47 ± 2 μm/s) 8.00 (Fig. 5.3).

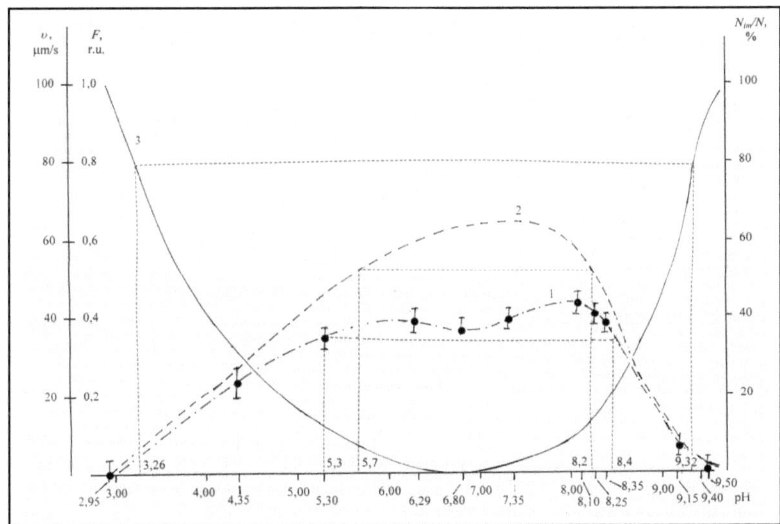

**Fig. 5.3.** Dependence of the linear velocity $\upsilon$ of movement (1), phototopotaxis $F$ (2), and relative quantity of immobile $N_{im}/N_0$ (3) cells of *Dunaliella salina* on the pH of the medium at the end of the first day of cultivation. The left vertical axis denotes values of linear velocity $\upsilon$ of movement (μm/s) and levels of phototopotaxis $F$ (r.u.); the right vertical axis denotes the relative quantity of immobile $N_{im}/N_0$ cells; the horizontal axis indicates the pH.

The relative quantity of immobile cells $N_{im}/N_0$ did not exceed 80 % of the maximum value between pH 3.26 to 9.32 (Fig. 5.3). Likewise, the linear velocity of cell movement did not exceed 80 % of the maximum value between pH 5.30 and 8.40 and phototopotaxis $F$ did not exceed 80 % of the maximum (Fig. 5.3) between pH 5.70 and 8.20. Both the linear velocity and phototopotaxis declined rapidly at a greater or lower pH.

Thus, various photomovement parameters possess either different pH optima or sensitivity pH extremes. Phototopotaxis $F$ was the most sensitive and motility the least. The linear velocity of movement $v$ was intermediate in sensitivity between the two parameters.

The results support our contention that the control of various photomovement parameters differ and are mediated by way of different mechanisms [Posudin et al., 1992, 1995, 2004; Massjuk et al., 2006]. The wide range in tolerance to medium pH (2.95 to 9.50) can be explained through the action of intracellular mechanisms whose homeostasis is dependent upon the concentration of $H^+$ ions in the cytosol at pH 7.1 (Pick et al., http://www.weizmann.ac.il/ Biological_Chemistry/ scientist /Pick/uri_pick.html). In addition, the optimum pH varies during algal development (Table 5.1).

**Table 5.1**. Changes of pH of the medium during cultivation of *Dunaliella salina* in 20-days experiment [Massjuk and Posudin, 2007].

| Days | *Values of pH* | | | | | | | | | | | |
|---|---|---|---|---|---|---|---|---|---|---|---|---|
| 1st | 2.95 | 4.30 | 5.30 | 6.29 | 6.80 | 7.35 | 8.10 | 8.25 | 8.35 | 9.15 | 9.40 | 9.50 |
| 7th | – | 5.10 | 6.00 | 6.70 | 7.20 | 7.70 | 8.10 | 8.20 | 8.30 | 8.25 | 9.10 | 9.20 |
| 20th | – | 6.50 | 7.25 | 7.70 | 7.87 | 8.07 | 8.15 | 8.15 | 8.17 | 8.35 | 8.47 | 8.42 |
| Difference between values of pH at 20th and 1st days | – | 2.20 | 1.95 | 1.50 | 1.07 | 0.72 | 0.05 | –0.10 | –0.18 | –0.8 | –0.93 | –1.08 |
| Foot-note: hyphens mean the absence of motile cells of alga. | | | | | | | | | | | | |

As illustrated in Table 5.1, over the course of the experiment there were changes in pH of the medium. In acidic, neutral and weak alkaline media, the pH increased with time, however, in alkaline and strong alkaline media the pH decreased. Maximum changes in pH occurred at pH 4.30 (+2.20) and 9.50 (–1.08). The lowest change in pH ($\Delta$pH = 0.05) during the course of the experiment was at a pH of 8.1 (Table 5.1). The pH interval between 6.50 and 8.47 is suitable for survival of this species; *D. salina* growing in nature has been observed at a similar pH range (pH 6.5-9.5) [Massjuk, 1973].

The optimum pH for motility and phototopotaxis of *D. salina* does not coincide with the optimum pH values for growth [Massjuk and Yurchenko, 1962] and catalase activity [Myronjuk et al., 1980] of the species. The pH optimum for linear velocity of cellular movement (8.0) was observed in a medium that underwent minimal changes in pH during cultivation and coincides with the pH for optimum growth (8-9) [Massjuk and Yurchenko, 1962]. The optimum pH for linear movement velocity and minimum change in pH during growth may be a useful criterion for determining optimum conditions for the commercial production of hyperhalophobous algae for carotenoids.

The pH of the medium changed during the course of the experiment (Table 5.1) with the range in pH becoming narrower between pH 5.10 and 9.20 by the end of the 7th day. D. salina displayed motility through out the entire initial pH range (2.95 to 9.50). With the exception of pH 2.5, the initial pH of the medium was reduced to 6.5 to 8.47 with pH 2.50 being the only one not altered. The adjusted pH range was more favorable for growth of the species.

The level of sensitivity of the cells to the hydrogen ion concentration and their pH optimum for various photomovement parameters differed indicating the possible existence of different mechanisms governing the parameters. Photomovement parameters may be used as an indicator aquatic media quality during the commercial production of algae for carotenoids [Massjuk & Posudin, 2007].


## 5.4. Simultaneous Effect of Several External Factors

In nature microorganisms that display photomovement are commonly exposed to multiple external factors that modulate their behavior. Ascertaining their joint effects on photomovement parameters is often a complex task. The simultaneous effect of external factors such as light, air temperature and electrical fields on different photomovement parameters (linear velocity $\upsilon$ and phototopotaxis $F$) were assessed in two species of *Dunaliella*. Samples were selected after a trice-repeated sequence of careful mixing of the algae suspensions. The study of the effect of an electrical field on photomovement was conducted in a plexiglass cuvette with two electrodes (see Section 5.2 for additional details). The cuvette was positioned on a microscope slide table in a controlled temperature chamber equipped with a transparent window for the introduction of light. A beam of white light was directed at a 30° angle onto the surface of the slide.

The effect of air temperature on photomovement parameters was studied in a thermostatically controlled chamber with a temperature precision of ± 1°C. The chamber was equipped with a TPK-ZP-128 electrocontact thermometer.

Two sets of external factors were tested: minimal $(l = 100$ lx, $t = 18$ °C and $e = 0$ V/cm) and maximal $(L = 500$ lx, $T = 30$ °C and $E = 2.4$ V/cm). Measurements of $\upsilon$ and $F$ were collected for all possible combinations of the external factors: *l-e-t, L-e-t, l-E-t, L-E-t, l-e-T, L-e-T, l-E-T, L-E-T*. The data was processed using plural regression [Martynenko et al, 1996; 2000] which allowed separating the effect of individual factors and their combinations on the cell photomovement parameters [Melnikov at al., 1972; Mosteller et al., 1978]. Using a linear regression model, the magnitude of model coefficients were assessed using Fisher's test and adequacy coefficients by Kohren criteria. Statistically insignificant coefficients were assessed by evaluating their magnitude. The regression equations contained only terms with statistically significant coefficients.

The effect on linear movement velocity $\upsilon$ and phototopotaxis $F$ in the two algae species in all possible combinations of the external factors (light, temperature and electrical field) are presented in Tables 5.2 and 5.3, where $X_1$ – level of illumination, $X_2$ – electrical field, $X_3$ – temperature, $<\upsilon>$ and $<F>$ are average values for photomovement parameters, and $S_\upsilon$ and $S_F$ are standard deviations.

Dependence of linear movement velocity $\upsilon$ in both *Dunaliella* species on external factors (level of illumination $L$, air temperature $T$ and electrical field $E$) is described by the following regression equations:

*D. salina:* $\upsilon = 34.8 + 0.6\,L - 0.62\,E - 1.66\,T + 1.3\;(L{\times}E) - 0.6(L{\times}E)$
$\qquad - 0.8\;(L{\times}E{\times}T);$ (5.1)

*D. viridis:* $\upsilon = 37.6 - 0.61 + 2.0\,T - 2.6\;(E{\times}T) + 0.53\;(L{\times}E{\times}T).$ (5.2)

The most important contributor in the equations was made by a free term (the uninfluenced movement velocity of the cells) that was independent of the external factors and was equal to 34.8 for *D. salina* and 37.6 for *D. viridis*. Weight coefficients of variables $L$, $E$ and $T$ define the contribution of each of the factors to the change in movement velocity $\upsilon$. In equation (5.1), the greatest effect on movement velocity of *D. salina* was due to temperature. Increases in temperature lead to deceleration (the coefficient was −1.66). Light and an electrical field modulated the other factors. *D. salina* accelerated when exposed to increased light (+0.6) while the electrical field decelerated movement (−0.62). The interaction of the two factors was also significant (+1.3) and resulted in acceleration. The interaction effects of temperature and light (−0.6) and the three-way interaction of temperature, light and an electrical field (−0.8) lead to a deceleration in movement. Thus, the effect of the interaction of the external factors on *D. salina* was statistically significant and commensurate with the influence of the separate factors.

In equation (5.2), the reactions of *D. viridis* to the same factors differed from *D. salina*. Increasing temperature (within the abovementioned limits) lead to an acceleration of cell movement while an increase in illuminance resulted in a deceleration of cell movement. The most significant effect on movement velocity was due to increasing temperature (coefficient +2.0) and the interaction of an electrical field and temperature (coefficient −2.6).

Increasing the level of illuminance resulted in an inhibiting effect on *D. viridis* with cell movement decelerating (weight coefficient −0.6). The electrical field, as a separate factor, did not have a significant affect on *D. viridis* cell movement velocity, but its interaction with illuminance and temperature significantly increased movement velocity (+ 0.5). Collectively the data indicate that under the same aquatic conditions the two species can occupy different ecological niches, a conclusion supported by observations in nature [Massjuk, 1973].

The dependence of cell phototopotaxis $F$ in both *Dunaliella* species on external factors is described by the following regression equations:

*D. salina:* $F = 0.07 - 0.07\,E + 0.09\;(E{\times}T);$ (5.3)

*D. viridis:* $F = -0.125\,E + 0.079\,T + 0.148\;(E{\times}T).$ (5.4)

The results of these experiments are presented in Tables 5.2-5.5.

**Table 5.2.** Dependence of linear movement velocity *(υ)* of cells *Dunaliells salina* Teod. on the effect of external factors: illuminance $(X_1)$, electrical field $(X_2)$ and temperature $(X_3)$

| $X_1$ | $X_2$ | $X_3$ | $υ_1$ | $υ_2$ | $υ_3$ | $<υ>$ | $S_υ$ |
|---|---|---|---|---|---|---|---|
| L | e | t | 38.200 | 39.900 | 37.300 | 38.467 | 1.743 |
| L | e | t | 45.800 | 35.700 | 37.100 | 39.533 | 29.943 |
| L | E | t | 40.900 | 31.700 | 32.700 | 35.100 | 25.480 |
| L | E | t | 40.400 | 37.800 | 38.000 | 38.733 | 2.093 |
| L | e | T | 35.300 | 33.200 | 33.000 | 33.833 | 1.623 |
| L | e | T | 33.800 | 32.000 | 32.700 | 32.833 | 0.823 |
| l | E | T | 33.700 | 32.300 | 31.200 | 32.400 | 1.570 |
| L | E | T | 33.200 | 33.600 | 33.700 | 33.500 | 0.070 |

*Note.* Here and in Table 5.3 $l$ = 100 lx; $t$ = 18 °C; $e$ = 0 V/cm; $L$ = 500 lx; $T$ = 30 °C and $E$ = 2.4 V/cm; $υ_1$, $υ_2$, $υ_3$ are absolute velocity significances measured in trifold sequence; $<υ>$ – average velocity significance; $S_υ$ – standard deviation.


**Table 5.3.** Dependence of linear movement velocity *(υ)* of *D. viridis* Teod. on the effect of external factors: illuminance $(X_1)$, electrical field $(X_2)$ and temperature $(X_3)$

| $X_1$ | $X_2$ | $X_3$ | $υ_1$ | $υ_2$ | $υ_3$ | $<υ>$ | $S_υ$ |
|---|---|---|---|---|---|---|---|
| L | e | T | 30.900 | 32.000 | 34.900 | 32.600 | 4.270 |
| L | e | T | 33.200 | 33.100 | 31.700 | 32.667 | 0.703 |
| L | E | T | 36.700 | 40.600 | 39.900 | 39.067 | 4.323 |
| L | E | T | 38.000 | 36.900 | 38.600 | 37.833 | 0.743 |
| L | e | T | 44.200 | 42.300 | 43.000 | 43.500 | 0.390 |
| L | e | T | 39.900 | 40.600 | 40.100 | 40.200 | 0.130 |
| L | E | T | 37.700 | 38.000 | 36.900 | 37.533 | 0.323 |
| L | E | T | 36.909 | 38.000 | 36.700 | 37.200 | 0.490 |


**Table 5.4.** Dependence of phototopotaxis *F* of *D. salina* Teod. on external factors effect: light $(X_1)$, electrical field $(X_2)$ and temperature $(X_3)$

| $X_1$ | $X_2$ | $X_3$ | $F_1$ | $F_2$ | $F_3$ | $<F>$ | $S_F$ |
|---|---|---|---|---|---|---|---|
| L | e | T | 0.110 | 0.190 | 0.240 | 0.180 | 0.004 |
| L | e | T | 0.310 | 0.360 | 0.270 | 0.313 | 0.002 |
| L | E | T | 0.160 | 0.030 | −0.180 | 0.003 | 0.029 |
| L | E | T | −0.140 | −0.250 | −0.097 | −0.162 | 0.006 |
| L | e | T | 0.140 | −0.150 | 0.200 | 0.063 | 0.035 |
| L | e | T | 0.000 | 0.060 | −0.090 | −0.010 | 0.006 |
| L | E | T | 0.020 | 0.120 | 0.160 | 0.100 | 0.005 |
| L | E | T | 0.100 | 0.050 | 0.060 | 0.030 | 0.007 |

*Note.* Here and in Table 5.5 $l$ = 100 lx; $t$ = 18 °C; $e$ = 0 V/cm; $L$ = 500 lx; $T$ = 30 °C and $E$ = 2.4 V/cm; $F_1$, $F_2$, $F_3$ are absolute values of the phototopotaxis; $<F>$ – average phototopotaxis value; $S_F$ – standard deviation

**Table 5.5.**Dependence of phototopotaxis $F$ of $D.$ *viridis* Teod. on external factors effect: light $(X_1)$, electrical field $(X_2)$ and temperature $(X_3)$

| $X_1$ | $X_2$ | $X_3$ | $F_1$ | $F_2$ | $F_3$ | $<F>$ | $S_F$ |
|---|---|---|---|---|---|---|---|
| L | e | t | 0.240 | 0.260 | 0.200 | 0.233 | 0.001 |
| L | e | t | 0.250 | 0.270 | 0.180 | 0.233 | 0.002 |
| L | E | t | −0.280 | −0.330 | 0.140 | −0.250 | 0.010 |
| L | E | T | −0.570 | −0.360 | −0.190 | −0.373 | 0.036 |
| L | e | T | 0.100 | 0.330 | −0.110 | 0.107 | 0.048 |
| L | e | T | −0.050 | 0.500 | 0.260 | 0.087 | 0.025 |
| L | E | T | 0.210 | 0.200 | 0200 | 0.203 | 0.000 |
| L | E | T | −0.200 | 0.180 | 0.260 | 0.080 | 0.060 |

In equation (5.3), the determining factor in $D.$ *salina* phototopotaxis was the effect of the electrical field which had an inhibiting effect (coefficient −0.07). Increasing the temperature canceled the inhibiting effect of the electrical field and stimulated phototopotaxis (+0.09). At the same time, the effect of temperature on phototopotaxis is not determinative.

Equation (5.4) for $D.$ *viridis* indicates that as with $D.$ *salina* the electrical field inhibited phototopotaxis (−0.125) while increased temperature increased phototopotaxis (+0.079). The greatest effect was due to the interaction of both factors (+0.148).

The rather complicated effect of external physical factors on photomovement parameters in both *Dunaliella* species can be explained in the following way. Since the cell movement velocity of both species depends on illumination by white light, maximum cell movement activity falls within a light intensity range of 150-550 lx and the maximum movement velocity between a temperature range of 20 to 25°C [Posudin et al., 1988]. Maximum positive phototopotaxis values were reached at a white light illuminance of 500 lx; at 1500 lx phototopotaxis is absent. With further increases in illumination, phototopotaxis becomes negative [Posudin et al., 1991]. The electrical field, applied to the sample (see Section 5.2) inhibited phototopotaxis.

The upper and lower limits for sample illuminance (100 lx and 500 lx) and temperature (18°C and 30 °C) used in the experiment were close to the outer limits for these external factors with regard to photomovement parameters and the range in which they can alter each other or the effect of an electrical field. The effect of external factors on photomovement parameters can be increased or decreased through the interaction of these factors.

Similar complementing and opposing effects caused by the external factors can be explained by differences in the ecology and behavior of these species under conditions found in nature [Massjuk, 1973].

## 5.5. Effect of Ultraviolet Radiation

Solar radiation is one of the important external factors that affect the viability and behaviour of plants. The spectral composition of solar radiation is characterized by the presence of ultraviolet (200-400 nm), visible (400-800 nm), and infrared (800 nm-50 μm) regions. Solar radiation is the primary factor influencing algal aquatic ecosystems and in particular, the algal photomovement parameters of motility, phototopotaxis, and movement velocity.

Ultraviolet radiation can be divided into three spectral groups depending on the effect of the radiation on biological objects [Forster and Lüning, 1996]: UV-A (320-400 nm), UV-B (280-320 nm) and UV-C (200-280 nm). Ultraviolet radiation in the UV-C region is characterized by the shortest wavelength and the highest energy. Under natural conditions this energy stimulates ionization processes in the upper atmosphere though little reaches the Earth's surface due to absorption by the ozone layer. However, UV-B radiation reaches the Earth's surface and its intensity depends on the latitude, solar elevation, cloud cover, reflectivity of the surface, and the thickness of the ozone layer. Ozone depletion has lead to increases in the UV-B radiation reaching the Earth's surface. UV-B causes damage to living organisms due to its absorption by nucleic acids, proteins and other labile molecules [Häder, 1996]. Absorption of UV-A radiation, in contrast, is due to its interaction with conjugated double bonds and with cyclic and polycyclic structures such as isoprenoids, flavines, quiniones, alkaloids, and photosynthetic pigments in phototrophic organisms [Garcia-Pichel, 1996].

Halldal (1976) has shown that near-UV radiation (310-390 nm) had little or no injurious effects, being similar to visible light, whereas far-UV (190-310 nm) produced immediate growth inhibition, delayed growth inhibition, or simulative growth effects. Halldal examined the action spectra for inhibition of chloroplast development, changes in diurnal rhythm, motility, phototopotaxis, and chloroplast movements. UV-A (320-400 nm) exposure caused an inhibition of photosynthesis, bleaching of photopigments, and a loss of biomass [Häder, 1991, 1995, 1996b; Häder et al., 1995; Ekelund, 1996].

Both natural and artificial ultraviolet radiation alters the behavioural strategy and productivity of algae [Häder, 1994; Huovinen et al., 2006]. A number of articles have assessed the interaction of natural and artificial ultraviolet radiation on algal photosynthetic activity and orientation, in particular, the motility and photoorientation of *Euglena gracilis* G.A. Klebs [Häder, 1985, 1986a; Häder and Häder, 1988]. In addition, pre- and post-treatment with thiourea, caffeine and cysteine on UV-induced damage in desmids [Sarma and Chowdhury, 1985], motility in *P. uncinatum* [Häder et al., 1986], photomovement and motility in *Astasia longa* [Häder and Häder, 1989a], photosynthesis, protein and pigment composition in *E. gracilis* [Gerber and Häder, 1992], gravitaxis in *Euglena gracilis* [Häder and Shi-Mei Liu, 1990], photoorientation, motility and pigmentation in *Peridinium gatunense* Nygaard [Häder et al., 1990] and *Cryptomonas* sp. [Häder and Häder, 1989b, 1990,1991], photomovement and pigmentation in *Gyrodinium dorsum* Kofoid & Swezy [Ekelund and Bjorn, 1990], positive phototaxis in *Volvox aureus* Ehrenberg [Blakefield and Calkins, 1992], damage of photoreceptor proteins in the paraflagellar body of *E. gracilis* [Brodhun and Häder, 1993], growth and motility of the flagellate, *E. gracilis* [Ekelund, 1993], photoorientation, motility, and chlorophyll photosynthesis in *Euglena sanguinea* Ehrenberg [Gerber and Häder, 1994], pigments and assimilation of [15]N ammonium and [15]N nitrate by macroalgae [Doehler et al., 1995], photosynthesis in *Laminaria digitata* (Hudson) J. V. Lamouroux [Forster and Lüning, 1996] and *Dictyota dichotoma* [Flores-Moya et al., 1999], motility in *Dunaliella bardawil* [Jimenez et al., 1996], flagellar apparatus in *C. reinhardtii* [Donk and Hessen, 1996], amino acids in macroalgae [Kusten et al., 1998], growth and pigment composition of *Ulva expansa* (Setch.) S. [Grobe and Murphy, 1998], photoinhibition of marine macrophytes [Aguirre-von-Wobeser et al., 2000], reproduction of *Enteromorpha intestinalis* E. [Cordi et al., 2001], fixation of inorganic nitrogen in the marine alga *Dunaliella tertiolecta* Butcher [Beardall et al., 2002], composition of photosynthetic and xanthophyll cycle pigments in *Ulva lactuca* L. [Bischof et al., 2002], photomovement of the swarmers of the brown algae *Scytosiphon lomentaria* (Lyngbye) Link and *Petalonia fascia* (O.F. Müller) Kützing [Flores-Moya et al., 2002], biochemical composition of *Ulva* sp. [2002], canopy structure of *Ulva* communities [Bischof et al., 2002], UV-induced biochemical processes in *Ulva* canopies [Bischof et al., 2003], photosynthetic UV responses in *Ulva* species [Figueroa et al., 2003; Posudin et al., 2004a,b], taxonomic composition of phytoplankton [Xenopoulos and Frost, 2003], variation in sunscreen compounds (mycosporine-like amino acids) in marine species [Lamare et al., 2004], increasing competition between marine macro-algae and micro-algae populations [Zhang et al., 2005], growth interactions between *Ulva pertusa* Kjellman and *Alexandrium tamarense* (Lebour, 1925) Balech, 1992 [Cai et al., 2005], oxidative stress and responses of the ascorbate-glutathione cycle in *Ulva fasciata* Delile [Shiu and Lee, 2005], UV-B protection in *U. pertusa* [Han and Han, 2005], carbon and nitrogen metabolism in *Fucus spiralis* L. and *Ulva olivascens* Dangeard [Vinegla et al., 2006], growth of *U. pertusa* and *Platymonas helgolandica* Kylin *var. tsingtaoensis* [Xu et al., 2006], vertical migration and photosynthesis in *E. gracilis* [Richter et al., 2007], effective quantum yield of *Ulva lactuca* L. [Xu and Gao, 2007], and photosynthesis in *U. lactuca* [Fredersdorf and Bischof, 2007] have been studied.

It has been assumed [Ghetti et al., 1992] that nucleic acids and proteins present in pigmented microorganisms can be damaged, not only by ultraviolet radiation, but also by near-UV and visible radiation through photosensitization reactions that result in the generation of singlet oxygen and/or other noxious oxygen species. To address this question, we studied the dependence of phototopotaxis in two species of *Dunaliella* Teod. and in *Tetraselmis viridis* (Rouch) Norris et al. (syn. *Platymonas viridis* Rouch), on the wavelength of lateral stimulating light within the ultraviolet region of the electromagnetic spectrum to compare differences in the action spectra between *Dunaliella* [Posudin et al., 1990] and *T. viridis* [Halldal, 1961].

The absence of phototopotaxis in *Dunaliella* within the ultraviolet region of the spectrum where flavins and rodopsin have maximum absorption supports the hypothesis that carotenes act as photoreceptor pigments and are responsible for phototopotaxis in the two species in *Dunaliella*. It is possible, therefore, that phototopotaxis in the ultraviolet portion of spectrum is also operative in the genus *Tetraselmis* [Posudin et al., 1990].

This section addresses the effect of ultraviolet irradiation on the two *Dunaliella* species using different intensities, wavelengths, and durations. Photomovement parameters were assessed in algae exposed to lateral white light irradiation (500 lx) and at a temperature of 18-20 °C [Posudin et al., 2004]. The application of artificial ultraviolet radiation allowed discerning its possible role on the viability and photo-behaviour of *Dunaliella* and identifying the optimal conditions for survival.

A mercury lamp with an emission spectrum in the 250-350 nm range was utilized as the ultraviolet radiation source. The dependence of photomovement parameters, in particular the linear velocity $v$ of movement, phototopotaxis $F$ and relative motility $N_m/N_0$ (where $N_m$– quantity of motile cells, and $N_0$ – total quantity of the cells) on the intensity of ultraviolet radiation were measured at 0.76 to 11 W/m$^2$ with the intensity monitored using a DAU-81 dosimeter. The duration of exposure was 5 and 10 minutes and the spectral sensitivity of the photomovement parameters determined using interference filters placed between the ultraviolet radiation source and the algal suspension. The filters had a maximum transmission at 248 nm, 280, 302, 313, 334 and 365 nm. Untreated algae illuminated laterally with white light (500 lx, 18-20 °C) were used as a control. Three replications of all measurements were used to calculate mean values and errors of measurements.

The dependence of the linear velocity $v$ of movement and phototopotaxis $F$ in both species on the intensity $I$ of unfiltered ultraviolet radiation is presented in Fig. 5.4. The values for $v$ did not change with increasing intensity of radiation, while phototopotaxis $F$ was inhibited by high intensity ultraviolet radiation (i.e., 2 to 11 W/m$^2$).

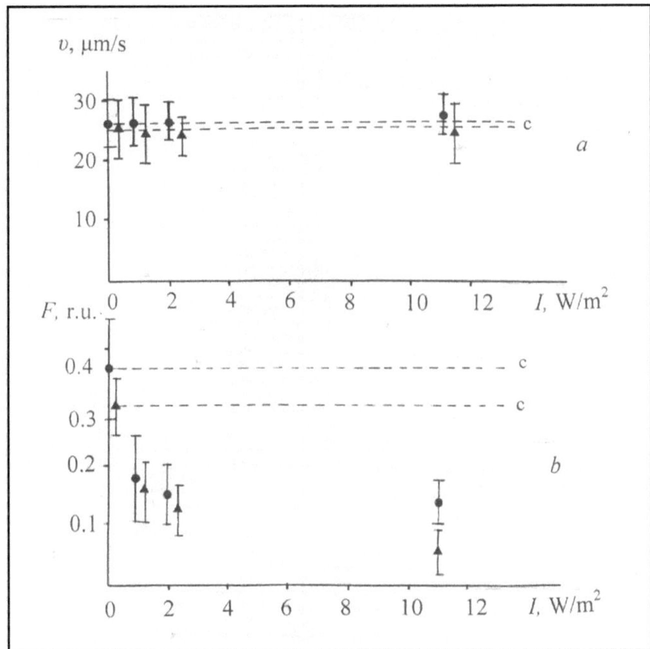


**Fig. 5.4.** Dependence of the linear velocity $v$ of movement  (*a*) and phototopotaxis $F$ (*b*) of the cells of two species of *Dunaliella* on the intensity (*I*) of preliminary exposure to nonfiltered ultraviolet radiation (wavelength range 250–350 nm, duration of irradiation 5 min)  ( –•– - *Dunaliella salina*; –▲– - *Dunaliella viridis*; c – control) [Posudin et al., 2004].

The dependence of the linear velocity $v$ of movement, phototopotaxis $F$ and relative motility $N_m/N_0$ of the cells on exposure duration $t$ to unfiltered ultraviolet irradiation (10 W/m²) is presented by Fig. 5.5.

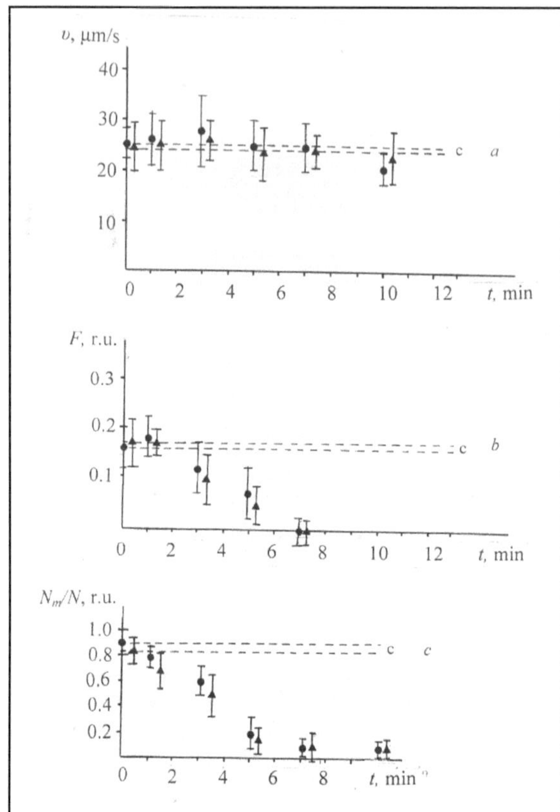

**Fig. 5.5.** Dependence of the linear velocity $v$ of movement  (*a*) and phototopotaxis $F$ (*b*) and relative motility $N_m/N_0$ (*c*) of the cells of two species of *Dunaliella* on the duration $t$ of preliminary exposure to nonfiltered ultraviolet radiation (wavelength range 250–350 nm; intensity of radiation  10 W/m$^2$ ( –•– - *Dunaliella salina*; –▲– - *Dunaliella viridis*; c – control) [Posudin et al., 2004].

The velocity $v$ did not differ from the control during 10 minutes of irradiation, while phototopotaxis $F$ and relative motility $N_m/N_0$ were inhibited by 7-10 minutes of irradiation. The results support the position that different mechanisms govern the velocity of linear movement (parameter $v$) versus photoorientation (parameter $F$) and motility (parameter $N_m/N_0$). It was not possible, however, to discern differences in the structure and size of the photoreceptor systems responsible for the velocity of linear movement and photoorientation of the cells. The results are in agreement with previous investigations assessing the effect of medium pH and ionizing radiation on photomovement parameters in *Dunaliella* [Posudin et al., 1992; Massjuk and Posudin, 2007] (see Sections 5.3 and 5.6).

The effect of natural solar and ultraviolet radiation on the velocity of linear movement and phototopotaxis in *E. gracilis* has been previously described in the literature [Häder, 1986*a*]. The mean value for linear movement was 120 µm/s; the velocity and phototopotaxis declined to zero after 1.5-2 hours exposure to solar irradiation. Use of an ozone filter (cuvette with 45 µg/mL ozone in air) decreased the intensity of the UV-B (290-320 nm) component of solar radiation (estimated to be 1.2 W/m$^2$) by 5.0 % which lead to an increase in cell viability (i.e., 50 % of motile cells remained after 3 hours of irradiation) and a decrease in the velocity of linear movement of up to 80 % of the initial value. The use of a glass lid that blocked ultraviolet radiation made it possible to reach a level of phototopotaxis comparable to control samples [Häder, 1986*a*]. This dependence of photomovement parameters in *E. gracilis* could not be explained by simply exceeding the light energy absorption capacity of chlorophyll since cells of *Astasia longa* Pringsheim devoid of pigment and bleached cells of *E. gracilis* demonstrate the same response to ultraviolet radiation [Häder, 1986*a*].

Inhibition of motility and linear velocity of movement due to unfiltered solar irradiation is also exhibited by the algal species *Peridinium gatunense* Nygaard, *Cryptomonas* spp., *Gyrodimium dorsum* Kofoid & Swezy, and *Cyanophora paradoxa* Korshikov [Häder, 1991]. The radiation leads to immobilization of *Cryptomonas maculata* Ehr. cells after 140 minutes exposure and by unfiltered artificial ultraviolet exposure after only 60 minutes [Häder et al., 1987; Häder and Häder, 1991].

The difference between our data and that of Häder appears to be due to the fact that we used a higher intensity of ultraviolet radiation (i.e., up to 10 W/m$^2$). Phototopotaxis $F$ and relative motility $N_m/N_0$ were inhibited by a 10 minute exposure indicating that irradiation is the primary factor affecting the photomovement parameters. The dose was calculated as $D=I{\cdot}t$, where $I$ is the intensity of ultraviolet radiation and $t$ the duration. The dosage used by Häder was $D = 1$ W/m$^2{\cdot}$120 min and in the same order of magnitude as in our experiments (i.e., $D = 10$ W/m$^2{\cdot}$10 min).

Differences in the velocity of linear movement in *Dunaliella* between our results and those of Häder and Häder (1991) are most likely due to differences in the way the measurements were made. We measured the velocity of individual cells that were mobile, not the mean velocity estimated by videomicrography. The cells maintained a constant velocity in our experiments even when there were only a small number of cells.

The dependence of velocity $\upsilon$ and phototopotaxis $F$ in the two *Dunaliella* species on the ultraviolet radiation wavelength (2 W/m$^2$ intensity) is presented in Fig. 5.6 (5 min exposure duration) and Fig. 5.7 (10 min exposure duration).

The effect of ultraviolet radiation wavelength on the linear velocity $\upsilon$ of *D. salina* (Fig. 5.6*a*) and *D. viridis* (Fig. 5.6*c*) was not established. The effect of monochromatic ultraviolet radiation on phototopotaxis $F$ in the two species was not straight forward. A 5 minute exposure decreased the level of photoorientation (parameter $F$) below control values until reaching negative values for phototopotaxis $F$ near 248-334 nm in *D. salina* (Fig. 5.6*b*) and near 248 nm in *D. viridis* (Fig. 5.6*d*). Prolonged exposure (i.e., 10 min) lead to a complete inhibition of phototopotaxis near 248-280 nm for *D. salina* (Fig. 5.7*b*) and 248-334 nm for *D. viridis* (Fig. 5.7*d*).

Ultraviolet irradiation in the 302-365 nm (*D. salina*) and 365 nm regions (*D. viridis*) for 10 minutes resulted in negative phototopotaxis in both species. The transition from positive to negative phototopotaxis in response to ultraviolet radiation with increased duration of exposure has not been previously reported. Exposure of *E. gracilis* to ultraviolet radiation at 295 nm and below, however, resulted in inhibitory effects on motility and photoorientation of the cells [Häder, 1985].

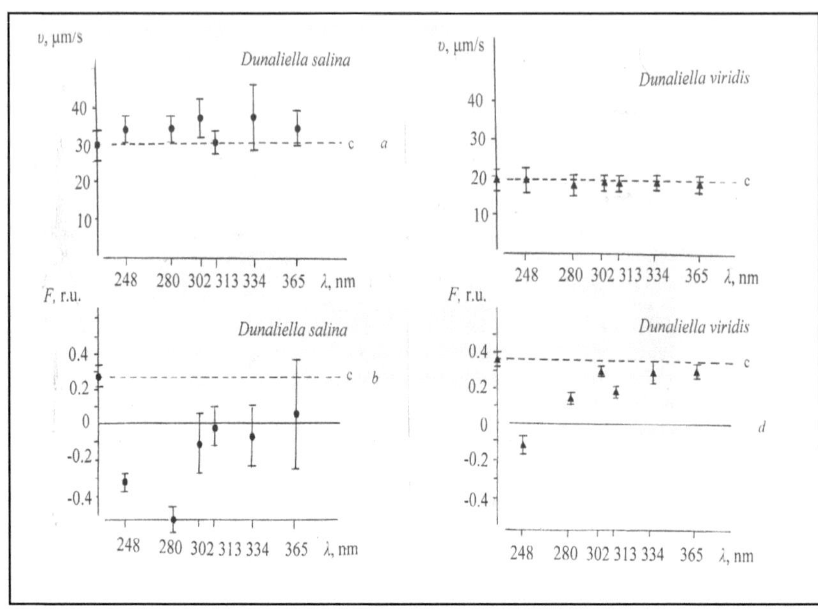

**Fig.5.6.** Dependence of the linear velocity $v$ of movement $(a,b)$ and phototopotaxis $F$ $(c,d)$ of the cells of two species of *Dunaliella* on the wavelength $\lambda$ of ultraviolet radiation (intensity of radiation is 2 W/m$^2$; duration of irradiation 5 min; c – control) [Posudin et al., 2004].

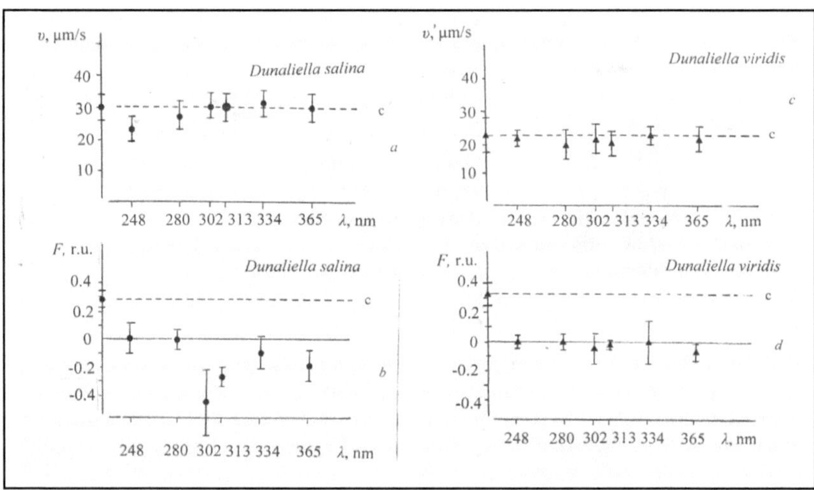

**Fig.5.7.** Dependence of the linear velocity $v$ of movement $(a,b)$ and phototopotaxis $F$ $(c,d)$ of the cells of two species of *Dunaliella* on the wavelength $\lambda$ of ultraviolet radiation (intensity of radiation is 2 W/m$^2$; duration of irradiation 10 min; c – control) [Posudin et al., 2004].

It is noteworthy that the effect of ultraviolet exposure on the parameter $F$ was reversible immediately after the ultraviolet treatment (2 W/m$^2$ intensity) and resulted in $F$ values reaching those of control at all wavelengths with the exception of 248 nm after a 2 hour exposure (Fig. 5.8). *Dunaliella bardawil* (=*D*. salina) cells recovered within 24 hours after the cessation of 10 hours of visible + UV-A radiation (26.01 W/m$^2$ intensity) and visible + UV-A + UV-B radiation (39.72 W/m$^2$) [Jimenez et al., 1996].

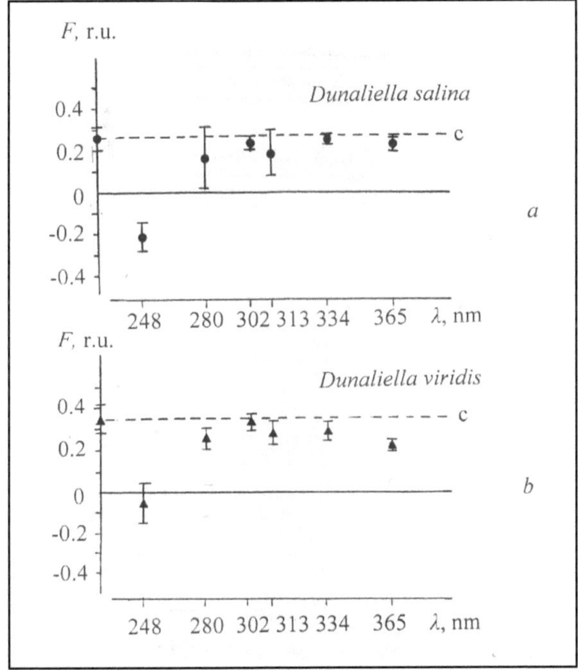



**Fig. 5.8.** Phototopotaxis of *Dunaliella salina (a)* and *D. viridis (b)* 2 hours after cessation of a 10 min pulse of ultraviolet radiation (intensity of radiation is 2 W/m$^2$) [Posudin et al., 2004].

There are several hypotheses concerning the mechanism of action of ultraviolet radiation on algae. The first proposes that the primary impact of ultraviolet irradiation is on DNA molecules and is based on the similarity in the absorption spectrum of DNA to the action spectrum for the inhibition of microorganisms [Yammamoto et al., 1983]. It was subsequently shown, however, that the fine structure of the action spectrum for the inhibition of motility in *E. gracilis* was characterised by a principal maximum near 270 nm (UV-C region), a smaller peak at 305 nm, and a shoulder at 290 nm (both in UV-B region) [Häder, 1991], a spectrum that does not resemble the absorption spectrum for DNA [Jagger, 1983]. In addition, the very fast effect of the radiation on motility and the absence of photo-repair, further argues against an effect on DNA being the primary avenue for the inhibition by ultraviolet radiation [Häder and Häder, 1988].

A second hypothesis is that ultraviolet damage of cells in through a photodynamic effect caused by the simultaneous action of ultraviolet radiation and chemical compounds. Ultraviolet radiation is thought to be absorbed by a photoreceptor molecule and if the excited molecule does not expend this additional energy via photochemical reactions or dissipative processes, the energy can be transferred to the triplet state and the formation of singlet oxygen $^1O_2$ [Maurette et al., 1983] or free radicals [Spikes, 1977]. Free radicals have highly reactive properties and can destroy membranes and other cellular components.

Arguments against this hypothesis are based on the application of specific diagnostic reagents and quenchers of singlet oxygen and free radicals and the absence of viability in algae exposed to radiation [Häder et al., 1986; Häder and Häder, 1988b]. Likewise, the addition of $D_2O$, which increases the half-life of singlet oxygen, does not inhibit the effect of ultraviolet radiation on the algae [Häder, 1991].

A more plausible explanation for the effect of ultraviolet radiation on photomovement parameters in algae may be through the effect of radiation on the proteins governing the activity of the flagellar apparatus or/and the photoreceptor system. Impairment of these proteins when detached from the paraflagellar body of *E. gracilis* by ultraviolet radiation supports this possibility [Häder, 1991]. The differential effect of varying levels of ultraviolet radiation on photomovement parameters in *Dunaliella* also supports this hypothesis. It is possible that the photoreceptor systems responsible for linear velocity and photoorientation differ in size and structure. The dependence of phototopotaxis and motility of algal cells on the intensity, wavelength, and duration of exposure to ultraviolet radiation can be used in biotesting of natural ultraviolet radiation.


## 5.6. Effect of Ionizing Radiation

Ionizing radiation results in morphological and behavioral changes in algae. For example, $\gamma$-irradiation of *D. bioculata* destroyed the cells within 24 hours [Saraiva, 1972]. As a consequence, ionizing radiation has been used to study the photoreceptor system in algae and to better understand the primary aspects of photoreception – localization, structure and function of the photoreceptor.

The influence of $\gamma$-radiation on the two species *Dunaliella* and an analysis of dose curves on velocity and direction of movement are addressed in this section. The range of $\gamma$-radiation was between 0–1000 Gr. Radiation was generated using a MPX-$\gamma$-25 "Investigator" system with 0.2 Gr/s from a $^{60}$Co source. Algal suspensions were placed in the radiation field for a given time interval and then removed and the linear velocity of movement and phototopotaxis of the cells measured. A Coulter Erics C cytofluorometer was used to evaluate cellular damage due to $\gamma$-irradiation. Control and irradiated suspensions were passed through the flow cuvette of the cytofluorometer to analyze cellular diffraction and fluorescence. Computer assessment of a large number of cells made it possible to construct mono- and biparametrical histograms of the number of cells of similar size. Amplitude of the signal was recorded on the abscissa axis and the number of the cells of the same size that gave similar amplitudes on the ordinate axis.

Three replications of the test on three flasks for each species were used to determine the effect of $\gamma$-irradiation on photomovement parameters $v$ and $F$. Doses of 30, 600 and 1000 Gr were assessed. The cells were illuminated with $1200 \pm 50$ lx; the temperature during irradiation was $23 \pm 1$ $^0$C, and during testing $17 \pm 1$ $^0$C. The velocity of cell movement $v$, phototopotaxis $F$, and the direction of movement were calculated by averaging the data from 9 samples (three flasks, three tests) for each irradiation regime. Measurements of $v$ and $F$ were made on the first, second, seventh and eleventh day after irradiation. The *Dunaliella* cultures were diluted 1:19 immediately after irradiation in order to study the effect of irradiation on the medium. Measurement of $v$ and $F$ were carried out on the first and 16$^{th}$ day after irradiation [Posudin et al., 1992].

The velocity of cell movement $v$ decreased linearly up to 600 Gr after the first day of irradiation, reaching 20 % (Fig. 5.9$a$) and decreased sharply at 1000 Gr. Phototopotaxis $F$ was also affected linearly up to 600 Gr, but exhibited a different slope in comparison with parameter $v$. Above 600 Gr (i.e., 1000 Gr) inhibited both velocity $v$ and phototopotaxis $F$.

The parameter $v$ decreased linearly up to 1000 Gr in $D.$ $viridis$ at a slope similar to $D.$ $salina$. Parameter $F$ decreased to the 10 % level at 30 Gr but exhibited considerable variation at higher levels (Fig. 5.9$b$). Differences in the dependence of parameter $F$ between species can be explained by differences in cell size. The character of the dose-response curves were essentially the same at the end of the first, seventh, and eleventh day indicating an irreversible effect of ionizing radiation. Dilution of the cultures with fresh medium did not appear to alter the affect of irradiation on parameters $v$ and $F$ in comparison with undiluted cultures on the 16$^{th}$ day, confirming the absence of an inhibitory effect by the medium on the photomovement parameters.

It is possible that there was a change in dimensions (or shape) of the cells due to the action of ionising radiation. Shifts in the maxima and changes in amplitudes of the histograms, characterized by either scattering of laser radiation by the cells or chlorophyll fluorescence due to the treatment, support this possibility (Fig. 5.10, 5.11). Thus, the histograms (Fig. 5.10, 5.11) indicate possible irreversible damage to the cells of both species ($D.$ $salina$ and $D.$ $viridis$) and their photosynthetic apparatus due to exposure to the above levels of $\gamma$-irradiation.

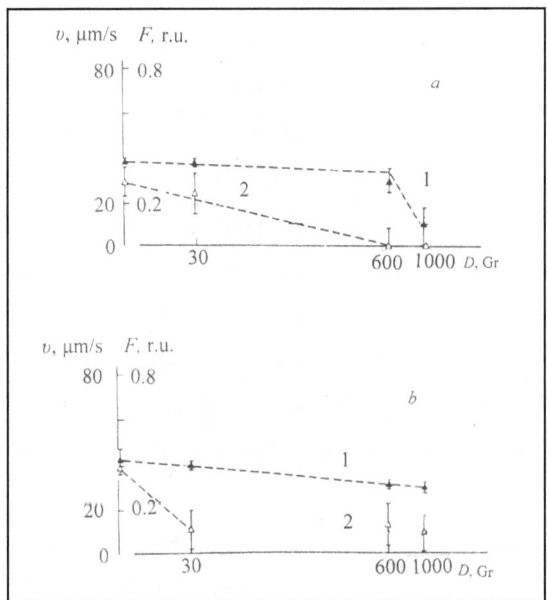

**Fig. 5.9.** Dependence of the linear velocity $v$ (*1*) and phototopotaxis $F$ (*2*) on the dose of ionizing radiation after one day of irradiation: $a$ – $D.$ $salina$; $b$ – $D.$ $viridis$. Axis of abscisses is the dose of irradiation in Gr [Posudin et al., 1992].

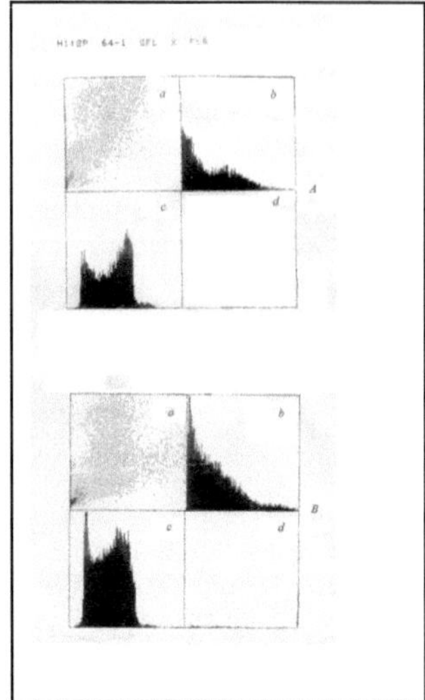

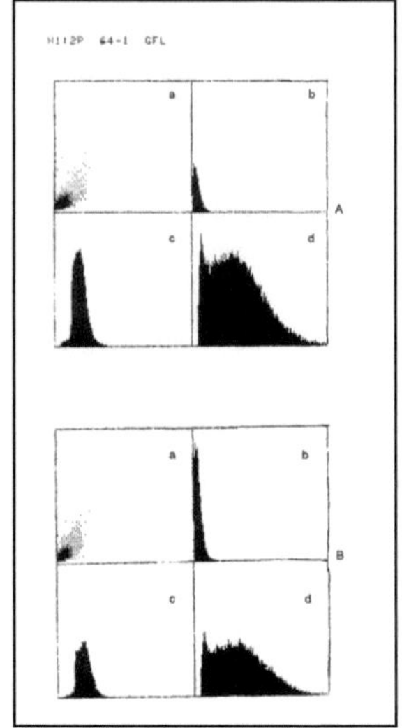

**Fig. 5.10.** Histograms which characterize relation between scattering and fluorescence ( *a* ) of the cells, fluorescence of the cells ( *b* ), scattering of laser radiation on the cells ( *c*), and scattering at an angle of $90^0$ ( *d* ) for *D. salina*: *A* – before $\gamma$-irradiation; *B* – after $\gamma$-irradiation [Posudin et al., 1992].

**Fig. 5.11.** Histograms that characterize the relation among scattering and fluorescence ( *a* ) of the cells, fluorescence of the cells ( *b* ), scattering of laser radiation on the cells ( *c*), and scattering at an angle of $90^0$ ( *d* ) for *D. viridis*: *A* – before $\gamma$-irradiation; *B* – after $\gamma$-irradiation [Posudin et al., 1992].

Ascertaining the location and the nature of photoreceptor that is responsible for light reception and the transduction of the signal controlling flagellar beating are critical questions in the study of photomovement. It is thought that in green algae such as *Chlamydomonas* the photoreceptor system is located near surface of the cell in the plasmalemma [Nultsch, 1983]. It is known that photoreceptor in *E. gracilis* is located inside the cell, at the basis of flagellum in the paraflagellar body [Colombetti et al., 1982].

In *Dunaliella,* there is at this time an absence of sufficiently precise data indicating the location of photoreceptor. One might assume that if the photoreceptor was located at the cell surface (as in *Chlamydomonas*) and the mechanisms responsible for controlling velocity and direction of movement were in different locations, there would be different characteristic dose-response curves for $v$ and $F$. Our results demonstrate that $v$ and $F$ in *Dunaliella* display different slopes for their dose-response curves (see Fig. 5.9) which supports the idea of different mechanism(s) govern the linear velocity of movement and phototopotaxis. The possible loss of protein conformation due to ionizing radiation can not be excluded. It is known that membrane proteins control ion transport, in particular, $Ca^{2+}$ ions are modulated by light conditions [Posudin and Suprun, 1992]. Photoreception can be viewed as an excitation of certain membrane proteins due to the absorption of light by a photoreceptor molecule and conformational alterations in the protein that lead to the activation of ion channels. Changes in the concentration of $Ca^{2+}$ ions around the flagella modulates their beating and in turn, the photoorientation of the cell.

The damage caused by $\gamma$-radiation exposure on the linear velocity of movement and phototopotaxis in *Dunaliella* differ. The photoreceptor that is responsible for photoorientation appears to be more sensitive to ionising radiation than the apparatus that controls the velocity of linear movement. There are, for example, several differences in the dose-response curves for $F$ between *D. salina* and *D. viridis* that may be related to differences in the dimensions of the respective photoreceptor systems or their sensitivity to ionising radiation. The linear character of the dose-response curves (i.e., for parameter $v$) indicates that algae might be useful as a biological sensor for ionizing radiation, at least up to 1000 Gr.


## 5.7. Summary

The effects of illuminance, temperature, pH, electrical field, ultraviolet and ionizing radiation on photomovement parameters such as motility [relative quantity of mobile cells $N_m/N_0$), velocity of linear movement ($v$, $\mu$m/s)], and phototopotaxis ($F$) in two species of *Dunaliella* and the interaction of several external factors (illuminance, temperature, electrical field) on photomovement were studied. Maximum values for motility, velocity, and positive and negative phototopotaxis were observed for the temperature range between 20 and 30°C and a pH range of 6.50 to 8.47. Optimum pH for the various photomovement parameters differed as did the sensitivity of these parameters to pH extremes.

Exposure of cells to an external electrical field results in suppression of phototopotaxis in *Dunaliella*, similar to other green algae (e.g., *Chlamydomonas reinhardtii* and *Haematococcus pluvialis*) underscoring the role of bioelectrical potentials in photomovement.

The species *D. salina* and *D. viridis* differed in their motile behavior in response to adverse external factors and combinations thereof. Increasing the temperature from 18 to 30 °C overcame the inhibition of the electrical field and stimulated phototopotaxis in both species. With an illumination of 100 and 500 lx and temperatures of 18 and 30 °C, there were levels in which photomovement parameters are highly dependent upon the external factors tested. The response may be positive or negative, altering the photomovement accordingly. In nature, differences in the ecology and behavior of *Dunaliella* species may be in part due to complimentary or opposing effects of these external factors [Massjuk, 1973].

Ionizing and UV irradiation inhibit phototopotaxis in both *D. salina* and *D. viridis*. The inhibiting effect depends on the dose of irradiation and in the case of UV-irradiation, wavelength. Ionizing radiation (up to 600 Gr) and UV-irradiation did not affect the velocity of movement. We found a previously unreported transformation of positive to negative phototopotaxis due to UV-irradiation, with an inhibition of phototopotaxis down to a zero value. Effects of UV radiation on photomovement of *Dunaliella* were reversible, while the effects of ionizing radiation were irreversible. The two species of *Dunaliella* displayed a differential in sensitivity to UV and ionizing radiation.

The relative number of motile cells ($N_m/N_0$) varied among populations of *Dunaliella* from 0 to 100 % and displayed the same dependence on characteristics of the light stimulus and environmental conditions as phototopotaxis. Phototopotaxis differs considerably, however, depending upon its presence or absence and by its degree of dependence on factors such as pH, intensity, wavelength, dose of preliminary ionizing and UV radiation. There also appears to be differences in the structure and dimensions of photoreceptor systems that are responsible for variation in the various photomovement parameters within and among algal species.

# Chapter 6

# Structure of the Photoreceptor System

## 6.1. Problems associated with Photoreception of Algae

Critical aspects in algal photoreception are: 1) the location and structure of the photoreceptor system; 2) the composition of the photoreceptor pigments; and 3) the mechanism(s) of photoreception. For nearly 100 years [Engelmann, 1982*a,b*], the photoreceptor was thought to be the stigma ("eyespot", "Augenflecke") [see reviews by Dodge, 1973; Sedova, 1977; Ettl, 1980; Massjuk and Posudin, 1991*b*]. This belief was supported indirectly by its presence in the coloured motile vegetative and reproductive cells of algae, its disappearance when the algae was kept for extended periods in darkness, and its restoration when exposed to light [Kivic and Vesk, 1972, 1974].

The conservation of the stigma in motionless cells of many tetrasporal algae is considered as example of atavism which is supported by the complete disappearance of the stigma at the coccoid level of organisation of the vegetative body in algae. At the same time, numerous cases of the absence of a stigma in motile cells of algae have been reported (e.g., *Raphidophyta*, many *Haptophyta*; zoospores *Chlorhormidium flaccidum* (Kütz.) Fott, *Coleochaete* Bréb., *Urospora penicilliformis* (Roth) Aresch., motile flagellar male gametes *Bacillariophyta*, and *Bryopsis hypnoides* Lamour.). As a consequence, it is difficult to accept a photoreceptor function for the stigma since species that are devoid of a stigma can respond the light.

Toward the beginning of the 19th century, photoreception was thought to be performed by a specialized region near the base of one of the flagella and that the region had the ability to receive signals from the stigma [Mast, 1927]. More recently, the stigma has been shown to function as a subsidiary device that is able to modulate the light during photoreception in flagellates. Therefore the photoreceptor system consists of a photoreceptor and a stigma that modulate the light.

A cross-section of flagellates with photoreceptor systems are found in various taxonomic divisions and classes and they vary in structure, location, and principle of action [Foster and Smyth, 1980; Kivic and Walne, 1983; Greuet et al., 1987; Ruediger and Lopez-Figueroa, 1992; Lenci et al., 1996; Barsanti et al., 2004]. Since a detailed assessment of these systems across a range of taxons has been given by Massjuk and Posudin [1991*b*], the following discussion focuses upon a comparative analysis of photoreceptor systems in green algae.

## 6.2. Structure of Photoreceptor Systems in Green Algae

The photoreceptor system in green algae consists of a distinct stigma and photoreceptor, however, the location and structure of the photoreceptor has not been adequately established. The stigma is part of the chloroplast and is located at its surface directly under the double membrane envelope contacting the plasmalemma. It consists of one or several (<9) layers of pigmented globules. The layers are found in a multi-layered stigma that is separated at intervals where the colour-less stroma of the chloroplast are located. Sometimes photosynthetic lamella or individual thylakoids deprived of pigments are found. The stigma of green algae are considered to be type A according to the classification of Dodge [1973].

Mast [1911] observed that eyespots of colonies of the green algae *Pandorina* Bory and *Eudorina* Ehrenb. reflect greenish blue light when illuminated with direct sunlight. The more anterior eyespots reflect more light than those posterior, however, the reflection is not visible when using a transmission light microscope since the reflected light does not reach the eye [Mast, 1927].

Mast first proposed the importance of this phenomenon in algal phototopotaxis, an effect that was later confirmed using modern techniques [Foster and Smyth, 1980]. The light reflecting eyespots in green algae consist of pigmented globules that are arranged in a hexagonal array. The distance between centers of the globules varies from 75 nm to 100 nm [Nakamura et al., 1973; Bray et al., 1974].

Using an electron microscope, the thickness of the four stigma layers in *Chlamydomonas reinhardtii* P.A. Dang. [Linder and Staehelin, 1979; Foster and Smyth, 1980] was measured. The pigmented and nonpigmented layers were 69.0 and 77.7 nm, respectively and the double membrane, found between the inner surface of each pigmented layer, was 15.4 nm thick. Based on the thickness and differences in the refraction indices between the pigmented and non-pigmented layers, Foster and Smyth [1980] concluded that the reflective properties of the stigma were very similar to the properties of a "quarter-wave-stack" (i.e., a stack of alternating layers with high and low refractive indices). The width of each layer and the distance between equalled $\lambda/4$ (where $\lambda$ is the wavelength). If the light reflects at the low- to high-refractive index interface, it changes phase by $\pi$ radians while the light that reflects at the high- to low-refractive index interface does not change phase. A constructive interference occurs as a result of the interaction of light with each layer (Fig. 6.1).

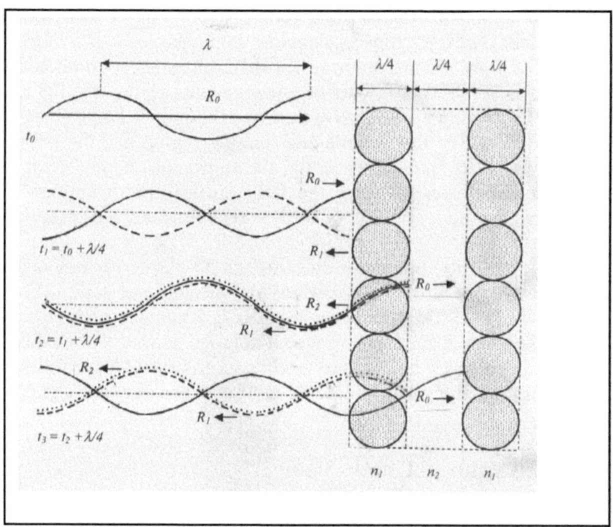


**Fig. 6.1.** Interaction of light with a quarter-wave stack of alternating layers of high and low refractive indices wavelength; light reflects from such a multilayer system without changing phase and producing an interference maximum at the site of possible photoreceptor [adapted from Foster and Smyth, 1980].

Based on microphotographies in the literature, the stigma of other species of green algae [e.g., *Volvulina pringsheimii* Starr, *Platydorina caudata* Kofoid, *Volvox aureus* Ehrenb., *V. tertius* Meyer, *Pteromonas tenuis* Belcher and Swale, *Pyramimonas montana* Geitl., *Eudorina illinoisensis* (Kofoid) Pascher] also function as a quarter-wave reflector [Foster and Smyth, 1980]. A mutation of *C. reinhardtii* without a cell wall displayed a high level of specificity

(Melkonian and Robenek, 1980; 1984). The plasmalemma near the stigma was characterized by a dense arrangement of intramembranous particles in comparison with other parts of its surface. Smaller particles (8-12 nm diameter) prevailed while larger particles (16-20 nm) were located outside the stigma area. The plasmalemma and chlorophyll envelope outer membrane adjacent to the stigma were only 20-30 nm from each other. It is thought that regions of both membranes take part in photoreception and the primary sensory transduction [Melkonian and Robenek, 1980]. Thus, the photoreceptor in green algae is thought to be located in the membrane between the stigma and the adjacent part of the chloroplast envelope.

The structural organization of intramembraneous particles on the cytoplasmic surface of the plasmalemma ["protoplasmic face" as per the terminology of Branton et al., 1975] near the eyespot was found in all species of algae under investigation. The plasmalemma of *C. reinhardtii* near the eyespot has a specific chemical composition of proteins and lipids [Melkonian, 1981; Melkonian and Robenek, 1980].

There are several variations in the outer membrane of the chloroplast envelope ranging from specialized to the absence thereof. Specialization of the intramembraneous particles, however, was not found in the inner membrane of the chloroplast envelope and thylakoid membranes near the eyespot [see review by Melkonian and Robenek, 1980]. If light beams enter the lateral side of the cell where the stigma is located, the photoreceptor receives a signal double in intensity due to an amplification that is equal to the sum of the incident and reflected intensities. If the opposite side of the cell is illuminated, the signal received by the photoreceptor is attenuated due to the absorption of light by internal cellular constituents and the stigma as well as due to reflection of light by the stigma. The stigma, therefore, modulates the light and acts like a directional antenna, determining the spatial positioning of the light. This effect is intensified due to the interchange of pigmented and non-pigmented layers in the stigma by about a quarter of a wavelength. If the cell is illuminated from the side of the outer surface of the stigma, a series of intensity maxima are produced. The location of these maxima coincides with the plasmalemma and thylakoid membranes inside the stigma. However, if the light strikes the stigma from the opposite side, a series of minima at the same position are produced within the stigma increasing the maximum contrast in the light received from both opposite sides. This suggests that the photoreceptor pigments are located on the outer membranes (plasmalemma and outer membrane of the chloroplast envelope) as well as the thylakoid membranes [Foster and Smyth, 1980]. Such a location for the photoreceptor (i.e., in the plasmalemma in the region of the stigma) was proposed earlier based on the fact that such a position would provide a direct linkage of the photoreceptor with the locomotor apparatus since the plasmalemma is continuous with the flagellar membrane [Arnott and Brown, 1967; Walne and Arnott, 1967]. The connection of internal membranes with the flagella is less evident. Therefore, the location of the photoreceptor pigments not only in outer membranes but inside the multilayer stigma on the surface of thylakoid membranes, appears plausable. Use of either outer membrane (plasmalemma and chloroplast envelope membrane) or the thylakoid membranes increases the surface area for photoreception and the intensification efficiency [Foster and Smyth, 1980].

The structure of the eyespot apparatus for almost 90 species of green algae has been studied (see review by Melkonian and Robenek, 1984) and categorized using 30 physical traits e.g., the shape of the eyespot (ellipsoid, egg-shaped, spherical, etc.), the shape of outer surface (flat, concave, salient), the size of outer surface of the eyespot (which ranged from 0.28 $\mu m^2$ in zoospores of *Chlorosarcinopsis gelatinosa* Chantanachat and Bold, to 9 $\mu m^2$ in vegetative cells of *Volvox* (L.) Ehrenb. and female gametes of *Bryopsis lyngbye* Lyngb. = *B. plumosa* (Hudson) C.Ag., and the size (80-130 nm, sometimes 200 nm), chemical composition, arrangement of packing (hexagonal or not), and electron density of the pigmented globules.

There are three main variations in the layers of pigmented globules and thylakoid membranes within the stigma.

1) *Chlamydomonas* Ehrenb. possesses a single thylakoid membrane between the layers of pigmented globules. The thylakoid is in close proximity to the back side of the globular layer. Between the thylakoid and the subsequent layer of globules there is an extensive space that is filled with a granulated substance. This variation is found in the multilayer stigma of *Chlamydomonadaceae* and the four-flagellar *Hafniomonas* Ettl and Moestrup [Foster and Smyth, 1980; Ettl and Moestrup, 1980; Melkonian and Robenek, 1984].

2) *Tetraselmis* Stein. possesses exaggerated thylakoids located between globular layers and contacting both adjacent globular layers. This variation is found in *Tetraselmis* Stein. and *Carteria* Dies., and zoospores of *Schizomeris leibleinii* Kütz., *Uronema belkae* G.M. Lokhorst, *Ulothrix zonata* Kütz., and gametes of *Acetabularia mediterranea* Lamour. [Manton and Parke, 1965; Parke and Manton, 1965, 1967; Crawley, 1966, 1970; McLachlan and Parke, 1967; Birkbeck et al., 1974; Melkonian and Robenek, 1979, 1984; Foster and Smyth, 1980; Hertz et al., 1981].

3) *Pyramimonas* Schmarda is characterised by the absence of thylakoids. The intervals between the pigmented layers are electron transparent and contain a fibrous substance that connects to the back surface of the globular layers. This type of structure is found only in species of a genus of coloured flagellates, except for the colourless representatives of *Polytoma* Ehrenb. and *Polytomella* Aragao. Their eyespots are located in leucoplasts that are devoid of thylakoids. The back of the second globular layer is connected to the thylakoids or with the opposite side of the chloroplast exterior in the two-layer eyes of *Pyramimonas orientalis* Butcher [Moestrup and Thomsen, 1974] located in the very thin anterior parts of the chloroplast. Even the most inner back globular layer does not contact the thylakoids within the two-layered eyes [Moestrup and Thomsen, 1974]. The various globular layers are connected to each other by a continuous transition at their edges. This results in a rigid design between layers with constant intervals between separate layers and the absence of any external supporting structures between these layers (e.g., thylakoids) [see review by Melkonian and Robenek, 1984].

One additional type of eyespot structure, where the globular layers are located inside the thylakoids, has been described in the literature, however, this variant is thought to be an erroneous [Foster and Smyth, 1980] interpretation of the micrographs of the eyes of *Carteria turfosa* Fott [Joyon and Fott, 1964], *Carteria crucifera* Korsch. [Lembi and Lang, 1965] and an unusual sea algae *Chlamydomonas reginae* Ettl and Green [Ettl and Green, 1973]. Based on their structure, the eyes of these species actually belong to *Tetraselmis* Stein. [Melkonian and Robenek, 1984].

The stigma of many green algae contains only one layer of pigmented globules that are located at the chloroplast surface and closely adjoin the plasmalemma. Such single-layered stigma are found in *Mantoniella squamata* (Manton and Parke) Desikach., *Monomastix* Scherff., *Nephroselmis* Stein., in some species of *Chlamydomonas*, and in mobile reproductive cells of many representatives of *Chlorococcales, Chlorosarcinales, Bryopsidales, Ulvales, Ulotrichales*, and *Chaetophorales* [reviews see: Foster and Smyth, 1980; Melkonian and Robenek, 1984]. The pigmented globules of single-layered stigma are separated by either the chloroplast lamella or a thylakoid-less granular substance. Despite differences in morphology and stigma structure in green algae, all appear to function based on the principle of a simple or quarter-wave interference reflector [reviews see: Foster and Smyth, 1980; Melkonian and Robenek, 1984].

The distance between the external chloroplast membrane and the plasmalemma of the stigma associated in a cross-section of species of algae (~50) [Melkonian and Robenek, 1984]) ranges from 10 to 53 nm (24.5 nm average). The interval is wider (30 to 53 nm) in single celled species of *Pedinomonas* Korsch., *Mesostigma* Lauterborn, and *Nephroselmis*, while the mobile reproductive cells of multicellular species *Microthamniales, Ulotrichales* and *Ulvales* are characterised by a narrow (10 to 17 nm) interval.

Species of algae can be separated into 5 principal categories base on the stigma location near the cell surface. These are type I (*Prasinophyceae* (*Mantoniella* Desikach., *Mamiella* Moestrup) [Melkonian and Robenek, 1984], type II (*Pedinophyceae* [Melkonian and Robenek, 1984; Ettl and Manton, 1964]), type III (*Chlorodendrophyceae* and *Prasinophyceae* with four-flaggelar cells of *Tetraselmis* and *Pyramimonas*) [Melkonian and Robenek, 1984], type IV (*Nephroselmis* and *Prasinophyceae*), and type V (many green algae of class *Chlorophyceae* sensu Mattox and Stewart, *Ulvophyceae* Stewart and Mattox, and *Microthamniales* sensu Melkonian [Watson, 1975; Melkonian, 1984; Melkonian and Robenek, 1984].

During reproduction, the photoreceptor apparatus in green algae can be transferred three ways [Barlow and Cattolico, 1980; Melkonian, 1981; Melkonian and Robenek 1984]. While a discussion of possible combinations of pigmented globules and thylakoids in the morphology, fine structure and location of the stigma of green algae, historical development of photoreceptor systems, and photobehaviour of green algae during evolution is not within the scope of this book, useful information can be found in the following references [Foster and Smyth, 1980; Ettl and Moestrup, 1980; Melkonian and Robenek, 1984; Massjuk et al., 2007].

The complexity and variety in the structure and location of photoreceptor systems in green algae allow their use in taxonomic systematic and phylogenetics. The following characteristics are the most useful for these purposes: the shape and a location of the stigma, quantity and the size of globules, character of their arrangement in layers, number of layers, specialization of photoreceptor membranes, and method of inheritance of the photoreceptor system by daughter cells [Melkonian and Robenek, 1984].

The greatest diversity in photoreceptor systems is found among representatives of the *Prasinophyceae* and correlates with other ultrastructural traits (e.g., structure of the locomotor apparatus, mechanisms of mitosis, cytokinesis) reiterating the heterogeneity of this taxon. Four different types of photoreceptor systems can actually be observed within the species of *Pyramimonas*. Such correlations underscore the high phylogenetic weight of the traits.

An absolutely unique position in the complex of morphological and ultrastructural traits (including peculiarities in structure, location, and photoreceptor inheritance) is held by the green alga *Pedinophyceae*. Likewise, there is a unique type of photoreceptor system in representatives of *Microthamniales* [Melkonian and Robenek, 1984]. Various types of photoreceptor systems are also found in species of *Carteria* and *Chlamydomonas*. Distinctions in the structure of stigma are found in species of the genus *Chlamydomonas* (e.g.. *C. reinhardtii* P.A. Dang, *C. moewusii* Gerloff), that are characterized by differences in the structure of a cellular envelope, flagellar apparatus, and reproductive behaviour [see review by Melkonian and Robenek, 1984].

The photophobic responses of the unicellular alga *Haematococcus pluvialis* Flotow and stimulus-response curves at four different wavelengths, were interpreted with regard to the structure of the photoreceptor apparatus [Cecconi et al., 1996]. The authors confirmed that the photoreceptors are located in the stigma region and suggested the presence of two photoreceptors.

Each of these examples is relevant with regard to the evolution and classification of green algae. Unfortunately, the functional value of the structural types of photoreceptor systems has not been elucidated until now, since information on the photobehaviour of green algae has been based on only few species. Of particular interest are small, asymmetric green flagellates that are considered an early evolutionary stage in green algae [Mattox and Stewart, 1984; Melkonian, 1984]. Despite a scarcity of data, an attempt has been made to construct the probable evolutionary sequence in the development of photoreceptor systems and photobehaviour in green algae and while speculative, it is of considerable interest [Melkonian and

Robenek, 1984]. Photoreceptor pigments (such as rhodopsin or bacteriorodopsin [Foster and Smyth, 1980; Foster et al., 1984]) that are spread chaotically in the plasmalemma were inherent to ancestors of green algae (bacteriorodopsin found in cytoplasmatic membranes was already present in the prokaryotic *Halobacteria* belonging to the *Archaebacteria* group that is connected to the predecessors of eukaryotes [Fox et al., 1980; George et al., 1983; Schnabel et al., 1983]). Such colourless flagellates, with one or two flagella that produce wavy movements and contain photoreceptor pigments within the plasmalemma, are capable of undergoing elementary photophobic reactions [Melkonian, 1983, 1984]. In addition, the concentrations of photoreceptor pigments led to strengthening of the light signal, and localization in the cell opposite the flagella attachment. The presence of a stigma that contains carotenoid pigments with absorption properties similar to rhodopsin and the phototopotaxis ability of the cell were thought to have evolved after the inclusion of chloroplasts. The photoreceptor system, with the arrangement of the photosynthetic organelle found under optimum light conditions, probably evolved earlier in the evolution of green algae. Originally it included one layer of pigmented globules that provided shading for the photoreceptor and increasing contrast in light illuminance at different positions of the organism relative to the light source. This layer was likely formed by chloroplast lipid globules (plastoglobules) and it did not have a fixed position. Subsequently the size of the globules and the distance between them and a photoreceptor became fixed forming the photoreceptor system.

Transfer of flagella from an apical position and changes of their movement character (from wavy to a rowing movement) resulted in a transfer of the photoreceptor system to a lateral position in relation to the attachment site of the flagella and perpendicular to the direction of movement of a cell.

The occurrence of communication between the microtubular system that provides a fixed position in relation to a plane of flagellar beatings and the subsequent occurrence of multilayered stigmata represent further developments in the evolution of photoreceptor systems in green algae.

There was a subsequent loss of the photoreceptor systems in mobile reproductive cells of many green algae (including representatives of the class *Charophyceae* sensu Stewart and Mattox), especially in connection with the transition to a terrestrial way of life [Melkonian and Robenek, 1984].

Many evolutionary schematics are based on hypothetical guesses as to the course of evolution and development of the photoreceptor systems in green algae that require verification and confirmation. It is impossible to reject the possibility of numerous occurrences of photoreceptor systems in green algae in various evolutionary branches of the monophylogenetic kingdoms of green plants *Viridiplantae*. The fact that photoreceptor systems similar to those of green algae were found within other lineages in eucaryotic algae (e.g., *Dinophyta*) argues in favour of such a position. The independent occurrence of such photoreceptor systems in these departments is quite evident based on differences in the structure of the chloroplasts and composition of the photosynthetic pigments. The scientific value of such speculation lies in the possibility of increasing the precision of selecting models for further investigation that may lead to a greater understanding of the basic biology operative.


## 6.3. Structure of the Photoreceptor System of Dunaliella

### 6.3.1. Stigma

The species of *Dunaliella*, as the majority of phytomonads (with the exception of *D. paupera* Pascher), possess a photoreceptor system that includes a stigma and photoreceptor, though the location of the latter has yet to be established. In spite of differences in the morphology and fine structure of the eyespots in green algae, it is believed that all of them function on the principle of a simple or quarter-wave reflector (see reviews by Foster and Smyth [1980] and Melkonian and Robenek [1984]).

Stigma of *Dunaliella* species that have been studied consists of one- or two-layer of pigmented globules between 100-200 nm in diameter. The number of globules can increase during ontogenesis. The stigma is located in the subapical, subplasmalemma portion of the chloroplast stroma, free from thylakoids, though sometimes part of the globules maybe located between lamella [Eyden, 1975; Hoshaw and Maluf, 1981]. Thus, based on the location of the stigma and its fine structure, *Dunaliella* does not differ from other green algae.


### 6.3.2. Structure of the Photoreceptor

Elucidation of the structure of the photoreceptor in algae is an important aspect in our understanding of photomovement. Photoreceptor molecules consist of structures that have opposite charges and can possess a dipole moment. If all of the molecules are oriented similarly relative to the longitudinal axis of the cell, it is possible to envision the *dichroism* of the photoreceptor system.

The dichroic nature of photoreceptor in *Dunaliella* can be studied by observing their photomovement in polarised light focused on the sample using a microscope condenser and polarizer. The appearance of dominating directions in histograms of the angular distribution of the cells and the dependence of these directions on the plane of polarization indicates the dichroism of the photoreceptor.

The effect of polarised light can be interpreted in the following manner. If the photoreceptor system has a dichroic structure, the absorption of light takes place only when the dipole moment is parallel to the plane of oscillation of the electric field vector of the stimulating light. If the moment is perpendicular to the orientation of the dipole moment, absorption is absent (Fig. 6.2).

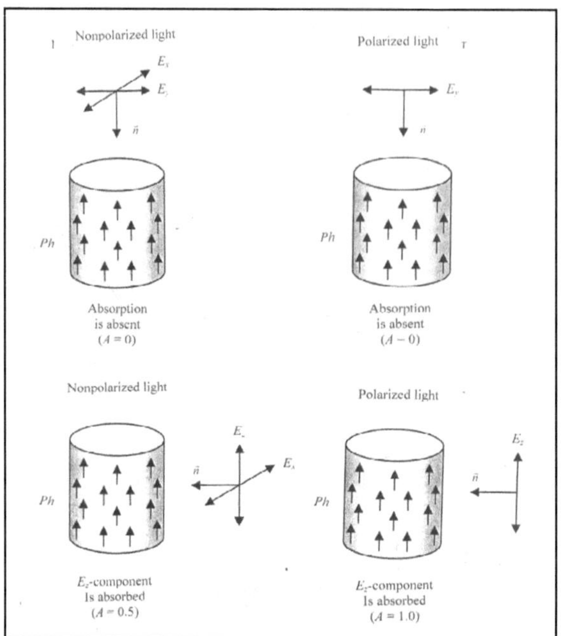


**Fig. 6.2.** Schematic of the relative orientation of dipole moments of photoreceptor molecules inside the photoreceptor *Ph* (vertical arrows) and direction of propagation $\vec{n}$ of stimulating light: *a* – non-polarised light (vectors $\vec{E}_x$ and $\vec{E}_y$ are perpendicular to dipole moments); *b* – non-polarised light (vector $\vec{E}_x$ is perpendicular to dipole moment and vector $\vec{E}_y$ is parallel to dipole moment); *c* – polarised light (vector $\vec{E}_y$ is perpendicular to dipole moments); *d* – polarized light (vector $\vec{E}_z$ is parallel to dipole moments); *a* and *c*: absorption is absent ($A = 0$); *b*: one component from both is absorbed ($A = 0.5$); *d*: the single component is absorbed ($A = 1$).

Maximal absorption of light takes place during movement of the cell parallel with the direction of light propagation [Creutz and Diehn, 1976]. Such a dichroism was found in *E. gracilis* [Häder, 1987*b*] and in *C. reinhardtii* [Yoshimura, 1994]. A simple mathematical model for the signal received by the dichroic photoreceptor molecules in *E. gracilis* when irradiated by polarized light is described by Hill and Plumpton [2000] and can be used to explain the experimental results of Häder [1987*b*].

Analysis of the angular distribution of *Dunaliella* cells moving under polarised light with an electric vector that is parallel to the plane of cell movement indicates that the distribution has a random character. The dominant directions of cell movement, which could be changed after rotation of the plane of polarisation, were absent [Posudin, 1992].

As mentioned earlier in Section 4.3.1, we did not find significant changes in cell linear velocity of either species in response to polarized and non-polarized white light at the same intensity (see Fig. 4.3). The same conclusion (i.e., the relative number $N_m/N_0$ of motile cells does not depend on the plane of polarization of the light) can be reached.

The absence of a dominating direction in histograms of angular distribution of the cells in polarised light and the dependence of the linear velocity of movement and the relative number of motile cells on the rotation plane of polarization also support the idea of a non-dichroic nature for the photoreceptor system. It is therefore evident that *Dunaliella* has a quite different photoreceptor system structure in comparison to *E. gracilis*.

### 6.3.3. Application of Two-Beam Irradiation to *Dunaliella* Cells

Irradiation of algae by two light beams, perpendicular to each other, is one of the methodo-logical approaches for investigating the photoreceptor's structure. Two sources of light producing light flows in two perpendicular directions and at an angle 30° to the surface of the slide with the algae suspension. The population of algae can move in the field of the two light flows in either direction or the resultant direction. The application of Fourier-transform to the effect of the two light flows of moderate illuminance ($E_1 = E_2 = 500$ lx) in perpendicular directions indicates the dominating maximum of angular distribution in sector 90°-180° (Fig. 6.3*a*). The first harmonic has a maximum amplitude in comparison with the others (Fig. 6.3*b*) and its phase corresponds to a direction of 135°. This along with an inverse Fourier-transform indicates movement of the population within one flow direction during positive phototopo-taxis (Fig. 6.3*c,d*).

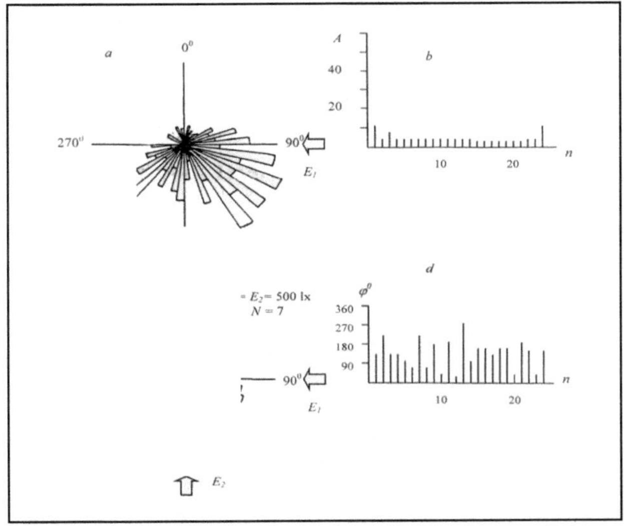


**Fig. 6.3.** Fourier-analysis of the angular distribution of *Dunaliella salina* cells due to two light flows of moderate illuminance ($E_1 = E_2 = 500$ lx) (see text for further explantion) [Posudin et al., 1991].

Increasing the illuminance to $E_1 = 10,000$ lx and $E_2 = 60,000$ lx leads to the appearance of a single maximum of angular distribution in sector 0°-45°. The first light beam is directed from 180° to 0° and the second – from 270° to 90° (Fig. 6.4*a*). The results of the Fourier-transform demonstrate the first four intense harmonics (Fig. 6.4*b*) that are characterised by phases that correspond to the direction of the 0°-45° sector (Fig. 6.4*c,d*). The population displays negative phototopotaxis (Fig. 6.4*c*) and orients away from both light sources.

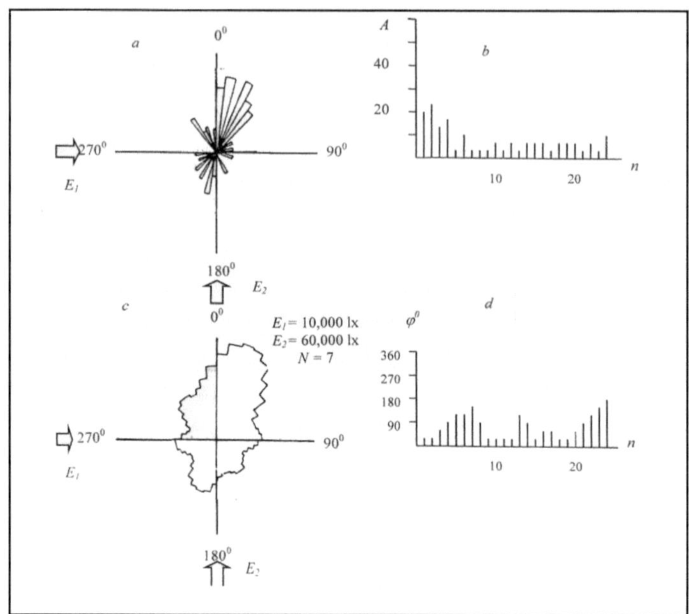

**Fig. 6.4.** Fourier-analysis of the angular distribution of *Dunaliella salina* cells due to two light flows of high illuminance ($E_1$ = 10,000 lx and $E_2$ = 60,000 lx) (see text for further explantion) [Posudin et al., 1991].

Comparison of our results with data that have been obtained for *E. gracilis* [Lebert and 1985; Häder, 1986b; Lebert and Häder, 2000] indicates a difference between *D. salina* and *E. gracilis* in photomovement when exposed to two perpendicular low illuminance beams. At low illuminance, the cells of *E. gracilis* display positive phototopotaxis and a bimodal distribution. The population splits into two fractions moving toward both light sources. In two strong light beams perpendicular to each other, the cells of both *D. salina* and *E. gracilis* move away from the light beams. These distinctions indicate two different mechanisms of photoreception at the moderate intensity. *D. salina* most likely exhibits periodical amplification or attenuation of the light signal by way of the stigma − a mechanism typical for green phytomonads. The photoreceptor system is located perpendicular to the surface of the cell and the direction of movement, and it has non-dichroic nature.

## 6.4. Summary

The photoreceptor system in the species of *Dunaliella,* as well as that of other green algae, consists of a photoreceptor, presumably located in the plasmalemma and in the chloroplast membranes near the stigma. The stigma, consisting of one-two layers of lipid globules in different species is located in the peripheral zone of the plastid. It has been shown that, in contrast to certain other algae species (e.g., *E. gracilis*), *Dunaliella* do not possess a photoreceptor with a dichroic structure.

As indicated in Chapter 5, the effect of ionizing and ultraviolet radiation indicates different mechanisms govern the linear velocity of movement and phototopotaxis in *Dunaliella* and the possible existence of two different photoreceptors that are responsible for different photomovement parameters.

# Chapter 7

# Identification of Photoreceptor Pigments

## 7.1. Characteristics of Photoreceptor Pigments

Rhodopsin, flavins, pterins and carotenoids are pigments that have been identified as participating in the photoreception and photomovement of eukaryotic algae [Lenci, 1975, 1995; Haupt and Häder, 1994; Kreimer, 1994; Lebert, 2001; Siebert, 2003]. Rhodopsin represents a protein (opsin), lipid and chromophore (retinal) complex. The absorption spectrum of rhodopsin has maxima at 231 and 278 nm (opsin), and at 350 and 500 nm (retinal). It fluoresces at 580 nm with a quantum yield $5 \cdot 10^{-3}$ in a digitonin solution [Konev and Volotovsky, 1979].

Flavins are isoalloxazine derivatives that can be presented by three forms – riboflavin (RF), flavin mononucleotide (FMN), and flavin adenine dinucleotide (FAD). The absorption spectrum of oxidized, non-ionized flavins in water is characterized by four bands found at 220, 265, 375 and 445 nm. The maximum fluorescence emission is at 520 nm with quantum yields of 0.29 (RF), 0.25 (FMN), and 0.038 (FAD) [Lenci, 1975].

Pterins are amphoteric molecules with weak acid and alkaline properties. The absorption spectrum of pterin has three (sometimes two) maxima. The position of absorption bands depends on the specific pigment, e.g., 240, 285, and 340 nm in leucopterin, 255 and 391 nm in xanthopterin, 252 and 385 nm in chrysopterin, and 240, 310 and 475 nm in erythropterin [Britton, 1986].

Carotenoids are tetraterpenes that are formed by eight isoprene subunits and are virtually ubiquitous in the plant kingdom. The spectrum of carotenoids is characterized by a wide absorption band in the 350-500 nm range with maxima at 425, 450 and 475 nm. Their fluorescence quantum yield is very small, i.e., less than $10^{-5}$ [Nobel, 1973].

The nature of alga photoreceptor pigments can be determined on isolated pigments using a cross-section of methods, e.g., microspectrofluorometry and microfluorometry, determining the action spectra of photobiological reactions, and investigating of the effects of specific chemicals and substitutes. These methods have resulted in the development of a fairly sound understanding of the pigments involved in the photomovement in the euglenophyte alga *Euglena gracilis* Klebs and green algae *Chlamydomonas reinhardtii* P.A. Dang. and *Haematococcus pluvialis* Flotow.

## 7.2. Identification of Photoreceptor Pigments in *Euglena gracilis*

### 7.2.1. *Euglena gracilis* Photoreceptor Pigments

Identification of the photoreceptor pigments in *E. gracilis* has not been without controversy. One group of scientists [Diehn and Kint, 1970; Lenci, 1975; Colombetti and Lenci, 1980; Doughty and Diehn, 1980; Lebert, 2001] consider flavins and pterins to be the primary photoreceptor pigments, while a second group [Gualtieri et al., 1989, 1992; Gualtieri, 1993; Sineshchekov and Spudich, 2005; Barsanti and Gualtieri, 2007;] favor rhodopsin. Opsin-like pigments have been identified in a wide range of vertebrates and invertebrates including unicellular organisms [Martin et al., 1986]. In spite of the variety of photoreceptor systems found, all have similar vision transformers – rhodopsin-like proteins that consist of seven transmembrane α-spiral receptors that are linked with retinal at the chromophore. Specific characteristics of retinal-opsin complexes include intense light absorption in the 380-640 nm range, the ability of retinal to isomerize in response to light, and structural changes (motion of the α-spiral) that are induced by retinal isomerization.

The arguments presented by both sides of the debate, those favoring one or another pigment system in *E. gracilis*, are based on a diverse range of methodical approaches and are of considerable historical interest.


### 7.2.2. Pigment Isolation

Batra and Tollin [1964] initially isolated the pigment from the stigma of *E. gracilis*, strain Z. The absorption spectrum of a suspension of the isolated stigma granules identified the presence of maxima at 490, 462, and 436 μm and a small shoulder at 387 μm. The action spectrum for phototaxis follows this spectrum very closely. All of the pigments appeared to be carotenoids. Subsequently cellular structures that contain (or that can contain) photoreceptor pigments were isolated. The paraflagellar body was first isolated by Rosenbaum and Child [1967].

Flagella were isolated from *E. gracilis* and from the closely related *Astasia longa* Pringsheim. The flagellar of the latter, which does not possess a paraflagellar body, was used as a control [Brodhun et al., 1994; Brodhun and Häder, 1990, 1995]. Using liquid chromatography, six major protein fractions were separated and identified. Fluorescence spectroscopy indicated the presence of flavin and pterin binding proteins in the paraflagellar body of the flagella sample while these chromoproteins were not been found in the *Astasia* control.

Several investigators [Gualtieri et al., 1986, 1988; Gualtieri, 2001] successfully extracted the paraflagellar body and separated the flagellar apparatus of *E. gracilis* using a high contentration $CaCl_2$ solution. Comparing the absorption spectra of the paraflagellar body and rhodopsin with maxima at 500 nm indicated the possible presence of a rhodopsin-like protein in the paraflagellar body. Extraction of retinal from intact and membrane-less cells further supported the presence of rhodopsin in the photoreceptor [Gualtieri et al., 1992].

A riboflavin-binding protein was also purified from isolated flagella of *E. gracilis* [Neumann and Hertel, 1994]. It was thought that the protein was part of the flagellar membrane and not within the paraflagellar body. It is possible that the flagellar riboflavin-binding protein has a functional role in the biochemical sequence between the receptor of the phototactic stimulus and the motile response.

A photoactive protein (*Erh*) was subsequently isolated from the photoreceptor of *E. gracilis* that had a photocycle resembling that of sensory rhodopsin but with at least one stable intermediate [Barsanti et al., 2000]. The absorption and fluorescence measurements suggested that *Erh* is a rhodopsin-like protein.


### 7.2.3. Microspectrophotometry and Microfluorometry of Pigments

Optical spectroscopy methods have proved to be an effective means of studying the primary characteristics of photoreceptor pigments [Lenci and Ghetti, 1989; Cubeddu et al., 1991]. Microspectrophotometry involves the transmission of a focused beam of light of variable wavelength through the organelle of interest. Microspectrofluorometry has been used to investigate *in vivo* the paraflagellar body of *E. gracilis* [Benedetti and Checcucci, 1975; Benedetti and Lenci, 1977; Ghetti et al., 1985] indicating the presence of flavins in the organelle. The presence of flavins and pterins in the paraflagellar body was subsequently confirmed by Galland et al. (1990). Preparations from the flagella of *E. gracilis* containing the paraflagellar body exhibited fluorescence maxima at 520 and 450 nm which is typical for flavins and pterins, respectively. The fluorescence emission spectra of pterins purified using thin-layer chromatography were similar.

Microspectrophotometry was used to investigate the paraflagellar bodies of isolated flagella from *E. gracilis* [Schmidt et al., 1990]. Flagella with attached paraflagellar bodies were separated from the cell bodies using a short exposure to near-UV light. Fluorescence emission spectra (excitation at 365 nm) of single paraflagellar bodies had maxima near 470 and 520 nm, indicating the presence of pterins and flavins. The occurrence of pterin-like fluorescence in the paraflagellar body lends further support to the earlier proposal that pterins as well as flavins may function as photoreceptor pigments for near-UV and blue light [Schmidt et al., 1990]. The fluorescence spectra of isolated paraflagellar bodies from *E. gracilis* excited at different wavelengths indicated the presence of more than one (flavin and pterin) photoreceptor pigment [Sineshchekov et al., 1994].

Fluorescence spectroscopy has indicated the presence of an energy transfer between pterins (pigments absorbing in the ultraviolet portion of the spectrum) and flavins [Sineshchekov et al., 1994]. With the exception of fluorescence at 440–520 nm, long-wave fluorescence at 580 nm (excitation wavelength 520 nm) and 620 nm (excitation wavelength 550 nm) was observed. The role of long-wave fluorescence is not clear. The presence of a third photoreceptor pigment was postulated by the authors. The participation of rhodopsin, however, was excluded due to the insignificant fluorescence quantum that is indicative of rhodopsins [Sineshchekov and Litvin, 1987]. Likewise, contrasting the temporal kinetics of rhodopsin fluorescence at 397 nm (excitation at 365 nm) that had a photocycle duration of about 18 s [Barsanti et al., 1997] with the real fluorescence kinetics at a frequency of 1-2 Hz induced by rotation of the cells [Ascoli et al., 1978] argues against this position.

Flavins were found not only in the paraflagellar body but also in the stigma of *E. gracilis* [Sperling et al., 1973]. The absence of stigma absorption dependence on the plane of stimulated light polarization argues against the participation of the stigma in photoreception [Benedetti et al., 1977]. In contrast, phototopotaxis is sensitive to changes in the polarization plane [Bound and Tollin, 1967].

The preceding experimental methods used by advocates of the flavin-pterin hypothesis supported the conclusions that: 1) the paraflagellar body of *E. gracilis* is the location of the photoreceptor; 2) the stigma is not a photoreceptor; 3) flavins are the primary chromophores for photoreception; and 4) the pterins function as a "light antenna" for the flavins.

High-performance liquid chromatography was used to identify the pigments associated with the flagellum of *E. gracilis*. A wild-type strain that contains FMN and FAD was compared with two mutant strains that did not display phototactic reactions and photoaccumulation in response to blue light. Flagella of the wild type contained flavin isoalloxazine derivatives, FMN and FAD. The mutants, which lacked the stigma but retained a small paraxonemal body, contained less flavins. The white mutant (FB), which retained a residual stigma and a paraxonemal body and the white phytoflagellate *A. longa*, a close relative of *Euglena*, had normal amounts of flagellar flavins. An unidentified pterin-like pigment (Pt16) was found in the cells and flagella of *Euglena*. Small amounts of the pigment were found in *Astasia* and the *Euglena* mutants. A third pigment (P528) was present mainly in the flagella. The three pigments probably participate in photoreception [Geiss et al., 1997].

Another group [Gualtieri et al., 1988] believed, based on microspectrofluorometry of the paraflagellar body (*E. gracilis*), that rhodopsin was responsible for photoreception. The close similarity of the absorption spectrum of the paraflagellar body with that of rhodopsin with a maxima at 500 nm supports this position [Gualtieri et al., 1989; James et al., 1992]. The intensity of the absorption band of the paraflagellar body depended on the relative orientation of the polarization plane and longitudinal axis of the cell. This suggested that a rhodopsin protein located in the quasi-crystalline structure of the paraflagellar body functions as the phototeceptor [Gualtieri et al., 1989; James et al., 1992].

The possible contribution of other cellular components that absorb light has been indicated as a possible weakness of the methods. In addition, the absorption bands of practically all of the pigments in the visible region of spectrum are superimposed to some degree (Fig. 7.1). However, extraction of retinal from intact cells and the photoreceptor after isolation from membrane-less cells, further supports rhodopsin as the photoreceptor of *E. gracilis* [Gualtieri et al., 1992]. A photochromic pigment was found in the paraflagellar body demonstrated a reversion of fluorescence [Barsanti et al., 1997].

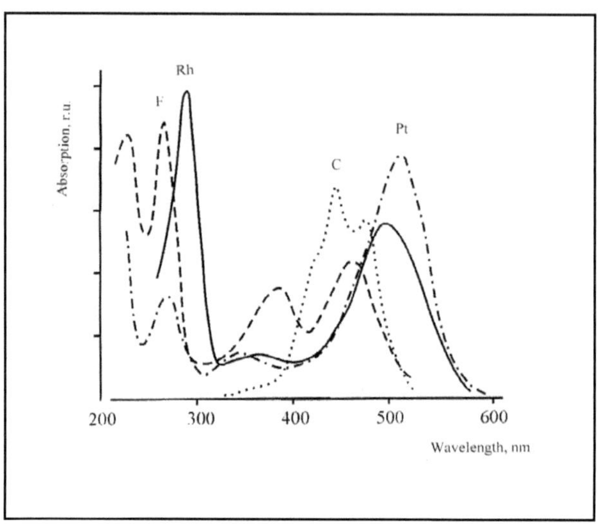


**Fig. 7.1.** Absorption spectra of the photoreceptor pigments: F – flavins [Britton, 1986], Ph – rhodopsin [Bensasson, 1980], C – carotenoids [Bensasson, 1975], and Pt – pterins [Britton, 1986].

Thus, the paraflagellar body is characterized by optical bistability – the primary nonfluorescent and secondary fluorescent forms appear after light excitation, further substantiating the paraflagellar body as photoreceptor in *E. gracilis*. However, since other fluorophores may be located in the photoreceptor or near organelles, the precise identification of the photoreceptor pigments is difficult.

## 7.2.4. Determination of the Action Spectra for Photobiological Reactions

Action spectra give an indication of the dependence of parameters characterizing photomovement (e.g., velocity, direction of movement) that are normalized to the intensity of the light stimulus or the number of incident photons, on the wavelength of the light. An action spectrum can be used as an indication of the functional pigments, such as those that are responsible for photomovement in microorganisms [Foster, 2001]. For example, the action spectrum for negative phototopotaxis in *E. gracilis* is characterized by primary maxima at 385 and 460 nm, and by small maxima at 410 and 490 nm [Häder and Reinecke, 1991]. Diehn [1969] found the action spectrum for the negative photophobic reaction, which should most closely resemble the absorption spectrum of the photoreceptor pigments, exhibits major sharp peaks at 450 nm and 480 nm, a minor peak at 412 nm, and a major broad peak around 365 nm. In contrast, the action spectrum for positive phototopotaxis and photophobic reactions in *E. gracilis* is within the 350 – 800 nm range [Diehn, 1969]. Positive photopotaxis and positive

86

photophobic reactions exhibited almost identical action spectra with peaks at 480 nm and 375 nm. Diehn [1969] supposed that these action spectra probably represent a composite of the absorption spectra for the photoreceptor and the screening pigments.

A number of authors indicate that the shape of the spectrum reflects the absorption properties of flavin-containing proteins. In general the experiments have focused on the measurement of action spectra of *E. gracilis* [Wolken, 1960, 1977; Bensasson, 1975; Lenci, 1975; Nultsch and Häder, 1979, 1988; Colombetti and Lenci, 1980; Galland et al., 1990] and have indicated the presence of more than one photoreceptor pigment (e.g, flavins, carotenoids, pterins).

A novel flavin-binding photoactivated adenylyl cyclase (EC 4.6.1.1, also known as adenylate cyclase and adenyl cyclase) consisting of two subunits has been isolated from the paraxonemal body [Ntefidou et al., 2003*a*]. This photoactivated adenylyl cyclase controls phototaxis in *E. gracilis* [Ntefidou et al., 2003*b*] and is the blue-light receptor flavoprotein identified as a sensor for photomovement in *E. gracilis* [Iseki and Watanabe, 2004; Iseki et al., 2006].

A chlorophyll-free strain of *E. gracilis* var. *bacillaris*, possessing stigma and a photoreceptor, displays negative phototopotaxis. The action spectrum for the reaction is characterized by a maxima at 410 nm and secondary peaks at 449 and 476 nm; above 530 nm no reaction was detected [Gossel,1957]. Another strain of *E. gracilis* with a photoreceptor but without a stigma displayed both positive and negative phototactic reactions. The maxima for positive phototopotaxis occurred at 410 and 425 nm, while negative phototopotaxis was at 449 and 476 nm [Gossel, I., 1957].

The colorless euglenoid flagellate, *Astasia fritschii* Pringsheim and Hovassee, exhibits a pronounced step-up photophobic response (i.e., elicited by a sudden increase in light intensity) and rather weak photokinesis. The action spectrum has a major peak at 450 nm suggesting that flavins may be the pigments responsible for the step-up response [Mikolajczyk and Walne, 1990].

In principle, the spectral sensitivity of photoreactions reflects the absorption properties of the key pigments operative, however, the analysis of pigments is much more complicated. For example, the possible participation of pigments that are present in shading structures (e.g., the stigma) or the presence more than one pigments in the photoreceptor structure confound the direct interpretation of the action spectrum obtained. In addition, the action spectra of photoreactions are seldom of high resolution and other pigments may have overlapping absorption bands in the visible region of the spectrum. As a consequence, precise correspondence in the patterns of the photoreceptor pigment and the action spectrum do not often occur. Finally, ascertaining the photoreceptor pigments by comparing the action spectra for photomovement parameters with the pigment absorption spectra is further complicated by alterations in the pigments due to their surroundings.


### 7.2.5. Biochemical Methods

Biochemical analysis of isolated paraflagellar bodies from *E. gracilis* has indicated the presence of four types of proteins, three of which contain pterins and the fourth, flavins [Brodhun and Häder, 1990, 1995]. This was further substantiated by the fact that ultraviolet radiation, where flavins and pterins demonstrate strong absorption, inhibits phototopotaxis in *E. gracilis*. The inhibition could be due to the high level of absorption of the shortwave ultraviolet region of the spectrum by flavins and pterins. In addition, the proteins of the paraflagellar bodies are destroyed by ultraviolet radiation. These proteins appear to play an important role in the reception of the light stimulus responsible for phototopotaxis.

## 7.2.6. Effect of Exogenous Chemicals on Photomovement

The application of compounds that prevent oxidative phosphorylation (e.g., cyanide, azide, rotenone, 2,4-dinitrophenol, 2,6-dichloroindophenol) in the photomovement of *E. gracilis* did not affect the parameters of motility and positive phototopotaxis equally. In contrast, compounds that prevent photophosphorylation (e.g., salicylaldoxime, tetramethylthiuram disulfide) inhibit both motility and phototopotaxis. Other compounds [e.g., ethanol, (dichlorophenyl)dimethylurea, methyl octanoate, carbonyl cyanide *p*-(trifluoromethoxy) phenylhydrazone)] have been shown to specifically inhibit positive phototopotaxis [Diehn and Tollin, 1967]. Based on these results, it was concluded that the primary energy source for positive photopotaxis is the photophosphorylation system of photosynthesis.

Additional confirmation of flavins functioning as the photoreceptor pigments was obtained using chemicals that affected their excitation state (e.g., potassium iodide). There are, however, some discrepancies in the results from different researchers. Several observed an effect of potassium iodide only on positive phototopotaxis in *E. gracilis* [Mikolajczyk, Diehn, 1975], while others found a strong effect on either positive or negative phototopotaxis [Lenci et al., 1983]. These contradictory results may be due to the effect of circadian rhythm and/or the age of the cultures on the response [Häder and Lebert, 2001].

The effect of nicotine, a highly biologically active compound, was tested at various concentrations on the photoreceptor in *E. gracilis* using phase-contrast microscopy, fluorescence microscopy, transmission electron microscopy and photobehaviour. Nicotine inhibited the photoaccumulation of the cells. The authors [Barsanti et al., 1993] concluded that the results strongly confirmed the presence of a carotenoid or carotenoid-like chromophore as the photoreceptor. However, the application of nicotine (that inhibits the biosynthesis of retinal) and hydroxylamine (that responds to retinal, either free or opsin-bonded forms) inhibits the formation of the paraflagellar body and photoorientation of the cells. Collectively the results suggest that retinal in the paraflagellar body acts as the photoreceptor pigment [Barsanti et al., 1992, 1993]. However, Foster [2001] argues that the inhibition of the formation of the paraflagellar body does not adequately support the participation of rhodopsin in photoreception, a position supported by the inhibition of phototopotaxis in *Chlamydomonas* by nicotine and norflurazon. Even though rhodopsin is present in *Chlamydomonas* cells, it does not substantiate its role as the photoreceptor pigment in that the effect of the action of specific inhibitors can be obtained by other means.

Formation of the paraflagellar body in *E. gracilis* can be dependent on rhodopsin. A more precise analysis demonstrated the absence of the effect of the inhibitor on both positive and negative phototopotaxis during extended exposure (>6 months) to nicotine, while retinal was identified in the cells using chromatography [Gualtieri et al., 1992].


## 7.2.7. Introduction of Alternative Photoreceptor Pigments

Another approach to identifying the photoreceptor pigment is by introducing substitutes for the pigment. For example, the introduction of roseoflavin which has a distinctive spectral shift in absorption maximum to the red region of the spectrum (i.e., up to 600 nm, in comparison to riboflavin, that has a maximum at 500 nm) into the paraflagellar body instead of the natural flavins (e.g., riboflavin). The action spectrum for phototopotaxis in wild-type cells of *E. gracilis* did not extend above 520 nm while the cells containing roseoflavin (i.e., cultured with roseoflavin for 30 days) displayed spectral sensitivity for phototopotaxis that shifted into the red region of the spectrum (550-580 nm) [Häder and Lebert, 1998].

Evidence for flavins and/or pterins as the photoreceptor pigments in *E. gracilis* includes:

1). The action spectra for phototopotaxis (i.e., wavelength dependence for response sensitivity) very closely resembles the absorption properties of flavins and there is a close relation between the absorption properties of the paraflagellar body and the known action spectra, thus further substantiating this structure as the location for the photoreceptor;

2) Fluorimetric measurements such as microspectrofluorometry of the pigments *in vivo* indicate an energy transfer from pterins to flavins in intact paraflagellar bodies; and 3). Several proteins containing flavins and pterins have been identified in the structure [Lenci and Ghetti, 1989; Cubedduet al., 1991; Lebert, 2001].

The following arguments are presented by supporters of the rhodopsin hypothesis [Gualtieri et al., 1989, 1992; Gualtieri, 1993; Barsanti and Gualtieri, 2007]. These include theoretical considerations, absorption microspectroscopic measurements, experiments inhibiting the biosynthesis of carotenoids, the photobehaviour in the presence of hydroxylamine hydrochloride, and the identification of all-*trans*-retinal by thin layer chromatography, high performance liquid chromatography (HPLC), absorption spectroscopy, and gas chromatography-mass spectrometry (GC-MS), each of which supports rhodopsin as the photoreceptive molecule in *E. gracilis*.

Neither supporters of flavin (pterins) nor rhodopsin as the photoreceptor believe the opposing group has adequately substantiated their case, regardless of the diverse range of analytical methods employed. It is evident that a lively scientific debate will continue until the question is finally resolved.


## 7.3. Identification of Photoreceptor Pigments in Green Algae

The action spectrum for positive phototopotaxis in *C. reinhardtii* displays a wide band in the 400-600 nm region with a maximum at 500 nm [Foster and Smyth, 1980] to which the authors believe indicates the participation of retinal in photoreception. Substitution of the natural photoreceptor with endogenous retinal analogues resulted in a shift in the action spectra for phototopotaxis pointing toward rhodopsin as the photoreceptor. The effective cross-section of absorption ($\sigma = 0{,}8 \cdot 10^{-20}$ m$^2$) supports this contention in that cross-sections of flavin and rhodopsin are known to be $0{,}48 \cdot 10^{-21}$ m$^2$ and $1{,}5 \cdot 10^{-20}$ m$^2$, respectively [Sineshchekov, 1991*b*]. Rhodopsin was also identified as the photoreceptor in the green algae species *H. pluvialis* Flotow [Sineshchekov et al., 1991*a,b*; 1994; Hegemann and Harz, 1993; Hegemann, 1994; Spudich et al., 1995; Ridge, 2002; Suzuki et al., 2003; Yoshikawa, 2005; Sineshchekov and Spudich, 2005].

An assessment of the photoreceptor in *C. reinhardtii* was made by comparing the photomovement of normal cells with those deprived of carotenoid pigments. In *H. pluvialis*, identification of the photoreceptor was based on the analysis of the action spectrum for photoreactions and the membrane potentials and electric currents passing through the membranes. In addition, a number of studies utilized a genetic approach, comparing mutants and retinal analogs.

The mutant strain CC-2359 of *C. reinhardtii* that has little pigmentation was found to be devoid of photophobic stop responses to photostimuli over a wide range of light intensities. The photophobic responses of the organism were restored by exogenous introduction of all-*trans* retinal [Lawson et al., 1991].

"Blind" *C. reinhardtii* cells of carotenoid-deficient mutants were studied through the addition of retinal analogs such as all-*trans*-retinal, 9-demethylretinal, or dimethyloctatrienal. These analogs restored the photoreceptor current and regenerative response of the alga indicating that both photoreceptor current and regenerative response were initiated by the same or similar rhodopsins that contain an archaebacterial-like chromophore(s) [Sineshchekov et al., 1994].

Recent advances in genetic techniques have facilitated the isolation of the genes causing mutations. A genetic approach was utilized for *C. reinhardtii* [Pazour et al, 1995] and seven new genes were identified (ptx2-ptx8). Some mutants had defects in the ability of their axonemes to respond to changes in $Ca^{2+}$ concentration, while the others exhibited a defective flagellar $Ca^{2+}$ channel. As a consequence, some mutants displayed one reaction (phototaxis) but not the other (photoshock) in response to $Ca^{2+}$ treatment.

Mutants ppr1, ppr2, ppr3, and ppr4 do not display a photophobic response but display phototactic activity at nearly the same level of sensitivity as wild type cells [Matsuda et al., 1998]. It was shown that with the all-or-nothing flagellar current, a $Ca^{2+}$ current generated by depolarization of the flagellar membrane was absent or seriously impaired in the mutants. These findings clearly indicated that the all-or-nothing flagellar current is necessary for the photophobic response but not for phototopotaxis. Similarly, a complementary DNA sequence from *C. reinhardtii* that encodes a microbial opsin-related protein was studied by Nagel et al. [2002]. This protein, termed channel rhodopsin-1, is believed to be involved in the phototactic activity of the algae.

In *C. reinhardtii*, the pigmented eye comprises the optical system and has at least five different rhodopsin photoreceptors [Kateriya et al., 2004]. Two of them, the channel rhodopsins, are rhodopsin-ion channel hybrids that switch between closed and open states by photoisomerization of the attached retinal chromophore. Channel rhodopsin-2, a light mediated gated cation channel, switches a cation-selective ion channel. The system employs 11-*cis* retinal as its chromophore [Flannery and Greenberg, 2006].

Rhodopsins, identified from cDNA sequences, function as low- and high-light-intensity phototactic receptors in *C. reinhardtii* [Ebrey, 2002; Sineshchekov et al., 2002; Govorunova et al., 2004]. The function of the two rhodopsins (CSRA and CSRB) phototactic receptors was demonstrated by *in vivo* analysis of photoreceptor electrical currents and motility responses in transformants using RNA interference (RNAi) directed against each of the rhodopsin genes. The kinetics, fluence dependencies, and action spectra of the photoreceptor currents differ greatly among transformants depending upon the relative amount of photoreceptor pigments that are expressed. The action spectrum of CSRA has an absorption maximum at 510 nm and mediates a fast photoreceptor current. The action spectrum of CSRB, in contrast, has a maximum at 470 nm and generates a slow photoreceptor current. The relative wavelength dependence of CSRA and CSRB activity for inducing phototactic responses precisely matches the wavelength dependence of the CSRA and CSRB generated currents, demonstrating that either receptor can mediate phototopotaxis. The discovery of two distinct sensory receptors (sensory rhodopsins A and B) responsible for phototactic activity of the alga was also reported [Ridge, 2004].

The effect of hydroxylamine on the phototopotactic activity of *C. reinhardtii* and the restoration of phototactic sensitivity after its removal appears to be due to the bleaching of rhodopsin and subsequent replacement via *de novo* synthesis or by added retinal [Hegemann et al., 1987, 1988].

Several researchers believe that the abundant retinal protein in the stigma of *Chlamydomonas* is not the photoreceptor for phototopotaxis and photophobic responses and that the responses are triggered by a yet to be identified rhodopsin species [Fuhrmann et al., 2001]. Incubation of the organism with retinal analogs shifts the phototopotaxis action spectral maximum in a manner similar to the shifts encountered in the bovine visual pigment [Nakanishi, 1985]. The author believes that the algal photoreceptor may be closely related to the bovine rhodopsin but not the bacterial rhodopsin.

The ability of different retinal isomers and analogs to restore photosensitivity in blind *Chamydomonas* cells (strain CC2359) was tested using flash-induced light scattering transients and by measuring phototaxis using a taxigraph [Hegemann et al., 1991]. They concluded that rhodopsin bears a closer resemblance to bacterial rhodopsins than to the visual rhodopsins in higher animals. The blind mutant FN68 lacks retinal and therefore cannot display phototopotaxis, however, with incubation in a media containing retinal analogs, phototopotaxis is restored [Nakanishi et al., 1989]

A real-time automated method was developed for simultaneously measuring phototactic orientation and the step-up photophobic response in flagellated microorganisms [Takahashi et al., 1991]. Addition of all-*trans* retinal restored both photoresponses in a carotenoid-deficient mutant strain of *C. reinhardtii* in a dose-dependent manner. The results strongly suggested that isomerization of the 13-14 double bond is important for photobehavioral signal transduction and that a single retinal-dependent photoreceptor controls both phototactic and photophobic responses. Zacks et al. [1993] concluded that although phototaxis and the photophobic response are mediated by retinal-containing receptors displaying a close similarity in chromophore structural requirements for both behaviors, the results indicate that differences exist between the two responses in their photoreceptor proteins and/or transduction processes.

Experiments with a *C. reinhardtii* mutant into which retinal and retinal analogs were incorporated [Takahashi et al., 1992] were interpreted as indicating negative and positive phototopotaxis are mediated by two rhodopsin species differing in their affinity for the exogenous chromophores. However, a more reasonable explanation, that requires fewer assumptions, is that the sign for phototaxis depends on a delay in intracellular photosignal transduction.

Around the same time, Kroeger and Hegemann [1994] proposed that photophobic responses and phototaxis in *C. reinhardtii* are triggered by a single rhodopsin photoreceptor. According to their hypothesis, chlamyrhodopsin triggers a photoreceptor current in the eyespot region, resulting in directional changes or phototopotaxis, however, when the light stimulus exceeds a critical level, flagellar currents appear and are the basis for the photophobic response.

The principal methodological approach that has been used in the investigation of photoreceptor systems in green algae was similar to studies on membrane potential or electric current that passed through cell membranes. Ions that pass through the membranes carry electrical charges that induce an electric current in the membrane of about $10^{-12}$ A. It is possible to measure these currents using microelectrodes within thin glass tubes. This technology is called the *patch clamp technique* and uses a micropipette attached to the cell membrane to allow recording a single ion channel. The micropipette has tip diameter about 0.5–1.0 μm which is placed on the surface of the cell membrane. Gentle suction is applied to the micropipette drawing a small area of the membrane (the "patch") into the microelectrode tip. The presence of the micropipette tip against the membrane and the suction forms a resistance "seal" between the glass and the cell membrane, electronically isolating the current across the membrane patch. The current measurements are expressed either for the entire protoplast or the portion that has been drawn into the microelectrode tip (Fig. 7.2). Thus, very small diameter micropipettes allow measuring the electric currents through separate membrane ion channels.

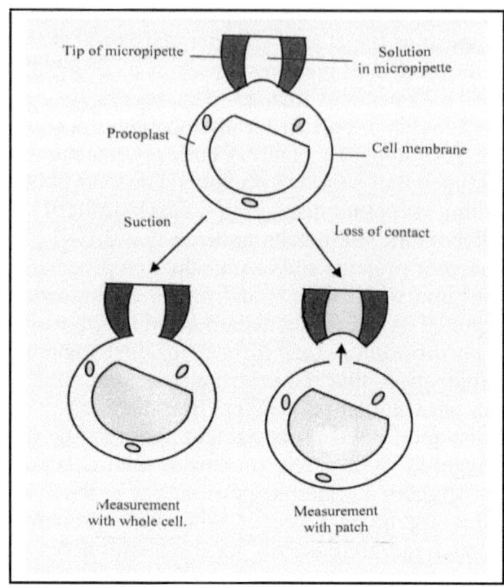


**Fig. 7.2.** Patch clamp technique for studying membrane potential

The specific membrane potential features in *C. reinhardtii* and *H. pluvialis* are the input resistance, 200–250 MΩ ($10^6$ Ω) for *C. reinhardtii* and several hundred MΩ for *H. pluvialis*, and input resistances of hundreds of GΩ (i.e., $10^9$ Ω). The clamp patch technique together with micro-irradiation of the cell has made it possible to establish that light-induced electric signals only appeared during irradiation of the stigma, further solidifying the presence of the photoreceptor in this area [Ristori et al., 1981; Sineshchekov et al., 1991*a*]. The technique has also allowed determining the action spectrum by monitoring the generation of the electric signal. Harz et al. [1992] concluded that the magnitude of the electric signal depends on the orientation of the cell and its photoreceptor relative to the light stimulus. The ratio of signals during irradiation of the photoreceptor and the posterior portion of the cell is approximately 8 for *C. reinhardtii* and 3 for *H. pluvialis* [Harz et al., 1992]. Differences in the ratio may be explained by structural differences in the stigma – four-layers in *C. reinhardtii* versus one in *H. pluvialis*. In addition, the stigma of *Chlamydomonas* functions as a quarter-wave plate that provides amplification of the light signal due to the reflection of the light waves from the layers and interference. Finally, the shape of the action spectrum for the generation of the electric signal is similar to the absorption spectrum for rhodopsin. The fine structure of the spectrum indicates a fixed location for the chromophore in a protein and the possible participation of more than one pigment in photoreception.

The microelectrode pipette technique was also applied to cell-wall-deficient *Chlamydomonas* cells [Holland et al., 1996] stimulated by short but intensive flashes of light. Three types of photocurrent were observed. 1) A transient current localized in the stigma that was identified as the photoreceptor current. The photoreceptor current consists of at least two components activated by different mechanisms. The first mechanism most likely involves translocation of ions across the membrane either by rhodopsin itself or through an ion channel directly coupled to it. 2) The mechanism of photoreceptor current generation appears to occur via a biochemical amplification cascade. Membrane depolarization, induced by the photoreceptor current, leads to an unbalanced motor response by the flagella, which is the basis for phototactic activity. This second transient current localized in the flagella is named the fast flagellar current. 3) The third current is a slow flagellar current, which has a very small amplitude. With an intense light stimulus, the photoreceptor current triggers a fast and a slow flagellar current that results in backward movement, stopping the organism. The photoreceptor current appears with a delay of less than 50 μs, suggesting that rhodopsin and the photoreceptor channel are closely coupled or form one ion channel complex.

It is believed that receptor photoexcitation in green flagellates triggers a cascade of rapid electric events in the cell membrane [Sineshchekov and Govorunova, 2001*a,b,c*]. The photoreceptor current has been the earliest process detectable thus far in the cascade. Measurement of the photoreceptor current at present has been the best approach for investigating the photoreceptor pigment, since the low receptor concentration in the cell makes existing optical and biochemical methods unsuitable. Existing physiological evidence indicates the photoreceptor in green flagellate algae is a unique rhodopsin-type protein.

Proteomic analysis (the study of the structure and function of proteins) of the stigma of *C. reinhardtii* identified 202 different proteins, in particular photoreceptors, retina(I)-related proteins, and members of putative signaling pathways for phototaxis such as casein kinase 1 [Schmidt et al., 2006].


## 7.4. Identification of the Photoreceptor Pigments in *Dunaliella*

### 7.4.1. Analysis of the Phototopotaxis Action Spectra in *Dunaliella*

The action spectra for phototopotaxis in both species of *Dunaliella* is within the 400–520 nm range and displays maxima at 410–415 nm and 465–475 nm [Posudin et al., 1991]. The action spectra for phototopotaxis in euglenoids (*E. gracilis* [Häder and Reinecke, 1991]), green algae (*C. reinhadtii* [Nultsch, 1971], *Platymonas subcordiformis* (Wille) Butcher (syn. *Tetraselmis subcordiformis*), *Stephanoptera gracilis* (Artari) G.M. Smith [Halldal, 1958]) and the dinophytes (*Peridinium trochoideum* (Stein) L. and *Goniaulax catenella* Whedon and Kofoid [Halldal, 1958]) have several general features. They have a maximum sensitivity in the blue-green (440–520 nm) region of the spectrum and very slight sensitivity (if at all) to light at wavelengths above 560 nm. The exception is *Prorocentrum micans* Ehrenb. (*Dinophyta*) which exhibits a maximum sensitivity in the red (640 nm) region of the spectrum. Representative cryptophytes (e.g., *Cryptomonas* sp. CR-1, *C. rostratiformis* Skuja, *C. nordstedtii* (Hansgirg) Senn) are characterized by an action spectrum for positive phototopotaxis with a maximum at 560 nm (yellow region of the spectrum) that is believed to be indicative of a photoreceptor of unknown nature [Watanabe, Erata, 2001].

Comparisons of the action spectra for phototopotaxis of the two *Dunaliella* species with the absorption spectra of a cross-section of well-known pigments, indicates that carotenoids, flavins, pterins or rhodopsin could function as photoreceptors. It is evident that more precise analytical techniques must be utilized to identify the photoreceptor pigment in these algae.


### 7.4.2. Application of Lateral Ultraviolet Irradiation

The photoreception response of algae to ultraviolet radiation makes it possible to determine the participation of the pigments (e.g., flavins) that absorb in this region of spectrum. Algae from the genus *Tetraselmis* Stein. – *Tetraselmis hazenii* ( = *Platymonas subcordiformis* (Wille) Hazen) respond to ultraviolet radiation [Halldal, 1961]. We have investigated photomovement of a representative of the same genus, *Tetraselmis viridis* (Rouch.) Norris et al.

(syn. *Platymonas viridis* Rouch.) strain N 68 from the collection at Institute of Botany of Ukrainian Academy of Science to compare *T. hazenii* and *T. viridis*.

An illuminator (LOS-2) with a 1 kW xenon lamp displaying a spectrum similar to solar radiation was used. The radiation was directed at the angle 30° to the plane of the cells positioned on a microscope stage. The radiation passed an infrared filter (a layer of distillated water) and interference filters with transmission maxima at 249, 279, 304, 310, 335, 364, 407, 434, 497, 541, and 655 nm. A wide-band ultraviolet filter (UVW) with a spectral band at 240-410 nm was also used. The spectral characteristics of these filters in comparison with absorption spectra of flavins and carotenoids are presented in Fig. 7.3. A unique feature of the experiments was the use of a 1.75 mm quartz plate (that does not filter out the ultraviolet region of the spectrum) as a cover slip [Posudin et al., 1990].

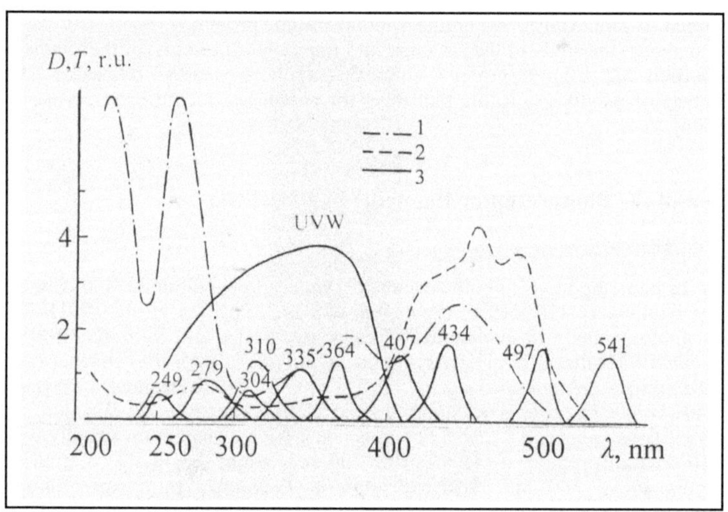



**Fig. 7.3.** Absorption spectra of pigments (*1* – flavins; *2* – carotenoids) and transmission spectra (*3*) of interference filters (figures correspond to maximum transmission) in the ultraviolet and visible portion of the spectrum; *D* – absorption of pigments; *T* – transmission of filters [Posudin et al., 1990].

Measurement of phototopotaxis in two species of *Dunaliella* and *T. viridis* (Table 7.1) established that the *F* parameter values for both *Dunaliella* species in response to a light stimulus at 249, 279, and 310 nm differed significantly from those of *T. viridis* [Posudin et al., 1990].

The values of this parameter are approximately the same for all three algae around 304 nm. Such a similarity can be explained by coincidence of absorption maxima of either flavins or carotenoids.

There were no significant differences in parameter *F* in the near ultraviolet (335 and 364 nm) and visible (407-655 nm) regions of the spectrum among the three species. The effect of the ultraviolet region on photomovement of *Dunaliella* indicates that both species do not display wavelength maxima for the dependence of parameter *F*. In the ultraviolet region of the spectrum there are absorption maxima for flavins and rhodopsin. At the same time, a significant increase in parameter *F* was observed in the ultraviolet region of the spectrum (i.e.,

249, 279, and 310 nm). The results obtained with a wide-band (240-410 nm) ultraviolet filter are similar (Table 7.1).

**Table 7.1.** Phototopotaxis of algae in the ultraviolet and visible portions of the electromagnetic spectrum.

| $\lambda$, nm | $I$, W/m$^2$ | $R(\lambda)$ D. s. | $R(\lambda)$ D. v. | $R(\lambda)$ T. v. | $F(\lambda)$ D. s. | $F(\lambda)$ D. v. | $F(\lambda)$ T. v. |
|---|---|---|---|---|---|---|---|
| **Ultraviolet part of spectrum** | | | | | | | |
| 249 | 0.17 | −0.05 | 0.05 | 0.35 | −11.81±0.01 | 11.81±0,01 | 82.68±0.02 |
| 279 | 0.19 | −0.08 | 0.01 | 0.19 | −15.09±0.01 | 1.89±0.02 | 35.84±0.02 |
| 304 | 0.21 | 0.07 | 0.08 | 0.08 | 10.96±0.05 | 12.53±0.10 | 12.53±0.15 |
| 310 | 0.18 | 0.02 | 0.05 | 0.12 | 3.58±0.15 | 8.24±0.07 | 21.50±0.02 |
| 335 | 0.27 | 0.03 | 0.01 | 0.05 | 3.32±0.02 | 0.66±0.03 | 5.53±0.10 |
| 364 | 0.28 | 0.08 | 0.07 | 0.07 | 7.85±0.10 | 6.87±0.10 | 6.87±0.13 |
| UVW | | 0.02 | 0.02 | 0.33 | - | - | - |
| **Visible part of spectrum** | | | | | | | |
| 407 | 0.37 | 0.15 | 0.20 | 0.16 | 9.96±0.13 | 13.28±0.05 | 10.62±0.08 |
| 434 | 0.37 | 0.10 | 0.12 | 0.29 | 6.23±0.07 | 7.47±0.02 | 23.04±0.02 |
| 497 | 0.41 | 0.42 | 0.37 | 0.45 | 20.61±0.09 | 18.15±0.10 | 22.08±0.04 |
| 541 | 0.44 | 0.07 | 0.06 | 0.02 | 2.94±0.01 | 2.52±0.03 | 0.84±0.04 |
| 655 | 0.47 | −0.09 | −0.07 | −0.09 | −2.87±0.01 | −2.24±0.03 | −2.88±0.03 |

N o t e : $R(\lambda)$ – relative number of cells, that are moving along direction of light stimulus; $F(\lambda)$ – parameter that characterizes phototopotaxis of the cells at given wavelength $\lambda$ of light stimulus; $I$ – intensity of light stimulus; $D$ .s. – *D. salina*; $D.$ $v.$ – *D. viridis*; $T.$ $v.$ – *Tetraselmis viridis*.

Comparison of the results with the action spectra for photomovement parameters of other algae (e.g., *C. reinhardtii* [Foster and Smyth, 1980], *H. pluvialis* [Sineshchekov, 1988], *T. hazenii* [Halldal, 1961] and *E. gracilis* [Foster and Smyth, 1980]) are presented in Fig 7.4.

Our results for parameter *F* in the ultraviolet region of the spectrum for both *Dunaliella* species and *T. viridis* in comparison with Halldal's [1961] indicates that the presence of phototopotaxis is inherent only to the latter species. Halldal believed that carotenoproteins function as the pigment. The complete absence of phototopotaxis in both species of *Dunaliella* in the ultraviolet region of the spectrum (i.e., 240-400 nm) suggests that flavins and rhodopsin cannot function in photoreception in these species [Posudin et al., 1990].

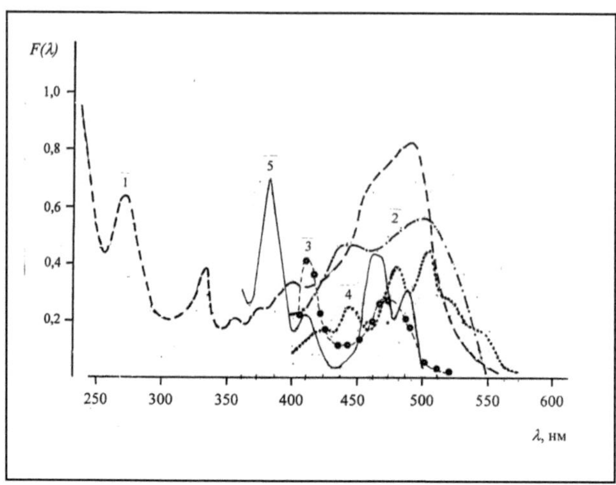

**Fig.7.4.** Action spectra of: *1* – positive phototopotaxis in *Platymonas subcordiformis* [Halldal, 1961]; *2* – phototopotaxis in *Chlamydomonas reinhardtii* [Nultsch et al., 1971]; *3* – phototopotaxis in *Dunaliella* spp. [Posudin et al., 1991]; *4* – photoinduction of phototopotaxis potential in *Haematococcus pluvialis* [Sineshchekov, Litvin, 1988]; *5* – phototopotaxis in *Euglena gracilis* [Foster, Smyth, 1980].

Existing data on the photomovement of *Dunaliella* [Posudin et al., 1991; Posudin et al., 1992] indicates that neither species displays a significant maxima for phototopotaxis in the ultraviolet region of spectrum where absorption maxima by flavins, pterins, and rhodopsin takes place. The absence of phototopotaxis in the ultraviolet region and the effect of the visible (i.e., 400-520 nm) region of the spectrum point toward carotenoids, that absorb between 350-500 nm, acting as the photoreceptor pigments. There is also evidence that carotenoproteins and rhodopsin participate in the photoreception of *D. salina* [Wayne et al., 1991].

## 7.5. Summary

The action spectrum for phototopotaxis in the two hyperhalobic species of *Dunaliella* is identical and within the 400–520 nm range, exhibiting maxima at 410–415 and 465–475 nm. The spectrum differs somewhat from those of *Chlamydomonas reinhardtii* and *Haematococcus pluvialis*, that display a wide band in the 400–600 nm region and a maximum at 500 nm. In contrast, in representatives of the class *Chlorophyceae*, such as *Tetraselmis viridis* (*Chlorodendrophyceae*), phototopotaxis occurs in the ultraviolet region of the spectrum. Phototopotaxis in *Euglena gracilis* (*Euglenophyta*) is in the 300–550 nm region with principal maxima at 385 and 460 nm and two smaller maxima at 410 and 490 nm. Thus, representatives of various genera, classes, and divisions of algae differ in the region of the spectrum that modulates phototopotaxis, indicating some degree of diversity in photoreceptor pigments among species.

# Chapter 8

# Mechanisms of Photoreception and Photoorientation in *Dunaliella*

## 8.1. Photoreception and Photoorientation Mechanisms in Algae

Common traits and specific peculiarities in the photoreceptor systems of flagellated algae have been previously reviewed [Massjuk and Posudin, 1991*b*], in particular from the standpoint of the structural and functional diversity among representatives of the various taxons. The objective of this section is to assess the evidence for the mechanisms controlling photoreception and photoorientation in *Dunaliella*. The following mechanisms of alga photoreception and photoorientation have been described in the literature [Foster and Smyth,1980; Kreimer,1994].

1) *Periodic shading and illumination of the photoreceptor by the stigma during rotation of the cell.* The rotational movement of the cell modulates light striking the photoreceptor. The amplitude of modulation depends on the direction of movement of the cell relative to the direction of propagation of the light (Fig.8.1).

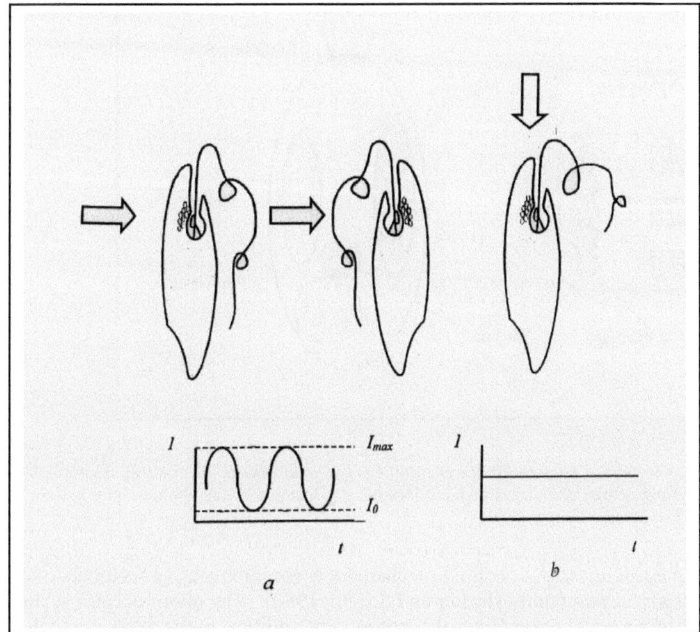


**Fig. 8.1.** Modulation mechanism for the photoorientation of *Euglena gracilis*. The rotational movement of the cell modulates the light striking the photoreceptor (*a*); the amplitude of modulation depends on the direction of movement of the cell relative to the direction of propagation of light and varies from maximal value ($I_{max}$) to zero ($I_0$) with subsequent flagellar beating. If the light is directed along the longitudinal axis of the cell (*b*), light modulation and flagella beating are absent (adapted from [Colombetti et al., 1982]).

Some algae possess a non-reflective stigma that acts as a shading device (*Euglenophyceae*) [Colombetti et al., 1982; Lebert and Häder, 2000] or a stigma which focuses the reflected light on the photoreceptor and acts as a shading device (e.g., *Chrysophyceae, Phaeophyceae,* some species of *Dinophyceae,* zoospores of *Eustigmatophyceae*) [Kreimer, 1994, 2001]. Normal movement of motile eukaryotic algae is accompanied by a rotational movement around the axis of the cell, a mechanism that is probably common to most algae. Therefore, it would appear that the mechanism was inherited from prokaryotic organisms and common ancestors. It is the primary event in photoreception in essentially all motile eukaryotic algae and their zoospores and it is possible to distinguish among various taxons secondary mechanisms that amplify the effect of the modulation.

2) *Waveguide mechanism.* Lateral light that is perpendicular to the longitudinal axis of the cell is believed to pass through coupled thylakoid disks. These disks contain photopigments and are perpendicular to the pigmented layer of the stigma and longitudinal axis of the cell. Thylakoids have a high refractive index in comparison to the light-coloured intervals between them. Therefore, the coupled thylakoids and intervals function as a waveguide for the light. The cell responds to the light stimulus due to its incidence on certain abdominal parts of the cell (Fig. 8.2). Such a mechanism is found in the Cryptophyta *Chroomonas* Hansg. and *Cryptomonas* Ehrenb. [Foster and Smyth, 1980].

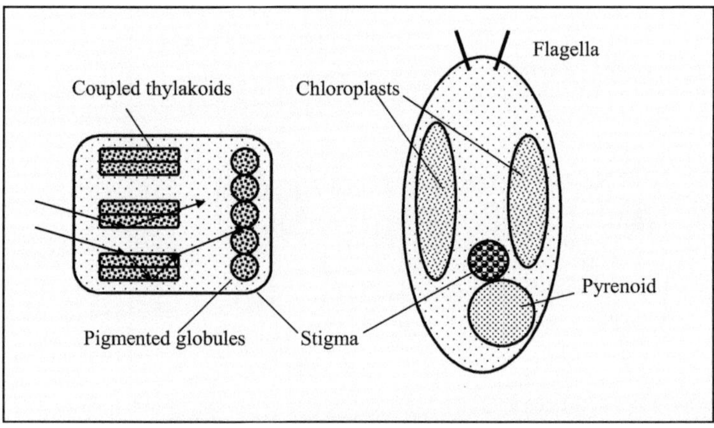


**Fig. 8.2.** Structure and location of photoreceptor system in *Chroomonas* Hansg.: *1* − part of chloroplast where coupled thylakoids and pigmented globular stigma are located; *2* − location of the photoreceptor system inside the cell (adapted from [Foster and Smyth, 1980]).

3) *Ocelloid mechanism.* An ocelloid mechanism is common among members of dinophyta from the *Warnowiaceae* family [Foster and Smyth, 1980]. The photoreceptor system of these alga is found in a special organelle − the *ocelloid,* in addition to the stigma, which possesses a refractive structure capable of acting as a focusing lens. The ocelloid is 20 μm in length and has a diameter of 6-15 μm. It consists of three principal parts: the *hyalosome* and the *melanosome* that are separated by the *ocelloid chamber.* The organelle is always lateral to the longitudinal axis of the cell body and can clearly function as a directional light antenna.

The ultrastructure of the ocelloid has been studied in three genera: *Nematodium, Warnowia* and *Erythropsidinium* [Francis, 1967; Mornin and Francis, 1967; Greuet, 1987].

The structure of the ocelloid in *Nematodinium armatum* (Dogiel, 1906) Kofoid & Swezy, 1921, is illustrated in Fig.8.3.

It has an egg-shaped structure with a *lens* that converge parallel beams of light deeply into the *retinoid* that is covered underneath by a layer of lipid globules. The *ocellar chamber* lies between the lens and the retinoid. Objects within about 50 μm are focused on the retinoid; therefore, as an aquatic object moves near the cell, the ocelloid produces a signal, the intensity of which depends on the distance, size, and direction of the object [Omodeo, 1975]. Optical processes in the ocelloid system include focusing the light by the lens, limiting light flux by structural elements in the ocelloid, refraction and interference.

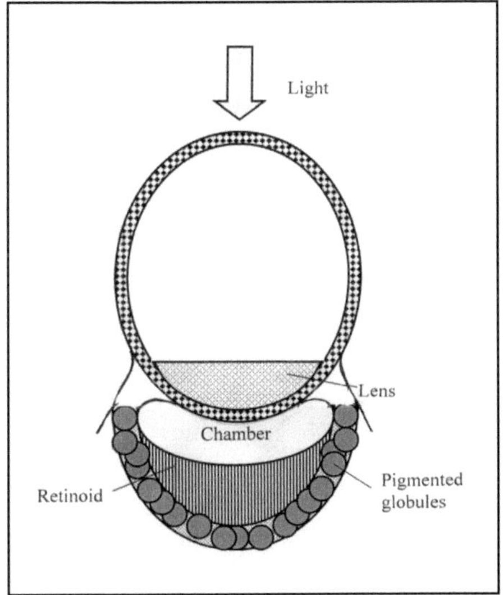


**Fig. 8.3.** Morphology of the ocelloid in *Nematodium armatum* (modified from [Omodeo, 1975]).

4) *Interference mechanisms.* A description of a four-layered stigma [Foster and Smith, 1980; Feinleib, 1985] is presented in Chapter 6 (see Fig. 6.1). According to Foster and Smith (1980), the stigma consists of several layers of pigmented and non-pigmented globules (some green alga have nine layers of pigmented globules). The stigma appears to be underneath the photoreceptor and acts as quarter-wave (interference) stack. The photoreceptor is located between the stigma and the adjacent cell surface.

Because of reflection and interference of the light, amplification or attenuation of the light that falls on the photoreceptor occurs. If the light shines on the surface where the photoreceptor is located, constructive interference takes place and the intensity at the photoreceptor is equal to the sum of the incident and reflected intensities. However, if the light falls on the opposite side of the cell, destructive interference occurs and there is an attenuation of the light by the cell and stigma, and reflection from the stigma. This allows the stigma to act as a directional antenna modulating the light. The antenna allows determining the location of the light source. The overall effect is strengthened by an alternation of pigmented and non-pigmented layers giving a periodicity of about ¼ of a light wavelength.

Light that falls on the outer surface of the stigma produces a series of interference maxima, the location of which coincides with the plasmalemma and thylakoid membrane inside the stigma. However, light falling on the opposite side of the cell results in several interference minima at the same location inside stigma. Thus, the contrast in the perception of light that falls on opposite sides of the cell is increased.

Assessment of the stigma in *Chlamydomonas reinhardtii* P.A. Dang. using confocal laser scanning microscopy and photoelectric measurements demonstrated the importance of an intact stigma for interference reflection and the absorption of phototactically active light, and therefore, the directional sensitivity of the stigma [Kreimer et al., 1992]. It is assumed that interference mechanism is present in the *Chlorophyceae, Prasinophyceae* and some species of *Dinophyceae* that have multilayered stigma [Foster and Smyth, 1980; Hegemann and Fischer, 2001]. If the thickness of the pigmented and non-pigmented layers equals one quarter of a wavelength, amplification of the light takes place [Hegemann and Harz, 1998]. Therefore, an interference quarter-wave mechanism is based on the assumption that there are several pigmented layers in the stigma. The presence of pigmented globules in continuous layers is a critical component of the hypothesis of Foster and Smyth (1980).


## 8.2. Diffractional Mechanisms of Photoreception and Photoorientation in *Dunaliella*

A criticism of the interference mechanism of Foster and Smyth is the fact that it is not possible to find stigma with continuous pigmented layers in nature. The stigma of green algae consists of either pigmented globules that are located separately in the layer (Fig. 8.4*a)* or globules of spherical (Fig. 8.4*b*) or hexagonal (Fig. 8.4*c*) form that are densely packed due to mutual compression. Stigma of many green algae consist of a single layer of pigmented globules. In a survey of 66 species, about 40 species of green algae had stigma with a single pigmented layer [Melkonian and Robenek, 1984].

For example, a monolayered stigma is found in the *Prymnesiophyceae*, (e.g., *Mantoniella squamata* (Manton and Parke) Desikach., *Monomastix* Scherff., *Nephroselmis* Stein) [Melkonian, Robenek, 1984] and in some species of *Chlamydomonas* Ehr., (e.g. *Chlamydomonas moewusii* Gerloff) [Walne and Arnott, 1967]. The number of pigmented globules varies from 18 (*Dunaliella salina* Teod.) up to 2000 *(Volvox* sp.) with the globule size generally ranging from 80 to 130 nm, though occasionally up to 200 nm [Massjuk and Posudin, 1991*b*].

We have proposed a *diffraction mechanism* for photoreception and photoorientation in unicellular green algae that have a stigma with a single layer or several layers of spherical or hexagonal pigmented globules that are densely packed [Posudin and Massjuk, 1996, 1997]. The diffraction mechanism for photoreception is based on stigma consisting of one (or more) layers of distinct spherical or hexagonal globules that are closely packed and are thought to act as a diffraction grating. The interaction of light with the structure results in diffraction of the light and the formation of diffraction maxima. The intensity and spatial location of these maxima depend on the geometry of the diffraction grating (i.e., the globule diameter, distance between, and number), the angles of incidence and diffraction, and the wavelength of the light. The intensity of light falling on the photoreceptor depends on the coincidence of diffraction maximum with the location of the photoreceptor (i.e., most probably in the plasmalemma above the stigma).

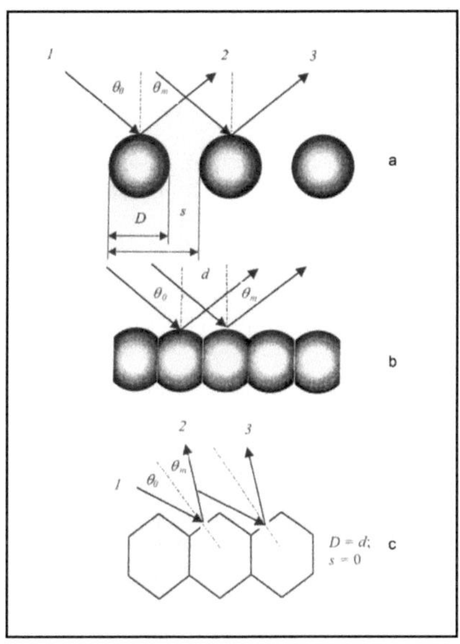

**Fig. 8.4.** Schematic of the optical phenomena that occur during the interaction of light with the structure formed by spherical (*a,b*) or hexagonal (*c*) globules that are densely packed due to mutual compression, where: $\theta_0$ and $\theta_m$ are the angles of incidence and diffraction, respectively; $d = D + s$ − period of diffraction grating (*D* is diameter of the globule and *s* is the interval between globules); *1* − incident beam of light; *2,3* − diffracted beams of light [Posudin and Massjuk, 1996].

Light interacts with periodically arranged pigmented globules in the stigma that form the diffraction grating which can be described by the following equation [Born and Wolf, 1968]:

$$p = d(\sin \theta_m - \sin \theta_0) = m\lambda \qquad (8.1)$$

where *p* is the parameter of diffraction; *d* − the period of diffraction grating *(d = D + s)*, where *D* is the diameter of the globules and *s* − the distance between them; $\theta_m$ − the angle of diffraction; $\theta_0$ − the angle of incidence of light on the plane of the diffraction grating; and *m* − an integer ( *m* = 0, ±1, ±2,....).

The intensity of diffraction maxima of *m*-order is described by the normalized function *F(p)* of light distribution during diffraction as:

$$F(p) = N^2 \, [\sin \, (Nkdp)^2 / N\sin \, (kdp/2)]^2 \cdot [\sin \, (ksp/2)/ksp/2]^2 \qquad (8.2)$$

where $k = 2\pi/\lambda$ is a wave number and *N* − the number of elements in the diffraction grating (in this case, the number of globules in the stigma). The function *F(p)* depends in a complex manner on the diffraction parameter *p*. A quantitative estimation of the diffraction mechanism was derived using electron microscopy to characterize the structure of the stigma monolayer of *D. salina* Teod. [Vladimirova, 1978]. The diameter of the pigmented globules (*D*) was 91.5 nm, the mean distance between globules (*s*) 3.5 nm, and the number of globules in the stigma (*N*) 18. These quantitative values are typical for all green algae [Melkonian and Robenek, 1984].

The diffraction parameters $D$, $s$, and $N$ were used to calculate the dependence of the intensity of diffraction maxima on the parameter of diffraction $p$ [Born and Wolf, 1968]. The dependence of the intensity $F(p)$ of diffraction maxima on parameter $p$ and, hence, on the angle of diffraction $\theta_m$ for the normal incidence of light ($\theta_0 = 0$) and wavelength $\lambda = 480$ nm is presented in Fig. 8.5.

The positions of diffraction maxima are determined by the values of parameter $p = \lambda/d$, $2\lambda/d$, $3\lambda/d$, etc. The dependence of the intensity $F(p)$ of diffraction maximum of the first $(m =1)$ order on the wavelength $\lambda$ of the light is presented in Fig. 8.6 (here $\theta_0 = 0$). It is clear that the positions of these maxima depend strongly on the wavelength $\lambda$ of the light.

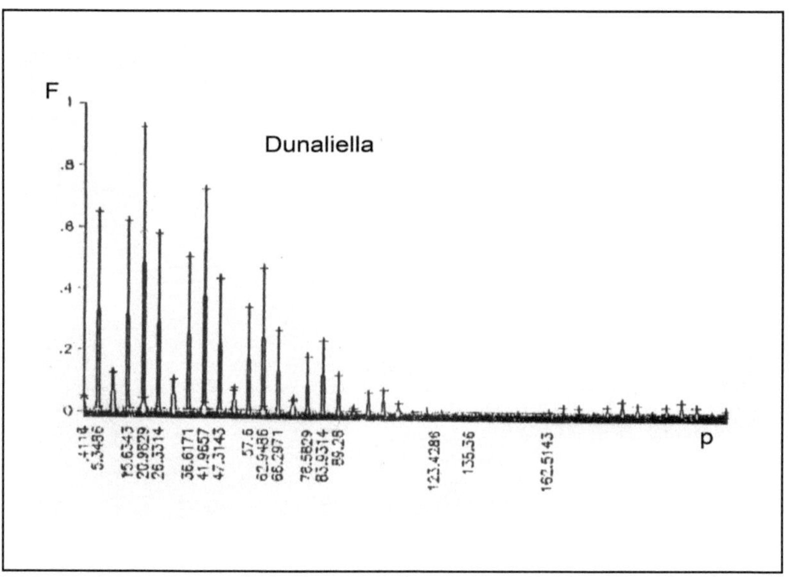


**Fig. 8.5.** The dependence of a function $F(p)$ of the light intensity diffracting on pigmented globules of *Dunaliella* on the parameter $p$ of diffraction. The normal incidence ($\theta_0 = 0$) of the light on the plane of the stigma is considered; the parameters of stigma as diffraction grating are: $D = 91.5$ nm; $s = 3.5$ nm; $N = 18$; the wavelength of light $\lambda = 480$.

The calculations support the hypothesis that there is a diffraction mechanism for photoreception in unicellular green algae containing a stigma with periodical structure. The mechanism allows the cells to respond to the angle of incidence and wavelength of the light. Changes in the direction of movement result in alterations in the angle of light incidence on the stigma along with spatial and temporal positions of diffraction maxima. This leads to changes in illuminance of the photoreceptor and a corresponding correction in the direction of movement (photoorientation).

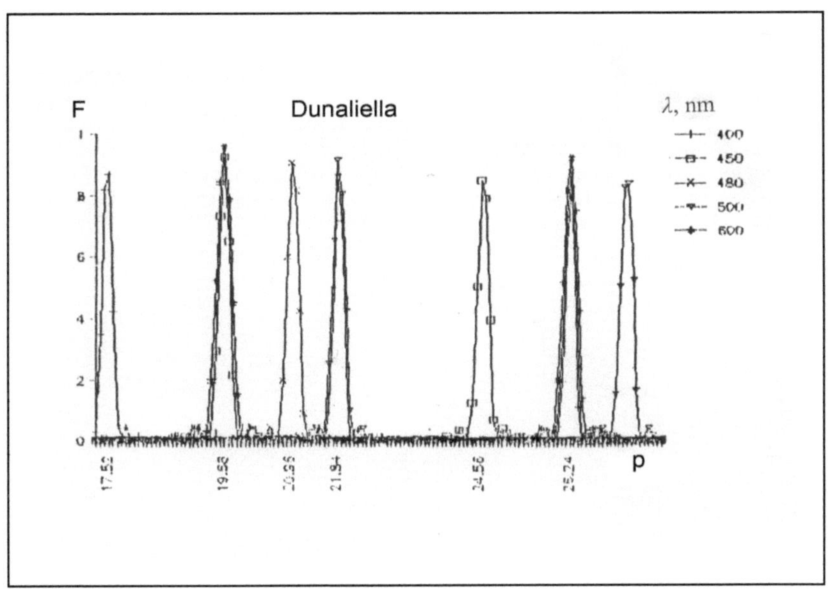


**Fig. 8.6.** The dependence of a function *F(p)* of the light intensity on the wavelength $\lambda$ of the light falling on stigma of *Dunaliella*, where: $p$ – parameter of diffraction; $\theta_0$ = 0; $D$ = 91.5 nm; $s$ = 3.5 nm; $N$ = 18; $m$ = 1 – the order of diffraction [Posudin and Masssjuk, 1996, 1997].

The proposed diffraction mechanism is believed to be present across all monad biological objects that possess a periodic stigma structure. This does not exclude the simultaneous action of other photoreception and photoorientation mechanisms that are based on modulation, interference, dichroism or other phenomena that provide light signal amplification and increased photoorientation accuracy.

## 8.3. Role of Proteins in Photoregulation Mechanisms in Flagellates

Proteins are components of photoreceptor systems and the flagellar apparatus. Those that undergo conformational changes are thought to play an important role in the photoregulation of movement. The processes involved in flagellar beating are related to the excitation dynamics of $\alpha$-spiral proteins [Kostuyk et al., 1988]. Two principal types of protein excitation occur, exciton and soliton [Davydov and Suprun, 1974]. Exciton is instigated by optical radiation in the visible or infrared region of spectrum. The process is characterized by a uniform distribution along the excited object. This type of excitation is the result of either membrane proteins that are reconstructing their spatial configuration under the scheme: light → exciton state in an $\alpha$-spiral → conformational reconstruction of the membrane protein, or due to photoreceptor under the scheme: light → exciton in photoreceptor → exciton transfer to an $\alpha$-spiral in the membrane protein and its excitation → conformational reconstruction of protein molecule.

Soliton excitations are interesting because they are localized along a small region of an $\alpha$-spiral and they can transfer excitation or an electric charge without dissipation. The velocity of transfer is close to that of sound and it can propagate a significant distance within the $\alpha$-spiral [Davydov and Suprun, 1974]. It is known that either exciton or soliton states are the result of symmetrical or anti-symmetrical excitations. Symmetrical excitation is characterized by the contraction of all three peptide chains in a $\alpha$-spiral, while the distance between the chains is increased symmetrically. Thus, the molecule contracts and thickens. These processes occur locally, within a small region (Fig. 8.7$a$). Anti-symmetrical excitation provides the contraction of a two peptide chains, while the third does not deform. This process leads to bending of the entire molecule in response to exciton excitation or local bending under soliton excitation (Fig. 8.7$b$).

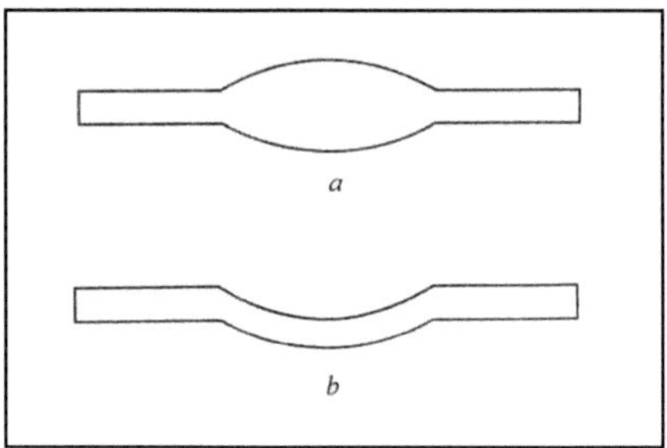


**Fig 8.7.** Deformation of peptide groups under $a$ – symmetrical; $b$ – antisymmetrical excitation. After Davydov and Suprun, 1974

The deformations result in conformational changes in the protein molecules that are accompanied by the activation of ion channels and flagellar beating. Thus, functioning of the membrane is tied to the activity of membrane proteins. Excitation of these proteins causes an alteration in their configuration and of ion channels that control the diffusion of calcium ions into the cell, stimulating motile activity of the algae [Posudin and Suprun, 1992].

## 8.4. Summary

The unique properties of the photoreceptor system of algae, in particular the structure of the stigma, suggest that all monad flagellates, with the exception of those devoid of a stigma, have possessed this mechanism for modulating photoreception since prehistoric times. It is proposed that the mechanism was inherited from prokaryotic organisms and their common ancestors. Certain green algae can simultaneously have three mechanisms (modulation, diffraction, and interference) that increase the intensity of the light signal absorbed by the photoreceptor. Proteins play a very important role in the photoregulation of movement in flagellates. They are a component of the photoreceptor system and flagellar apparatus and conformational changes in these proteins occur in response to a light stimulus.

# Chapter 9

# Sensory Transduction

## 9.1. Methods for Investigation of Sensory Transduction

Investigation of the photoregulation of movement in algae involves the transformation of light absorbed by photoreceptor molecules into a signal that governs the activity of the motor apparatus. Collectively, the molecular pathway linking photoreception with the activity of the motor apparatus is called *sensory transduction*. Familiarity with the methods used to study sensory transduction helps in understanding the mechanisms involved. These include: 1) analysis of the effects of calcium ions on photomovement and the application of chemicals that change membrane permeability to calcium ions (blockers of calcium channels, ionophores); 2) application of chemicals that impede specific steps in photomovement (e.g., ouabain which inhibits a $Na^+$-$K^+$-ATPase and ions that involve recording electrical phenomena at the cell level or through altering the cell with external electric fields). The following sections critique the principal methodical and experimental approaches used to study sensory transduction in algae.

## 9.2. Sensory Transduction in *Euglena gracilis*

A hypothesis by Tollin [1969, 1973] is based on the effect of changes in the light intensity striking the photoreceptor in *Euglena gracilis* G.A. Klebs which activates the flow of ATP, formed in the photosynthetic pathways, to the flagellar apparatus, thereby inducing flagellar beating and changes in the direction of movement. However, the hypothesis does not explain the presence of photomotile reactions in cells devoid of pigmentation (i.e., etiolated in the dark) and lacking chloroplasts and photosynthetic ability [Checcucci et al., 1976].

Jahn and Bovee [Jahn and Bovee, 1968; Bovee and Jahn, 1972] proposed that the axoneme paraflagellar rod and paraflagellar body function as a piezoelectric source. Piezoelectric materials can generate an electric charge with the application of mechanical pressure. Conversely, they can change physical dimensions with the application of an electric field. They believe [Jahn and Bovee, 1968] that the paraflagellar rod and paraflagellar body in *E. gracilis* are quasicrystals that under compression or stretching in certain directions, induce electrical polarization. The piezoelectric activity displaces cations along the flagellum, producing a sequential mechanical bending of the flagellum from base to tip resulting in its beating. The photoreceptor of the paraflagellar body acts as a capacitor that discharges with changes in illumination. To date, the hypothesis has received little experimental support.

Piccini and Omodeo [1975] proposed that signals from the photoreceptor on the paraflagellar body were proportional to the light-induced bleaching of the photoreceptor pigments. An electric signal from the photoreceptor travels to the base of both flagella, similar to a signal delayed by a synaptic junction (located between plasma membrane of the main and the short flagellum). After comparison, a signal is sent to an effector with longitudinal actomyosin-like fibrils that contract and change the flagellar position.

The effect of chemicals that alter the metabolic system [e.g., 1,1-dimethyl urea (DCMU), 2,4-dinitrophenol (DNP), sodium azide ($NaN_3$)] have been studied to elucidate the possible connection between photomovement and photosynthetic activity in *E. gracilis* [Barghigiani et al., 1979]. The chemicals did not affect the photophobic response at concentrations that impaired cell motility or induced serious morphological alterations. Therefore a connection between photomovement and photosynthetic activity was not established.

Sodium azide had no effect on positive phototopotaxis in *E. gracilis* but inhibited negative phototopotaxis at a concentration of $5 \cdot 10^{-5}$ M [Colombetti et al., 1982]. At $3 \cdot 10^{-5}$ M it induced decreased cell motility up to 40 % and velocity of movement up to 86 % [Barghigiani et al., 1979]. It was concluded that there was a connection between phototopotaxis and photophobic reactions though both reactions are through separate sensory transduction pathways.

Ion transport may also be involved in sensory transduction in *E. gracilis*. The effect of ions of certain elements on photomovement supports the involvement of ion transport [Doughty and Diehn, 1979; Doughty et al., 1980]. Cells display maximum motility in the presence of $Mg^{2+}$, $Ca^{2+}$ and $K^+$, while $Ni^{2+}$ results in immobilization. Increasing concentrations of ions such as $Ca^{2+}$, $Mn^{2+}$, and $Ba^{2+}$ increased the frequency of directional change in the cells. The duration of the photophobic response was enhanced by divalent ions in the following sequence: $Ca^{2+} > Ba^{2+} > Mn^{2+} > Co^{2+} > Mg^{2+} > Ni^{2+}$.

The effect of potential inhibitors of flavin photochemistry on negative phototopotaxis in *E. gracilis* was studied by Lenci et al. [1983]. KI and $MnCl_2$ were found to react with the excited states of flavins, impairing the negative phototopotaxis response. At high concentrations, these substances completely inhibited phototactic orientation. Collectively the results supported the hypothesis that a flavin-type chromophore acts as a photoreceptor in phototototaxis.

A cross-section of biologically active chemicals has been tested to ascertain their effect on photomovement and thereby broaden our understanding of the mechanisms involved. Ammonium ions specifically enhanced the step-down photophobic response in *E. gracilis* [Matsunaga et al., 1999]. Conversely, L-methionine-DL-sulfoximine (L-MSO), an inhibitor of ammonium assimilation, specifically enhanced the step-up photophobic response. The duration of photophobic reaction in *E. gracilis* was increased with application of NaCl and ouabain   (3-[(6-deoxy-α-L-mannopyranosyl)oxy]-1,5,11,14,19-pentahydroxy-card-20(22)-enolide), the latter being an inhibitor of $Na^+$-$K^+$ ion membrane transport. The $Ca^{2+}$ ionophore A23187 (4-benzoxazolecarboxylic acid) induces a specific light-independent but concentration-dependent response in *E. gracilis* that was expressed as discontinuous tumbling of the cells. In contrast, application of gramicidin *D* and carbonycyanidechlorophenyl-hydrazone, chemicals that specifically affect proton transport, did not effect photosensory transduction in *E. gracilis* [Castiello, et al., 1980].

Using electrical fields to alter membrane potential did not alter photoorientation in *E. gracilis* [Häder, 1986*b*]. Addition of the lipophylic cation methyltriphenylphosphonium that penetrates the membrane and dissipates its potential, likewise, did not affect photoorientation of the cells [Nultsch and Häder, 1988].

A model of the molecular processes involved in sensory transduction in *E. gracilis* has been proposed [Doughty and Diehn, 1979; Doughty et al., 1980]. Flagellar reorientation is governed by a transient increase of $Ca^{2+}$ ions in the intraflagellar space. As light is absorbed by the chromophore molecule (flavin) located in the paraflagellar body, the excitation energy of the chromophore activates a $Na^+$-$K^+$-pump in the flagellar membrane that controls the flux of monovalent ($K^+$ and $Na^+$) and divalent ($Ca^{2+}$) cations ions across the membrane. The $Na^+$-$K^+$-pump stimulates a high concentration of $K^+$ ions and low concentration of $Na^+$ ions inside the cell in comparison with the external medium. Active transport is necessary for the transfer of $K^+$ and $Na^+$ ions across plasma membrane. $Na^+$ ions are believed to be the driving force that induces the net efflux of $Ca^{2+}$ across the plasma membrane [Kostyuk et al., 1988]. Activity of the $Na^+$-$K^+$-pump is controlled by light and pharmacologically inhibited by ouabain [Colombetti et al., 1982]. Thus, changes of photomovement and photosensitivity in *E. gracilis* are induced by the flux of mono- and divalent cations, controlled by a membrane $Na^+$-$K^+$-pump that is triggered by light and inhibited by ouabain.

At this time, the very limited amount of data on the control of the primary stages in sensory transduction in *E. gracilis* has limited our understanding such that the process remains a proverbial "black box" [Lebert and Häder, 2000].


## 9.3. Sensory Transduction in Green Algae

The role of calcium ions and membrane phenomena controlling their transport in sensory transduction in *Chlamydomonas reinhardtii* P.A. Dang. has been established. Photostimulation is the result of $Ca^{2+}$ ion flux across the cell membrane, altering the intracellular concentration [Halldal, 1957; Marbach and Mayer, 1971; Stavis and Hirshberg, 1973; Stavis, 1974, 1975; Schmidt and Eckert, 1976; Nichols and Rikmenspoel, 1978; Hyams and Borisy, 1978; Schmidt, 1978; Nultsch, 1979; Kamiya and Witman, 1984; Merten et al., 1995; Marangoni et al., 1996]. Experimental measurement of photomotile reactions or light-induced electric currents in the presence of various levels of $Ca^{2+}$ ions in the medium [Hegemann et al., 1990] or in response to different inhibitors of calcium channels [Nultsch et al., 1986; Hegemann et al., 1990] supported a principal role for $Ca^{2+}$ ions in the photoresponses of *C. reinhardtii* and *Haematococcus pluvialis* Flotow.

Phototopotaxis and photophobic reactions in algae are gradually inhibited in the presence of omega conotoxin and pimozide [Hegemann et al., 1990], that selectively inhibit calcium channels. There is most likely another type of calcium channel that participates in phototopotaxis but is not linked to photophobic reactions. This channel is inhibited by verapamil [Hegemann et al., 1990]. The electrical signal that is generated by a cell is dependent on the extracellular concentration of calcium ions [Sineshchekov, 1991*a*]. In spite of the evident participation of calcium ions in sensory transduction in both species, the processes involved in transduction remain unclear. The effect of specific drugs on phototopotaxis in *C. reinhardtii* has shown the absence of a link between photomovement and photosynthesis [Stavis and Hirshberg, 1973; Stavis, 1974]. Motility and phototactic rate in *C. reinhardtii* were measured in the presence of isobutylmethylxanthine (IBMX), 3',5'-cyclic AMP dibutyrate (db-cAMP) and neomycin [Korol'kov and Rychkova, 1996]. No evidence was found for the involvement of cyclic nucleotide phosphodiesterases or inositol phosphates on the phototactic signalling pathway.

The effect of sodium azide, a respiration inhibitor, on the rate of respiration of *Chlamydomonas snowiae* Printz and *Dunaliella salina* Teod. was studied by Myroniuk [2000]. The complete suppression of phototaxis in *C. reinhardtii* occurred at a sodium azide concentration of $3.5 \cdot 10^{-4}$ M; the number of motile cells decreased to 62 % of the total number of the cells and the velocity of movement declined to 93 % [Stavis, 1974]. At $10^{-5}$ M it decreased phototopotaxis up to 80 % and motility up to 30 % [Pfau et al., 1983].

Both *C. reinhardtii* and *H. pluvialis* display light-induced membrane potentials that can be measured using microelectrodes. Two types of potential were identified. A positive potential reflects the surface properties of the membrane and a negative potential transmembrane properties. In addition, strictly periodic changes in positive potential in response to light and fast reverse changes in level were found [Sineshchekov et al., 1976].

An electrophysiological approach made it possible to establish that both phototactic and photophobic responses in *Chlamydomonas* were mediated by a rhodopsin-like photoreceptor [Holland et al., 1997]. The $Ca^{2+}$ currents, measured using the pipette electrode system, probably trigger all of the behavioral light responses in the cell.

The chlorpromazine-HCl results in a light intensity-dependent reversal of phototopotaxis in *C. reinhardtii* [Hirschberg and Hutchinson, 1980]. At moderate light intensities, treated cells swam away from the light (negative phototopotaxis), while untreated cells swam toward it (positive phototopotaxis). At low light, both treated and untreated cells exhibited normal positive phototopotaxis.

Sineshchekov et al. [1989] established that phototactically inactive red light induces a fast change in phototopotaxis from positive to negative in *Chlamydomas* cells exposed to short-wave irradiation (450-500 nm). The stimulation of negative phototopotaxis was readily reversible in the dark thereby excluding the participation of phytochrome. The long-wave boundary was near 700 nm and was inhibited with diuron (N'-(3,4-dichlorophenyl)-N,N-dimethyl-urea). These facts indicate the possible existence of a fast (i.e., seconds) control of the positive or negative phototaxis sign by photosynthesis. The spectral sensitivity of photo-topotaxis was determined by the collective absorption spectra of the photoreceptor, photosyn-thetic pigments, and stigma.

Takahashi and Watanabe [1993] confirmed that photosynthesis modulates the sign of phototopotaxis in wild-type *C. reinhardtii*. This conclusion was based on: 1). The transient nature of phototopotaxis was preferentially observed in blue-green actinic light rather than green actinic light; 2). Red background lighting induced negative phototopotaxis under actin-ic-light conditions, however, without background light, the cells exclusively display positive phototaxis; and 3). Both the effect of red background light and the transient change in the sign of phototaxis were inhibited by 3-(3',4'-dichlorophenyl)-1,1-dimethylurea, a relatively specific inhibitor of photosynthesis. Their conclusion altered the accepted view in the early 1970s that photosynthesis was not linked to phototaxis (e.g., see Stavis and Hirshberg, 1973; Stavis, 1974).

Application of electrophysiological methods (microelectrode recording of electric sig-nals on the protoplast's surface) made it possible to identify high-frequency rhythmic processes that are related to changes in electric potential of *H. pluvialis*. The duration of an oscillation was several tenths of a second. The processes pointed toward the existence of two independent oscillators in the cell and that the frequency change and phase shift between the two parallel rhythms probably determined the phototopotaxis sign [Sineshchekov et al., 2001; Sineshchekov and Govorunova, 2001*a*].

It is possible that there is a link between periodic processes and the functioning of con-tractile vacuoles. There are two contractile vacuoles in *C. reinhardtii* that are located near the basilar bodies of the flagella. Their behaviour has been analyzed on micropipette tips using videomicrography. The time interval between two contractions of the vacuole was 30 s. Con-traction of both vacuoles occurred with frequencies that were close in magnitude but shifted in phase. The magnitude of the shift changed periodically [Sineshchekov and Govorunova, 2001*a*].

These processes play a role in the regulation of cell movement. It is possible to ob-serve periodic spontaneous changes of the direction of cell movement during prolonged moni-toring of the movement trajectories of individual cells of *H. pluvialis* and *C. reinhardtii*. These changes are similar to periodic changes in electric potential at the cell surface and in the micromovement of chloroplasts, indicating the possibility of a common origin [Sinesh-chekov and Govorunova, 2001*a*]. The interaction of both oscillators in the cell, the function of which is controlled by the contractile vacuoles, is believed to be the basis for the mechanism regulating the phototopotaxis sign [Sineshchekov and Govorunova, 2001*a*].

It was shown that the two flagella display different levels of response to light (i.e., changes in frequency and beating plane) [Sineshchekov, 1991*a, b*]. Free-motile cells were characterized by the preferential reaction of the *cis*-flagellum, located on the side of the pho-toreceptor that leads to rotation away from the source of light (negative phototopotaxis), while the preferential reaction of the *trans*-flagellum, located on the opposite side from the photo-receptor, lead to rotation of the cell toward the source of light (positive phototopotaxis).

The trajectories of movement of individual cells under extended illumination change depending upon the source of the light. The frequency of such changes can be compared with the frequency of electric pulses on the cell surface. The sign for phototopotaxis is determined by the level of the phase shift between the rhythms of two oscillators – the contractile vacuoles. The level depends on many factors, such as the intensity of illumination, aeration, ion composition of medium, and age of the culture. Differences in the sensitivity of the oscillators to these factors are thought to be the probable cause [Sineshchekov and, Govorunova, 2001$a$]. An electrochemical system using dual electrodes allowed the simultaneous monitoring of algal motility and phototopotaxis in an investigation of photomovement in *C. reinhardtii*. The effect of diltiazem (3-(acetyloxy)-5-[2-(dimethylamino)ethyl]-2,3-dihydro-2-(4-methoxy-phenyl)-1,5-benzothiazepin-4(5H)-one hydrochloride), sodium azide, or ethanol on the redox currents were assessed as indices of photoinduced behavior [Shitanda and Tatsuma, 2006].

*C. reinhardtii* mutants with defects in thier dynein arm structure (ida4 and oda22), ptx mutants deficient in axonemal sensitivity to calcium ions, and ppr mutants lacking an ionic channel specific for the photophobic response were analyzed [Ermilova et al., 2000]. The mechanism of cell orientation in phototactic and chemotactic responses depended on functional differences between *cis*- and *trans*-flagella that differed in beat frequency and in the sensitivity of their axonemes to submicromolar calcium concentration. These responses are distinct from the photophobic response.

Using DNA sequencing, overlapping segments of cDNA have been identified in *C. reinhardtii* that encode two channelopsins proteins [Nagel et al., 2005]. Nagel et al. proposed that phototopotaxis and photophobic responses in green algae were mediated by rhodopsins with microbial type chromophores, i.e. all-*trans*-retinal in the ground state.

The effect of copper ions on phototactic orientation in *H. pluvialis* (syn. *H. lacustris* (Girod) Rostafinski) [Braune et al., 1994] indicated that phototopotaxis was inhibited at concentrations that did not impair the velocity of cell movement. The effects of chromium ($Cr^{6+}$, the more toxic species) on the photoreceptive apparatus in *C. reinhardtii* [Rodriguez et al. 2007] was studied by *in vivo* absorption microspectroscopy of both the thylakoid compartments and the stigma. Decomposition of the overall absorption spectra of the pigment constituents indicated that $Cr^{6+}$ induced a modification of the carotenoids present in the stigma. It is therefore possible that C. *reinhardtii* might be useful as a bioindicator of $Cr^{6+}$.


## 9.4. Sensory Transduction in *Dunaliella*

### 9.4.1. Methods of Investigation

Sensory transduction in *Dunaliella* was studied through the application of calcium ions ($10^{-6}$ M $- 10^{-2}$ M $CaCl_2 \cdot 6H_2O$), ionophore A23187 ($10^{-5}$ M), sodium azide ($10^{-7} - 10^{-3}$ M $NaN_3$), cobalt ions ($10^{-6} - 10^{-3}$ M $CoCl_2$), and the calcium channel blockers cinnarizine ($10^{-6} - 10^{-3}$ M 1-(diphenylmethyl)-4-(3-phenyl-2-propen-1-yl)-piperazine) and isoptin ($10^{-7} - 10^{-4}$ M benzeneacetonitrile). Individual dots on the figure indicate the dependence of the photomovement parameter on the concentration of a chemical and represents the average of several ($\geq 3$) measurements for each sample. Measurements were conducted one hour after exposure of the cells to the chemical. Positive phototopotaxis was assessed at an illuminance of 500 lx and negative phototopotaxis at 40,000 lx. The objective of the investigation was to compare the effect of the chemicals on photomovement in two species of *Dunaliella* [Posudin et al., 1993].


### 9.4.2. Effect of Calcium Ions

Dependence of the photomovement parameters linear velocity $\upsilon$ and factor $F$ on the concentration of calcium ions between $10^{-6}$ M to $10^{-2}$ M is presented in Fig. 9.1 for *D. salina* and in Fig. 9.2 for *Dunaliella viridis*.

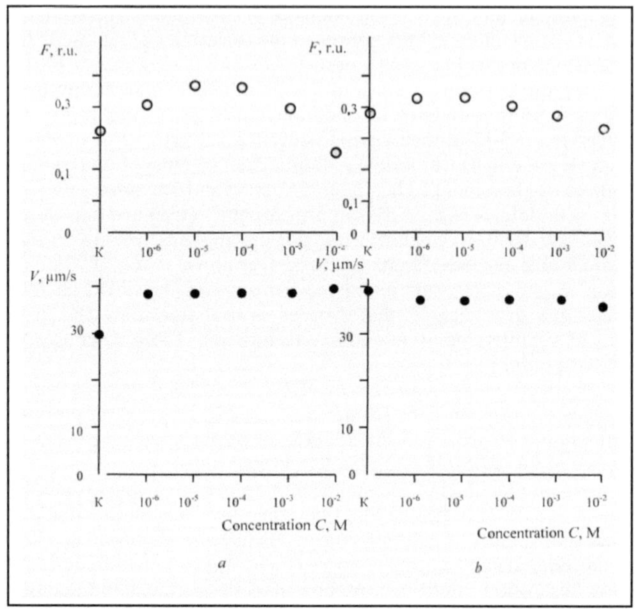


**Fig. 9.1.** Dependence of photomovement parameters $F$ and $U$ in $a$ – *Dunaliella salina* and $b$ – *Dunaliella viridis* on the concentration of $CaCl_2{\cdot}6H_2O$ in the water [Posudin et al., 1993].

Maximum values for parameter $F$ were between $10^{-6}$–$10^{-4}$ M for both species; the value of $F$ decreased to 80-90 % at a higher ($\sim 10^{-2}$ M) concentration in comparison with control values (without calcium ions in the medium). There was not a significant difference among the Ca concentrations on the velocity of movement $U$ and control values.

In spite of the evident participation of calcium in algal photomovement (i.e., activation of calcium channels, triggering the ion pumps, changes in membrane permeability), a number of aspects of its role remains to be determined. The dependence of photomovement parameters in the two species was characterized by a maximum for parameter $F$ in the concentration range of $10^{-6}$–$10^{-4}$ M. Velocity $U$, however, did not appear to be affected by calcium ions in the concentration range tested. These results are similar to those reported by Avron and Ben-Amotz (1992), who assessed motility, velocity of movement, and linearity of trajectory in *D. salina* and *Dunaliella bioculata* Butcher and found that Ca had little or no effect. Our results in part coincided with the those obtained by a number of authors for *C. reinhardtii*.

The effect of the pesticide lindane (1,2,3,4,5,6-hexachloro-cyclohexane) on motility in *D. bioculata* is related to a specific interaction between the chemical and Ca transport that results in an increase in cytoplasmic Ca. The pronounced effect of lindane on ciliary beating is probably related to a modification of the Ca balance within the cell [Marano et al., 1988].

Maximum phototopotaxis occurs at a Ca concentration of $5{\cdot}10^{-5}$ M [Dolle et al., 1987; Nultsch, 1979, 1983]; the level of phototopotaxis decreases to 50 % at $10^{-5}$ M and to 25 % at $2{\cdot}10^{-4}$ M in comparison to control values. In contrast, phototopotaxis in *Chlamydomonas* is completely inhibited at $10^{-3}$ M. Phototopotaxis in *Chlamydomonas* is very sensitive to the intensity of the light stimulus [Dolle et al., 1987; Morel-Laurens, 1987]. In contrast, the velocity of movement was dependent upon the concentration of calcium ions and the intensity of light stimulus [Morel-Laurens, 1987].

The *Dunaliella* species differ from *Chlamydomonas* in that the velocity of movement of the cells is not dependent upon the concentration of calcium ions.


## 9.4.3. Effect of Ionophore A23187

Ionophores are the compounds that facilitate the transport of ions (such as calcium) across the cell membrane by binding with the ion or by increasing the permeability of the membrane to the bound ions. The effect of the application of ionophore A23187, that increases the permeability of the membrane to calcium ions, is presented in Fig. 9.2*a,b* for both species of *Dunaliella* ($CaCl_2 \cdot 6H_2O$ concentration in the medium was $10^{-4}$ M).

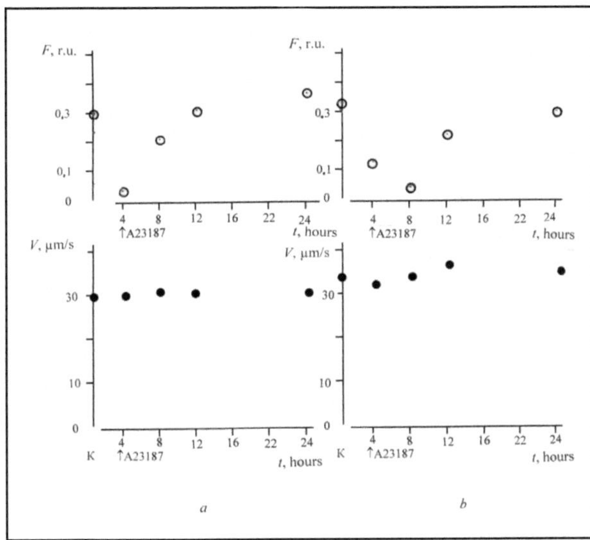


**Fig. 9.2.** Temporal dependence of photomovement parameters $F$ and $v$ in two species of *Dunaliella* (*a* − *D. salina* and *b* − *D. viridis*) on the addition of ionophore A23187 to the medium (moment of addition indicated by arrow) [Posudin et al., 1993].


Insertion of $10^{-5}$ M ionophore into the medium induced practically complete inhibition of photopotaxis in *D. salina* and *D. viridis*. Recovery of parameter $F$ to the initial control value was reached after 12 hours for *D. salina* and 24 hours *D. viridis*. Addition of calcium ($10^{-3}$ M) 24 hours after introduction of the ionophore did not change parameter $F$ and there were no changes in the velocity of movement of the cells. The addition of the ionophore at $10^{-5}$ M supports the role of calcium entering the cell in that phototopotaxis of *Dunaliella* was inhibited practically instantly. The absence of its effect on the velocity of movement can be explained by the increased permeability of the membrane to calcium ions.

Calcium ions can be present in the water used for the medium without the addition of calcium salts. Likewise, microscope slides and cover glasses can be a source of calcium ions [Dolle et al., 1987]). The fact that the velocity of movement in *Dunaliella* did not change in response to the ionophore in comparison with other green *Chlamydomonas* species [Pfau et al., 1983] demonstrated inhibition of either phototopotaxis or velocity of movement by the chemical ($10^{-5}$ M). The inhibitory effect of the ionophore on the velocity of movement in *Chlamydomonas* was thought to be due to contraction (or detachment) of the flagella and their subsequent restoration (regeneration).

The effect of the pesticide lindane on *D. bioculata* was tested with an ionophore (A23187) that functions as a mobile ion carrier. The results were compared with those of *Dunaliella* at 15 ppm lindane and a combination of lindane (15 ppm) and ionophore A23187 ($10^{-5}$ M). The ionophore enhanced the effect of lindane on the motility indicating that the chemical may interfere with intracellular calcium flux [Krishnaswamy-Chang, 1997]. Experiments with inhibitors, ionophores, and drugs support the conclusion that *Chlamydomonas* cells have an energy-dependent, outward-oriented $Ca^{2+}$ pump [Hutchinson and Hirschberg, 1985].

Comparing the distinctions and similarities of the action of the ionophore on photomovement parameters of representatives of two genera (*Dunaliella* and *Chlamydomonas*) underscores the greater adaptive ability of *Dunaliella* for survival under extreme conditions. Across taxons the chemical induces similar photomovement responses (e.g., direction of movement in *Dunaliella*, direction and velocity of movement in *Chlamydomonas*, frequency and duration of spatial tumblings in *E. gracilis* [Doughty and Diehn, 1979]) indicating it affects a fundamental control mechanism.


9.4.4. Effect of Ouabain

Inoculation of *Dunaliella* with ouabain at concentrations ranging from $10^{-7}$ to $10^{-4}$ M did not result in notable changes of photomovement parameters $\upsilon$ and $F$. Ouabain, an ionotropic chemical that inhibits $Na^+$-$K^+$-ATPase, increased either the tumbling of *E. gracilis* cells during a change in light intensity or in the velocity of cell accumulation in the illuminated area [Doughty et al., 1980]. The effect of ouabain appeared to be due to an inhibition of the flux of monovalent ions of sodium out of the cell and potassium into the cell. Ouabain therefore changes the electrical gradient across the cell membrane affecting the influx of divalent ions of calcium into the cell and causing a reorientation of the flagella [Meyer, Hildebrand, 1988]. The fact that ouabain did not affect photomovement parameters in *Dunaliella* suggests that the influx of calcium ions into the cells is triggered not by a $Na^+$-$K^+$-pump (as in *Euglena*) but probably through a direct light-enhanced entry of calcium ions (as in *Chlamydomonas* [Nultsch,1983]).


9.4.5. Effect of Cobalt Ions

We have shown that the addition of cobalt to the medium at concentrations from $10^{-6}$ to $10^{-3}$ M ($CoCl_2$) affects phototopotaxis in *Dunaliella*, while the velocity of movement essentially did not change (Fig. 9.3*a,b*). The inhibitory effect of Co remained more than 5 days after its introduction. Addition of calcium ions to the medium restored parameter $F$ to control values in cells grown without addition of $CoCl_2$ and $CaCl_2 \cdot 6H_2O$.

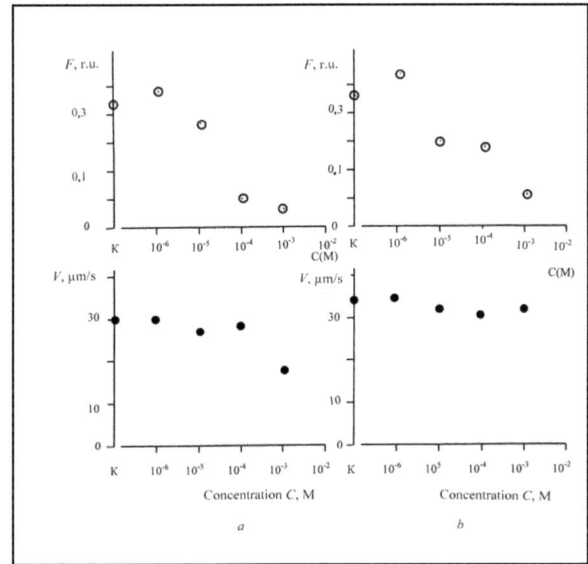

**Fig. 9.3.** Dependence of photomovement parameters $F$ and $\upsilon$ in two species of *Dunaliella* (*a* – *D. salina* and *b* – *D. viridis* ) on the concentration of $CoCl_2$ [Posudin et al., 1993].

Ions of cobalt, lanthanum, manganese and nickel are known to block membrane calcium channels inducing reversible flagellar beating, whereby changing the direction of movement in *Chlamydomonas* [Schmidt and Eckert, 1976; Doughty and Diehn, 1982]. The effect of the presence of blockers on photomovement indicates the participation of calcium ions in the sensory transduction of alga.

Differences among species and temporal changes in behaviour (i.e., photophobic reactions) of the cells were observed in *E. gracilis* due to the presence of calcium, barium, cobalt, and magnesium ions [Colombetti et al., 1982]. The inhibitory action of cobalt ions on phototopotaxis in *Dunaliella* suggests the transport of Ca across the membrane and therefore, the ionic nature of sensory transduction in *Dunaliella*.

### 9.4.6. Effect of Cinnarizine and Isoptin

Application of cinnarizine which blocks calcium channels results in the inhibition of both photomovement parameters ($\upsilon$ and $F$) in *Dunaliella* at a concentration range of $10^{-6}$–$10^{-4}$ M. The two parameters were completely inhibited at $10^{-3}$ M (Fig. 9.4 *a,b*). We did not observe a recovery in the photomovement parameters within 24 hours after exposure to cinnarizine.

Isoptin ($\alpha$-[3-[[2-(3,4-dimethoxyphenyl) ethyl]methylamino]propyl]-3,4-dimethoxy-$\alpha$-(1-methylethyl)-benzeneacetonitrile), also a calcium channel blocker, inhibits both photomovement parameters. Phototopotaxis of both species was decreased to 60-70 % and velocity of movement to 75–90 % in comparison with control cells (Fig. 9.5*a,b*).

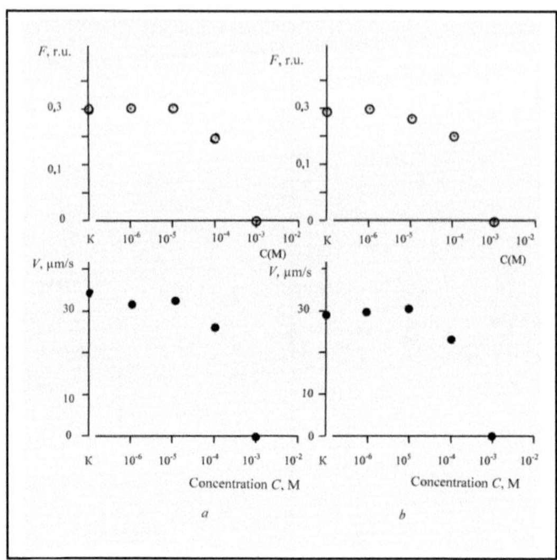

**Fig. 9.4.** Dependence of photomovement parameters $F$ and $v$ in two species of *Dunaliella* (*a* – *D. salina* and *b* – *D. viridis* ) on the concentration of cinnarizine [Posudin et al., 1993].

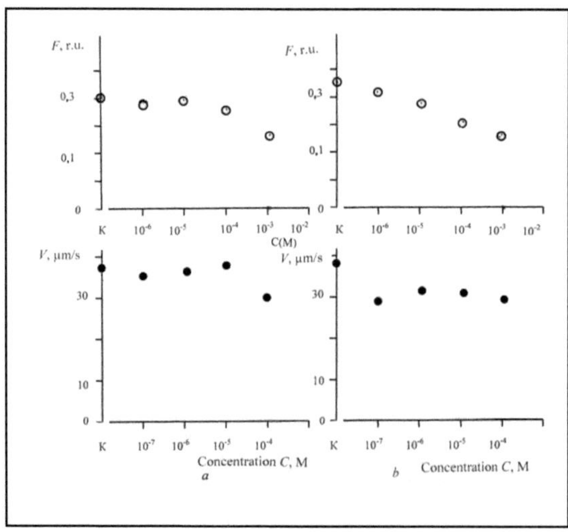

**Fig. 9.5.** Dependence of photomovement parameters $F$ and $v$ of two species of *Dunaliella* (*a* – *D. salina* and *b* – *D. viridis* ) on the concentration of isoptin [Posudin et al., 1993].

Application of calcium channel blockers demonstrated an inhibition of both photomovement parameters ($\upsilon$ and $F$) in the two species of *Dunaliella*. Comparing these results with those using calcium channel blockers such as flunarizine (1-[bis(4-fluorophenyl)methyl]-4-[(2E)-3-phenyl-2-propenyl]-piperazine), verapamil ($\alpha$-[3-[[2-(3,4-dimethoxyphenyl)ethyl] methylamino] propyl]-3,4-dimethoxy-$\alpha$-(1-methylethyl)-benzene-acetonitrile), diltiazem, and nimodipine (1,4-dihydro-2,6-dimethyl-4-(3-nitrophenyl)-3-(2-methoxyethyl) 5-(1-methylethyl) ester 3,5-pyridinedicarboxylic acid) [Nultsch et al., 1986] provides additional insight into the mechanisms involved. For example, increasing the concentration of flunarizine from an initial level of $10^{-6}$ M decreased phototopotaxis and motility in *Chlamydomonas* [Nultsch et al., 1986]. At a concentration of $5 \cdot 10^{-5}$ M both parameters were completely inhibited. The authors considered the effect specific (i.e., induced the inhibition of one parameter only).

Microscopic analysis indicated that the loss of motility was related to the contraction or detachment of the flagella as is exhibited by sodium azide or ionophore A23187 [Pfau et al., 1983]. Phototopotaxis and motility were recovered 6 hours after the application of cinnarizine indicating a regeneration of the flagella.

Verapamil inhibits phototopotaxis and motility at a concentration $2 \cdot 10^{-5}$ M in varying degrees (e.g., phototopotaxis up to 40 %, and motility up to 50-60 %) indicating a gradual detachment of the flagella [Nultsch et al., 1986]. Up to 20 % motility was recovered from its initial level, while phototopotaxis was not recovered. The results suggest a specific action for verapamil on phototopotaxis in *Chlamydomonas*.

Diltiazem and nimodipine, in contrast, affect phototopotaxis in *Chlamydomonas* without an effect on motility. Both chemicals inhibited phototopotaxis up to 30–35 %. Recovery was observed 6 hours after inoculation with diltiazem and 10 hours after inoculation with nimodipine. Collectively, the results indicate that the affect of these blockers (verapamil, diltiazem, and nimodipine) on photomovement parameters in *Chlamydomonas* are specific in that the chemicals acted on phototopotaxis and motility to a different degree.

In our experiments with *Dunaliella*, cinnarizine and isoptin both acted on photomovement parameters of *Dunaliella*, inhibiting completely the velocity of movement and phototopotaxis at a concentration $10^{-3}$ M.


## 9.4.7. Effect of Sodium Azide

The effect of sodium azide on $\upsilon$ and $F$ at a concentration range of $10^{-7}$ to $10^{-3}$ M (Fig. 9.6) was assessed one hour after its introduction into the algal suspension. Positive phototopotaxis occurred when *Dunaliella* cells were laterally illuminated with white light at 500 lx. In contrast, negative phototopotaxis occurred at 40,000 lx [Posudin et al., 1995].

As indicated in Fig. 9.6 *1,2*, sodium azide did not effect the velocity of movement in the two species of *Dunaliella* under both moderate (500 lx) and intense (40,000 lx) illumination with white light. Sodium azide completely inhibited positive phototopotaxis at concentrations $\geq 10^{-4}$ M and negative phototopotaxis at $\geq 10^{-5}$ M in *D. salina* and $\geq 10^{-4}$ M in *D. viridis* (Fig.9.6, *3,4*). Thus, while sodium azide selectively affected phototopotaxis, it did not alter the velocity of cell movement.

No significant differences were found in the response to sodium azide by the two *Dunaliella* species. The absence of a response was not due to light intensity which ranged from 500 to 40,000 lx.

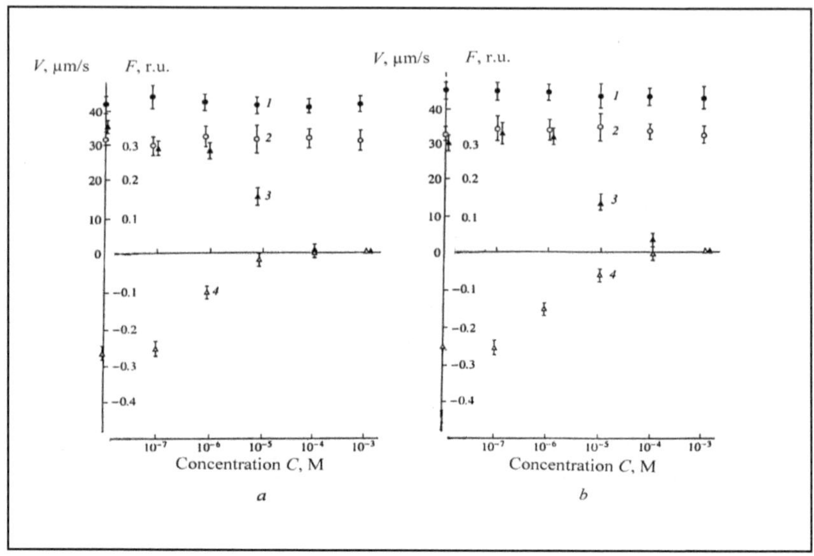



**Fig. 9.6.** Effect of sodium azide on the velocity ($v$) of movement and positive and negative phototopotaxis ($F$) in *D. salina* (*a*) and *D. viridis* (*b*) illuminated with white light: *1* and *2* − $v$ at 500 and 40,000 lx, respectively; *3* and *4* − *F* at 500 and 40,000 lx, respectively. Vertical bars indicate the standard error [Posudin et al., 1995].

In comparision of the effect of sodium azide on photomovement in other species, it inhibited positive phototopotaxis in *C. reinhardtii* at concentrations in the same order of magnitude (i.e., $3{,}5{\cdot}10^{-4}$–$10^{-4}$ M) as those tested with *Dunaliella* [Stavis and Hirschberg, 1973; Pfau et al., 1983]. It did not effect positive phototopotaxis in *E. gracilis* and *Anabatna variabilis* Kütx. but inhibited negative phototopotaxis of *E. gracilis* ($5{\cdot}10^{-5}$ M) [Colombetti et al., 1982] and *A. variabilis* ($10^{-3}$ M) [Nultsch et al., 1983].

Sodium azide did not change the velocity of movement in the two *Dunaliella* species nor *A. variabilis* under intense illumination [Nultsch et al., 1983]. However, it did affect the velocity of movement of *E. gracilis* [Barghigiani et al., 1979], *A. variabilis* under weak illumination [Nultsch et al., 1983], and *Phormidium uncinatum* (Ag.) Gom. [Nultsch and Häder, 1979].

The sodium azide effect is thought to be due to intracellular structural changes in *E. gracilis* [Barghigiani et al., 1979] and by disturbance of noncyclic photosynthetic electron transport in *A. variabilis* [Nultsch et al., 1983]. Effect of the drug on motility in *C. reinhardtii* may be due to shortening or detachment of the flagella [Pfau et al., 1983].

The sodium azide selectively affects the ability *Dunaliella* spp. cells to orient their movement relative to the direction of the light. The absence of an effect on the velocity of movement indicates the occurrence of two separate pathways in the sensory transduction chain. The first is sensitive to sodium azide and transmits the signal responsible for the orientation of cell movement relative to the direction of the light. In the second, the signal governs cell velocity. These conclusions are in agreement with our previous results [Posudin et al., 1992] where there were differences between phototopotactic and photokinetic responses of *Dunaliella* to $\gamma$-radiation level.

116

## 9.5. Summary

Distinctions were found in response to certain biologically active chemicals (e.g., ionophore A23187) between two species *Dunaliella*. In contrast, *C. reinhardtii* often reacts *non-specifically* (autotomy of flagella) to chemicals that stimulate or block ionic channels while *Dunaliella* cells did not lose their flagella. As a consequence, their locomotory reactions can be regarded as *specific* in response to the blocking/stimulating of ionic processes. Likewise, distinctions in photomovement responses to specific substances on hyperhalobic species of *Dunaliella* and fresh-water *E. gracilis* and *C. reinhardtii* indicate a higher tolerance in the hyperhalobic species to external inhibitors.

Maximum values for phototopotaxis in the hyperhalobic species of *Dunaliella* are observed at $10^{-5}$–$10^{-3}$ M $CaCl_2 \cdot 6H_2O$. Increasing the calcium chloride concentration to $10^{-2}$ M suppressed phototopotaxis by 10–20 %. The addition of the ionophore A23187, that increases the permeability of the cellular membrane to calcium ions, completely inhibited phototopotaxis in both *D. salina* and *D. viridis*. The addition of $CoCl_2$, that blocks membrane calcium channels at $10^{-6}$–$10^{-3}$ M, suppressed phototopotaxis in both species. Other calcium channel blockers (e.g., cinnarizine, isoptin, and sodium azide) similarly inhibited phototopotaxis. In contrast, ouabain, which stimulates $Na^+$-$K^+$-ATPase, did not influence phototopotaxis.

The effect of substances that act specifically on photomovement parameters in *Dunaliella* indicates that sensory transduction in the two *Dunaliella* species, as well as in *Chlamydomonas*, is of ionic nature and that photoregulation of the processes governing the velocity of movement of the cells and phototopotaxis occurs through separate sensory transduction pathways in response to light that triggers flagella beatings. The absence of an effect by ouabain on photomovement in *Dunaliella* indicates the absence of a $Na^+$-$K^+$-ATPase system participating in the photoregulation of movement.

# Chapter 10

# Flagella Apparatus

## 10.1. Structure

Flagella are organelle that are typically located at the apical end of the cell. Their lengths are equal and they are a slightly shorter or longer than the length of the cell [Massjuk, 1973]. Each flagellum presents a whip-like structure that is 0.2 μm in diameter. The flagella provide the cell with linear mobility through a pulling or pushing action mediated by active bending and a simultaneous rotation around its longitudinal axis. The flagellum that is closest to the stigma is called *cis*-flagellum, while the more distant one is the *trans*-flagellum.

The flagella apparatus consists of three main parts: the flagellum proper, the basal body, and structures that are associated with basal body (i.e., connecting fibers and flagellar roots). Each flagellum in turn consists of three parts: the flagellar top, shaft, and transition region [Melkonian, 1984]. The flagellum is enclosed within a membrane that is continuous with the plasmalemma. The *axoneme* is the central core of the flagellum and consists of two central microtubules surrounded by nine peripheral microtubule pairs. The microtubules are arranged in a 9+2 pattern and are immersed in an amorphous matrix. The basal bodies of both flagella form a V-like figure that is connected by distal and proximal striated fibers. The basal bodies are connected with the flagellar roots by a system of microtubules and microfibrils [Ringo, 1967]. The fine structure of flagellur apparatus has been described by Ringo [1967], Kvitko et al. [1978], and Melkonian [1982, 1984], while the flagellar apparatus, from the point of view of the phylogeny of green algae, is described by O'Kelly and Floyd [1983-1984].

Flagellar mediated algal motility in response to light (e.g., phototopotaxis and other photoresponses, the role of calcium channels, and photoreceptors) has been reviewed by Kreimer (1995).

$Ca^{2+}$-dependent flagellar dominance in *Chlamydomonas* has been studied in a newly isolated mutant (lsp1) that displays weak phototopotaxis. The *trans*-flagellum in the mutant beats more strongly than the *cis*-flagellum in a $10^{-9}-10^{-6}$ M $Ca^{2+}$ concentration range, a range in which wild-type cells display a switching of flagellar dominance. When compared with other mutants, $Ca^{2+}$-dependent flagellar dominance control and inner-arm dynein subspecies appear to be important for phototopotaxis, but not absolutely necessary [Okita et al., 2005].

## 10.2. Peculiarities of Flagellar Beating

### 10.2.1. Flagella Beating in *Euglena gracilis*

Flagellar bending in *Euglena gracilis* G.A. Klebs is characterized by a helical form in which each of a series of bendings represents part of the spiral. The parts are separated by rectilinear sections in the flagellum. This form of flagellum beating is called a "broken (or interrupted) helix" [Jahn and Bovee, 1968]. Neither the typical ciliary nor undulate types of flagellar beating are observed in euglenids.

Flagellum beating in *E. gracilis* provides cell movement along a helicoidal trajectory. The cell can be characterized by the frequency of the cell body rotation, 2 Hz or about 0.32 revolutions per second (19 r/min).

## 10.2.2. Flagella Beating in Green Algae

Flagella of *Chlamydomonas reinhardtii* P.A. Dang. display a synchronous beating of symmetrical character with regard to the longitudinal axis of the body in one plane. The process of flagella movement consists of two stages. The first stage involves movement of the flagella from front to back in the straightened state that yields a "power stroke". The second involves the recovery of the flagella to the initial state mediated by a smooth bending from the base to the tip − the "return stroke" [Ringo, 1967]. The cell moves forwards due to the power stroke, while during the reverse stroke it moves slightly backward. This type of movement was referred to by Ringo (1967) as "swimming by style of breast stroke" and the flagellar beating a cilia-like type [Ringo, 1967; Kvitko et al., 1978]. The movement of the cell backwards during photophobic response is mediated by undulatory waves (flagellum-like beating) that propagate from the base to the tip [Colombetti and Marangoni, 1991]. The principal types of flagellar beating in *Chlamydomonas* are depicted in Fig. 10.1.

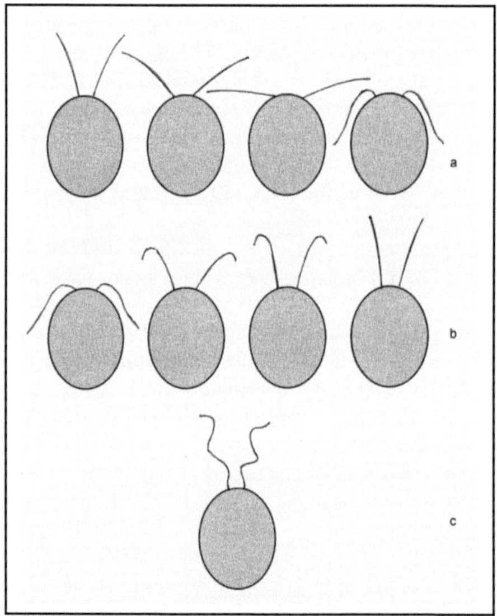


**Fig. 10.1.** Flagellar beatings in *Chlamydomonas*; *a* − forward motion (breast stroke, ciliary-like style); *b* − backward motion (return stroke, ciliary-like style) according to [Ringo, 1967]; *c* − reverse beatings (flagellar-like style) according to [Colombetti and Marangoni, 1991].

The velocity of rotation of the cell around the longitudinal axis during its linear movement is 2 Hz [Rüffer and Nultsch, 1985]. Turning of *C. reinhardtii* cells relative to the source of light occurs in the following manner [Rüffer and Nultsch, 1990; Nultsch, 1991]. Light is detected by a photoreceptor that is located in the plasmalemma opposite the stigma and asymmetrically in relation to the longitudinal axis of the cell. Illumination of the photoreceptor is accompanied by a flux of calcium ions across the flagellar membrane. The *cis*-flagellum (closest to the stigma) increases the amplitude of its beating in response to the influx of calcium ions relative to the beating of *trans*-flagellum. The differential in response between the two flagella to changes in the calcium ion concentration within the intracellular space results in the change in direction of movement toward the light source (i.e., phototopopotaxis).

Analysis of the movement of *Chlamydomonas* flagella using the radial-spoke system between wild-type and mutant flagella was described by Brokaw et al. (1982). The flagellar beat frequency of the biflagellated green alga *C. reinhardtii* was measured using fast Fourier transform analysis of light intensity fluctuation in microscope images of swimming cells. Live cells had a mean beat frequency of 48-53 Hz at 20 °C. However, detergent-extracted "cell models", when reactivated in the presence of 1 mM ATP, appeared to have two different beat frequencies of about 30 and 45 Hz. These observations suggest that the two flagella of *Chlamydomonas* have different intrinsic beat frequencies but that they are somehow synchronized and beat together in swimming cells [Kamiya and Hasegawa, 1987].

Although the two flagella of *Chlamydomonas* appear similar to each other, they differ in the beat frequency. The *trans*-flagellum beats at a 30–40 % higher frequency than the *cis*-flagellum in demembranated and reactivated cell models. Experiments with a set of mutants (*oda*) suggested that the attachment site for the outer dynein arm is important in determining the flagellar beat frequency and the basal portion of the outer arm dynein is important in regulating the flagellar activity and therefore the behavior of the cell [Saeko and Ritsu, 1997].

*C. reinhardtii* cells are able to change the beating frequency, pattern, and synchrony of the *trans*- and *cis*- flagella in response to light stimulation and the response of each flagellum is quite different. The *trans*-flagellum responds with less delay than the *cis* for both beating frequency and stroke velocity. With light stimulation at 2 Hz for the critical cell-rotation frequency, the *trans*- and *cis*-flagella responses are about 180 degrees out of phase. The *trans*-flagellar beating frequency peaks at a stimulus frequency of 5-6 Hz, higher than the *cis* at 1-2 Hz. The stroke velocities of the *trans*-flagella and *cis*-flagella have the same stimulus-frequency response (2 Hz), however, the *trans*-flagellum has a shorter delay than the *cis*. The length of time to reach a maximum response is much shorter than the time required for a single rotation of the cell. The use of two different mechanisms to enable the *trans*-flagellum to respond ahead of the *cis*-flagellum in both the beating frequency and stroke velocity responses suggests the importance of both responses in phototopotaxis [Josef et al., 2006]. The same peculiarity in flagellar function that is characterized by a differential in the influx of calcium ions between the *cis*- versus the *trans*-flagellum is found in *Haematococcus pluvialis* Flotow [Sineshchekov, 1991*a*].

It is possible to observe a transfer in the undulation from cilia-type to a flagella-type. Such changes of character of flagellar beating are observed during photophobic reactions in response to an electrical reaction triggered by the opening of potential-dependent calcium channels in the flagellar membrane [Sineshchekov et al., 1978; Beck and Uhl, 1994; Holland et al., 1997].

Using electro optic monitoring of flagella beating in *C. reinhardtii,* it was possible to record the beating frequency, stroke velocity, and stroke duration of each flagellum and the relative phase of the *cis*- and *trans*- flagella. Each beat cycle was resolved such that each asynchronous beat was detected [ Josef et al., 2005].


### 10.2.3. Flagella Beating in *Dunaliella*

The cells of *Dunaliella bioculata* Butcher move with a sinusoidal trajectory and rotate around their longitudinal axis [Marano, 1992]. Each flagellum moves from front to back and settles along its longitudinal axis. Reverse beating restores the initial position of the flagellum due to bending that propagates along its length.

There is a unique function of the flagellar apparatus in comparison with *Chlamydomonas* in that just before a change in movement direction of the cell, one of the two flagella remains transiently immobile while the other rotates the cell via ciliary beating. As a result, the cell moves in the new direction with both flagella active. In addition, there is a discrepancy in flagella beating frequency in *D. bioculata*. One flagellum beats at 60 Hz and the other at 50 Hz. This changes the angle between the planes of beating causing rotation of the cell and its helicoidal movement [Shoevaert et al., 1988; Marano, 1992]. Analysis of the flagellar assembly in *D. bioculata* is discussed in the work Marano et al. [1988] and Schoppmeier and Lechtreck [2002].

## 10.3. Analysis of Flagellar Beating

### 10.3.1. High-Speed Microcinematography

High-speed microcinematography (100-500 frames/s) makes it possible to analyze frame by frame the movement of algae cells. The parameters of movement in *C. reinhardtii* cells were thereby estimated. The linear velocity at room temperature was 100-200 µm/s (maximum value 240 µm/s); the velocity of rotational movement was 1.4-2 Hz or 0.22-0.32 r/s (maximum value 2.5 Hz or 0.4 r/s), and the flagellar beating frequency was from 45 to 62-70 Hz for flagellum located on the external side of helicoid and 45 Hz for flagellum located on the internal side [Rüffer and Nultsch, 1985, 1998]. The velocity of linear movement in *D. bioculata*, determined by microcinematography was $105 \pm 10$ µm/s [Shoevaert et al., 1988].

The motility of *D. bioculata* cells in response to pesticides was also quantified using microcinematography [Marano et al., 1988; Krishnaswamy-Chang, 1997]. The flagella of the marine algae *D. bioculata* and the freshwater *Chlamydomonas* resemble very much the cilia lining the bronchial tubes in human lungs. A commercial formulation of lindane was tested at various concentrations on the two algae. Concentrations of 5 ppm to 30 ppm were cytotoxic and had a remarkable effect on the motility of both *D. bioculata* and *Chlamydomonas*. Isomers of lindane instigated similar effects on *D. bioculata*. Cytotoxicity was determined using growth curves and motility quantified using microcinematography and Doppler laser velocimetry.

### 10.3.2. Laser Light Scattering

Several light scattering methods have been utilized to study swimming organisms. The first method assessed the Doppler shift of the scattered light to measure the velocity of movement and flagellar beating frequency [Ascoli, 1975; Ascoli et al., 1978; Ascoli and Frediani, 1980; Angelicini et al., 1986]. Laser Doppler spectrometry has also been used to study the rate and energy of cell mobility in *E. gracilis* in response to the duration of exposure to potassium bichromate $K_2Cr_2O_7$ (i.e., 1, 4 and 7 days). The degree of chromium toxicity depended upon the concentration and length of exposure to the chemical [Novikova et al., 2007].

The *Doppler effect* is the change in frequency (wavelength) of a wave for an observer moving relative to the source of the wave. When an object is moving at a constant velocity $v$ and is irradiated with light of a certain wavelength $\lambda$, the scattered light undergoes a Doppler frequency shift $\Delta f$. This shift depends on the velocity $v$ of movement of the object, the angle $\theta$ of scattering of the light, and the angle $\varphi$ between the direction of movement of the object and the direction of light propagation [Ascoli et al., 1980]. The Doppler effect can be calculated as:

$$\Delta f = \frac{2\upsilon}{\lambda}\sin\frac{\theta}{2}\cos\varphi.$$

( 10.1 )

Investigation of the interaction of laser radiation with cells of *E. gracilis* using Doppler frequency shifts allowed estimating the velocity of linear movement (100 μm/s), the frequency of the cell rotation (~2 Hz), and the flagellar beating frequency (about 30-50 Hz) [Ascoli et al., 1978; Ascoli and Frediani, 1980; Ascoli and Petracchi, 1991]. Similarly, Doppler shifts in laser light scattered by the cells of *Haematococcus pluvialis* Flotow using heterodyne detection techniques allowed measuring the swimming velocities of cells along the light stimulus axis [Cantatoreet al., 1989]. The time taken by a cell population to change orientation was about 1 s.

Using laser Doppler spectroscopy, the linear velocity of *D. bioculata* cells was determined to be 109 ± 5 μm/s [Marano, 1992] and the frequency of flagellar beating in *D. salina* approximately 25 Hz [Ascoli et al., 1980]. The method identified two different frequencies for *E. gracilis*; 30-50 Hz that corresponded to the flagella beating and 2 Hz that was related to the cell body rotation [Angelini et al., 1986; Ascoli and Petracchi, 1991]. The motility of *D. bioculata* under the influence of pesticides was also quantified by Doppler laser velocimetry [Krishnaswamy-Chang, 1997].

### 10.3.3. Method of Microphotometry

Microphotometry involves recording the absolute or relative values of radiation flowing through the base of the flagella. Spatial changes in the position of the bases modulates the light which when measured estimates the frequency of flagella beating. The process of microphotometry is depicted in Fig. 10.2.

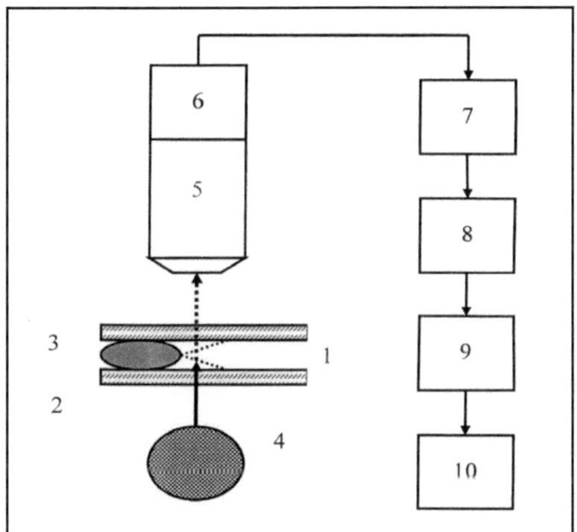

**Fig. 10.2.** Principle of microphotometry for the study of flagella beating [Posudin, 2009, Patent of Ukraine № 45376, Registration N U200905299]

The algal cell *1* is placed between the slide *2* and cover glass *3*. Optical radiation of the source *4* passes though the bases of the flagella thus modulating the intensity of light. This modulated light enters the objective of microscope *5* and photodetector *6*. The electric output signal of the photodetector then enters the amplifier-converter that amplifies the signal and filters it in the 3-100 Hz range. The signal is elaborated due to the Schmidt trigger which produces a standard pulse of certain duration if the intensity of signal exceeds a certain level. The pulse is indicated by the pulse counter *9*. Individual pulses corresponding to flagellar beating enter the analog-digital transformer *10* that estimates the frequency of flagellar beating. We have used microphotometry to measure the frequency of flagella beating in *D. salina* which was found to be in the 20-50 Hz range [Posudin, 2009].

## 10.4. Summary

Green algae such *Chlamydomonas* and *Dunaliella* differ during the process of photomovement in the character of their flagellar beating. Progressive movement of *Chlamydomonas* is due to ciliary beating of the flagella and backward movement by undulate beating. Since cells of *Dunaliella* are not capable of photophobic reactions, the undulate mode of flagella beating is not found in the genus. Turning of *Chlamydomonas* cells occurs due to an unequal frequency in beating between the *cis-* and *trans-*flagella. In contrast, the cells of *Dunaliella* temporarily cease beating in one flagellum, thereby initiating a turn.

The frequency of flagellar beating in seawater and hyperhalobic species of *Dunaliella* (20-50 Hz) is of the same order as in freshwater species of *Haematococcus* and *Chlamydomonas*.

# Chapter 11

# Applied Aspects of Aquatic Biomonitoring Using the Photomovement of *Dunaliella*

Observation of the state of the Earth's biotic component, its response to anthropogenic effects, and deviations from a *normal* state assessed at various organizational levels (e.g., molecular, cellular, organismal, populational, biocenotical) is called *biological monitoring*. A *test-object* (biomonitor) is an organism, part of an organism, or a community of organisms that contains quantitative information on the quality of the environment [Butterworth et al., 1995]. A number of organisms (e.g., bacteria, fungi, actinomyces, algae, protozoa, invertebrates, fish, amphibians, aquatic plants) can be used as test-objects [Shubert, 1984, 1984; Krajnyukova, 1988; Rosenberg and Resh, 1993; Markert et al., 2003].

A *test-function* is a physiological or behavioral response of an organism to changes in the quality of the environment. The intensity of reproduction, immobilization of cells, motility, photosynthetic activity, membrane permeability, bioluminescence, impedance of cell suspension, bioelectrical reaction [Barenboim and Malenkov, 1986; Krajnyukova, 1988]; growth, reproduction, mobility and energy potential of the cells [Parshikova, 2003; Parshikovaet al., 2004]; protoplasmic streaming [Mustacich and Ware, 1976; Evdokimov et al., 1982; Pileri, 1987], and phototopotaxis of organisms [Simone et al., 1978; Wang et al., 2001; Wu et al., 2006] are examples of typical test-functions that can be used for biomonitoring.

The proposed use of photomovement of high density algae (*Euglena gracilis* G.A. Klebs) for the removal of nutritive substances in photo-bioreactors is of considerable interest [Nakajima and Takahashi, 1991]. A number of articles and reviews have focused on the use algae as test-objects. For example, a review on the development of a biosensor for monitoring flagellar movement and phototopotaxis in algal cells for the detection of chemical toxicity in wastewater was proposed by Shitanda et al. [2006]. Waste water toxicity monitoring is also critiqued by Diao [2007].

The principal advantages of biomonitoring are high sensitivity, rapidity, reliability and the possibility of creating automated systems for collecting and processing environmental information. The very limited availability of useful biomonitors is related to the absence of adequate quantitative measures of individual toxicants present in aquatic media and the possible interaction between separate toxic components that are present in a mixture. Test-objects used under natural conditions are termed passive while in the laboratory they are considered active.

## 11.1. Algae of Genus *Dunaliella* as Test-Objects

Monocultures of unicellular algae represent a special segment among living organisms used as test-objects in that they provide a very uniform population that is suitable for relatively precise quantitative estimations of various influences. Monocultures of green algae of the genus *Dunaliella* (*Chlorophyta*) are highly desirable in that they are close relatives of higher plants. Their ability to reflect conditions modulating the plant kingdom (*Viridiplantae*) is extremely important since plants provide the essential requisites for the survival of mankind. Their microscopic size, high rate of reproduction, and active movement are peculiar to species of this genus and represent essential advantages of these organisms as model test-objects. Representatives of the genus change their behaviour and development in response to wide fluctuations in water salinity, light exposure, and a diverse range of toxicants. Current applications of algae in the genus *Dunaliella* are presented in Table 11.1.

**Table 11.1.** *Dunaliella* as test-object during biomonitoring of aquatic medium

**Heavy Metals**

| Pollutant | Test-Object | Test-Function | Reference |
|---|---|---|---|
| copper and lead | *Dunaliella salina, D. bioculata* and *D. tertiolecta* | growth and pigment content | [Pace et al., 1977] |
| copper, lead, cadmium and mercury | *D. salina* | cell growth | [Barghigiani et al., 1981, 1983; Serritti et al., 1981] |
| Hg, Cu, Cd, Pb | *D. viridis, D. tertiolecta* | uptake and intracellular accumulation of heavy metals | [Tzvylev and Tkachenko, 1981] |
| orthovanadate | *D. parva* | motility | [Gilmouret al., 1985] |
| copper | *Dunaliella salina, D. tertiolecta, and D. viridis* | absorption of copper | [Lustigman et al., 1985] |
| copper | *D. salina* | glycerol production | [Lustigman et al., 1987] |
| copper | *D. tertiolecta* | fluorescence induction | [Samson et al., 1988] |
| mercury, copper | *D. tertiolecta* | fluorescence induction | [Samson and Popovic, 1988] |
| boron | *D. tertiolecta* | growth | [Ahmed et al., 1988]. |
| copper | *D. salina* | thermostability, photosynthesis rate, motility | [Veselova et al., 1990] |
| Al, La, Cu, Cd, Hg, W | *D. acidophila, D. parva* | photosynthesis and growth | [Gimmler et al., 1991] |
| heavy metals, aromatic hydrocarbons and salts | *D. tertiolecta* | analitical and biotoxicity tests | [Baldi et al., 1993] |
| copper | *D. tertiolecta* | loss of flagella, changes in cell shape, lack of motility, and collapse | [Khristoforova et al., 1996] |
| Cd, Cr, Cu, Pb, Ni and Zn | *D. tertiolecta* | algal inhibition | [Pun et al., 1995] |
| Cd, Cr, Cu, Pb in coastal sediments | *D. tertiolecta* | inhibition of growth | [Wong et al., 1999] |
| Zn | *D. tertiolecta* | synthesis of phytochelatins | [Tsujiet al., 2002] |
| copper | *D. salina* and *D. tertiolecta* | total chlorophyll and carotenoids | [Nikookar et al., 2005] |
| lead and aluminium | *D. tertiolecta* | growth response and ultrastructure | [Sacan et al.,2007] ] |
| copper | *D. tertiolecta* | inhibition of growth rate | [Levy et al., 2007] |

126

**Pesticides**

| Pollutant | Test-Object | Test-Function | Reference |
|---|---|---|---|
| organophosphates Baytex and Abate; carbamate Baygon; and the chlorinated hydrocarbon; DDT | *D. euchlora* | photosynthesis | [Derby and Ruber, 1970] |
| chlorinated hydro-carbons (DDT, dieldrin, and en-drin) | *Dunaliella* | photosynthesis and growth | [Menzel, et al., 1970] |
| organochlorine insecticide lindane | *D. bioculata* | growth and structure | [Levain and Mara-no-Le Baron, 1973] |
| lindane | *D. bioculata* | division, cell cycles, and biosyn-thesis | [Jeanne, 1979] |
| hexachlorocyclo-pentadiene, EPN, chlorpyrifos, car-bophenothion, atrazine | *D. tertiolecta* | population growth, death of cells | [Walsh, 1983] |
| atrazine, DCMU, Dutox, and Soil-gard | *D. tertiolecta* | fluorescence induction | [Samson and Po-povic, 1988] |
| pesticide residues | *D. salina* | | [Yarden et al., 1993] |
| lindane | *D. bioculata* | motility | [Krishnaswamy-Chang, S., 1997] |
| organophosphorus pesticide | *D. salina* | photosynthesis, growth, biochem-ical compounds | [Cai et al., 1999] |
| organophosphorus pesticide methami-dophos | *D. salina* | growth, photosynthetic rate and biochemical compounds | [Xie et al., 1999] |
| profenofos | *D. salina* | tolerance to pesticide | [Xie et al., 1999] |
| pyrethroids (cy-permethrin and fenvalerate) and organophosphorus insecticides (diazi-non and cyano-phos) | *D. salina* | growth and some cellular ma-cromolecules (chlorophyll *a*, proteins, carbohydrates, RNA and DNA) | [Noaman et al., 2002] |
| herbicide atrazine, the insecticide chlorpyrifos, and the fungicide chlo-rothaloni | *D. tertiolecta* | population growth rate | [DeLorenzo and Serrano, 2003] |
| atrazine, chlorpyri-fos, and chlorotha-lonil | *D. tertiolecta* | population growth rate | [DeLorenzo and Serrano, 2003] |

| pesticides | D. salina | motility, biochemical parameters, cell size, chlorophyll content, population growth rate, biomass | [Orme, Kegley, 2004] |
| irgarol, fungicide chlorothalonil, and herbicides atrazine and 2,4-D | D. tertiolecta | growth rate | [DeLorenzo and Serrano, 2006] |

**Other toxicants, pollutants and chemicals**

| Pollutant | Test-Object | Test-Function | Reference |
|---|---|---|---|
| cigarette smoke (solid and gas phases) | D. bioculata | ciliostatic activity | [Izard et al., 1967, a; Izard and Testa, 1968] |
| sodium chloride, sodium sulfate, sulfuric acid and glucose | D. acidophila | growth, photosynthesis, and respiration | [Fuggi et al., 1988] |
| phenols: hydroquinone, pyrocatechol, phenol, guaiacol, resorcinol | D. salina | motility | [Stoma and Roth, 1981] |
| pyrocatechol and p-benzoquinone | D. salina | loss of motility | [Stom et al., 1984] |
| porphyrin derivatives, furcoumarins, acridines | Dunaliella | motor inactivation and loss of viability | [Posudin and Repetskii, 1988] |
| salt | D. salina | photosynthetic pigments and proteins | [Heidari et al., 2000] |
| salt | D. salina | photosynthetic pigments and proteins | [Heidari et al., 2000] |
| surface-active substances | D. salina, D. minuta | chlorophyll fluorescence, photomovement | [Parshikova, 2004] |
| nitrobenzenes | D. salina | grow inhibition | [Shen et al., 2006] |
| pharmaceuticals and personal care products (PPCPs) | D. tertiolecta | toxicity threshold | [DeLorenzo et al., 2008] |

There are also reports detailing the possible use of *Dunaliella* species in the phytoremediation of aquatic media. The effectiveness of *Dunaliella* in removing copper and nickel from solutions was demonstrated by Abdel-Raouf and Ibraheem [2001]. The highest concentration of the tested substances that did not inhibit the growth rate of the organism was 10 ppm of copper and 20 ppm of nickel. The bioaccumulation and toxicity of germanium in *Dunaliella salina* Teod. were investigated by Zhu and Wang (2001). The authors believed that most of the germanium in the algae was associated with proteins and amino acids-dissolved carbohydrate that may represent detoxicated storage forms of germanium. Several research groups have described the accumulation of copper, lead, selenium [Sacan et al., 2000], and aluminum [Sacan et al., 2001] by *Dunaliella tertiolecta* Butcher and chlororganic substances and oil products from water [Tzvylev and Tkachenko, 1981] by *Dunaliella viridis* (Snow) Printz and *D. tertiolecta*.

At present, there is considerable interest in developing new diagnostic methods for indicating the need for surgery in humans that require only a small amount of a test organism and could provide a rapid assessment of the state of the patient. The cells of *Dunaliella bioculata* Butcher have been used as cellular models and test-objects for monitoring the efficiency of gossypol, a chemical known for its contraceptive properties in humans and several other mammals, through its induction of spermatogenesis disorders and the inhibition of spermatozoal motility [Druez et al., 1989]. Inhibition of *D. bioculata* motility by gossypol was observed at the same concentrations modulating spermatozoa.

Another possible medical application of *Dunaliella* is the introduction of aqueous extracts from *D. tertiolecta* into a subject and assessing the central nervous system, spontaneous motor activity, rectal temperature, exploratory behaviour, muscle relaxation, catalepsy, and conditioned avoidance responses. The extracts can be used as a central nervous system depressant and a potential muscle relaxant [Villaret al., 1992]. Monocultures of *D. viridis* have also been used for estimating the level of cytotoxical compounds in the blood using changes in the relative motility of their cells as a rapid indicator [Dmitriev et al., 2005].

## 11.2. Photomovement Parameters of *Dunaliella* as Test-Functions

The majority of methods for biomonitoring the impact of chemicals on aquatic environments are through assessment of only one parameter that is modulated by the compound of interest. Assessment of only one test-function significantly limits the effectiveness of biomonitoring in that other chemicals in an aquatic medium may produce the same effect. Increasing the number of test-functions monitored significantly increases the level of qualitative and quantitative precision for assessing the toxicants present in the medium.

We have previously used several (i.e., $\geq 2$) photomovement parameters as test-functions and assessed them simultaneously. Toxicants present in an aquatic medium can produce specific responses in several photomovement parameters (e.g., linear and rotational velocities, relative number of motile cells, phototopotaxis, frequency of flagellar beating) depending on the type and concentration of the toxicant and its mechanism of interaction with the cells. We believe that the simultaneous recording of several photomovement parameters significantly increases the sensitivity of the biomonitor. Therefore, we investigated the possible use of green algae of *Dunaliella* species as test-objects and the simultaneous analyses of several photomovement parameters (e.g., linear and rotational velocities, phototopotaxis and motility) as test-functions. In addition, the vector method for estimating the effects of different pollutants (e.g., surface-active substances, pesticides, heavy metals) on photomovement of the microorganism was used.

Unialgal cultures *Dunaliella salina* Teod., strain N 10 and *Dunaliella viridis* Teod. (strain N 42) from the N.G. Kholodny Institute of Botany algal collection at the Ukrainian Academy of Sciences [Massjuk and Tereshchuk, 1983] were used in the study. Linear velocity $v$, velocity $n$ of rotation, phototopotaxis $F$ and relative motility $N_m/N_0$ of the cells ($N_m$ is the number of motile cells, $N_0$ – total number of the cells) were the photomovement parameters monitored. A description of the experimental set-up is given in Section 4.1. A cross-section of toxicants was tested (e.g., surface-active substances, salts of heavy metals, pesticides) [Parshikova et al., 1990; Posudin et al., 1996]. Surface-active substances tested were: cation surface-active substance (CSAS) – catamine or cationic surfactant (alkyldimethylbensylammonium chloride), anion surface-active substance (ASAS) or sodium salt of dodecyl sulphoacid (NaSDS), non-ionogenic surface-active substance (NSAS) or hydropol (from Collection of Institute of Colloidal Chemistry and Chemistry of Water of National Academy of Sciences of Ukraine) and natural surface-active substances of polysaccharide origin (PSAS) extracted from cyanobacteria [Peskov, 1979]. The concentrations of these surface-active substances varied from 1 to 40 mg/1.

The effect of various types of surface-active substances, their combinations and the duration of action on the velocity of movement of the species was determined using dispersive analysis of three-factorial non-orthogonal complexes. Data on the velocity of movement of the cells (from 0 to 55 $\mu$m/s) were grouped in 11 categories and statistically analyzed using biometric methods [Lakin, 1973].

Two three-factorial complexes were investigated. The application of the first complex made it possible to study the effect of such factors as type of surface-active substances (CSAS, ASAS, NSAS), their combinations (CA – cation-active-anion-active surface-active substances, cation-active-nonion-active surface-active substances, AN – anion-active-nonion-active surface-active substances, CAN – cation-active-anion-active-nonion-active surface-active substances), type of algae (*D. salina* and *D. viridis*) and duration of surface-active substances action (in 0.5; 1; 2; 3 and 4 hours after inoculation) on the velocity of cell movement.

The second complex was used to study the effect of different concentrations of surface-active substances (1, 5, 10, 20, 30, and 40 mg/l), types of surface-active substances (CSAS, ASAS, NSAS, PSAS), and type of algae (*D. salina* and *D. viridis*) on the velocity of cell movement.

The effects of the following pesticides were studied: acetal (1,1-diethoxy-ethane) (55 %), acetazine (1-[10-[3-(dimethylamino)propyl]-10H-phenothiazin-2-yl]-ethanone) (50 %), alachlor (2-chloro-N-(2,6-diethylphenyl)-N-(methoxymethyl)-acetamide) (45 %), arylon (75 %), basta (2-amino-4-(hydroxymethylphosphinyl)-butanoic acid) (20 %), dual (2-chloro-N-(2-ethyl-6-methylphenyl)-N-(2-methoxy-1-methylethyl)-acetamide) (96 %), DPC (20%), harmoni (75 %) and tecto (2-(4-thiazolyl)-1H-benzimidazole) (45 %). The concentrations of the pesticides ranged from $10^{-7}$ to $10^{-2}$ M. The effect of heavy metals was also determined using the salts of copper ($CuSO_4 \cdot 5H_2O$), cadmium ($CdCl_2$) and lead ($Pb(NO_3)_2$) in a concentration range of $10^{-7}$ to $10^{-2}$ M.


## 11.3. Effect of Surface-Active Substances on Photomovement of *Dunaliella*

### 11.3.1. Characteristics of Surface-Active Substances

The surface-active substances tested were the same or similar to chemicals commonly encountered in water reservoirs. The accumulation of these surface-active substances occurs through the action of either anthropogenic factors or natural processes [Parshikova, 2004].

*Synthetic surface-active substances* enter the environment as sanitary and domestic sewage, and as municipal and industrial wastewater. The level of surface-active substances in the sewage from textile plants can reach 2500 mg/l and in products of organic synthesis, as much as 10,000 mg/l [Stavskaya, 1981]. Effective removal of surface-active substances from sewage is difficult and often insufficient, leading to the introduction of these chemicals into the environment. For instance in 1980, 23,000 tons of surface-active substances (from an initial 27,000 tons) entered water reservoirs in Germany [Taranova, 1988]. Surface-active substances of varying chemistries were found in practically all regions of the world and at concentrations reaching nearly 5 mg/l in some reservoirs [Filenko, 1988].

*Natural surface-active substances* are produced during the metabolism of many organisms (e.g., some bacteria [Margaritis et al., 1979; Cooper and Zajic, 1980; Duvnjak et al., 1982], green algae, cyanobacteria and diatoms [Chamberlain, 1976; Sirenko and Kositskaya, 1988], and cell cultures of higher plants [Vakhmistrov and Bogorov, 1987]). Unfortunately, the chemical nature of biological surface-active substances and their role in metabolism have not been adequately elucidated.

There is considerable interest in assessing the membranotropic action of surface-active substances that produce bactericidal [Kalinichenko et al., 1986; En-Zanfeily and Nawar, 1980] and fungicidal [Zlochevskaya et al., 1981] effects, in addition to the negative effects of surface-active substances on culture growth, pigment composition, and photosynthetic activity in algae [Braginsky, 1986; Parshikova, 1988; Parshikova and Pakhomovas, 1988]. Lenova et al. [1989] reported the toxic effect of anion-active surface-active substances (sodium dodecyl sulfate) on *D. viridis* and its concentration dependant effect on cell size distribution.

Some surface-active substances are also known to stimulate the growth of algae-macrophytes in aquaculture [Kalugina-Gutnik and Belyayev, 1987] which is of significant practical interest with regard to the industrial production of various macroalgae species. As a consequence, it is essential to determine the interaction of various surface-active substances on the growth and behavior of algae in both natural and cultivated conditions.

## 11.3.2. Effect of Various Types of Surface-Active Substances, their Combinations and Duration of Action on the Velocity of Movement in *Dunaliella*

Using a three-factorial dispersive analysis, the effects of various surface-active substances on the velocity of movement of different species of *Dunaliella* are presented in Tables 11.2 and 11.3. The results identify statistically significant independent and total effects of such factors as type and concentration of surface-active substances and the duration of action using the velocity of movement of the algae. The effects of species and combinations of concentration, type of substance, and algal species were not statistically different.

**Table 11.2.** A three-factorial dispersive analysis of the effect of type and duration of action of SAS on the velocity of movement by different species of *Dunaliella*.

| Factor | Degrees of freedom | Variance, $\sigma^2$ | Criterion of significance, $F_\Phi$ | $p = 0.05$ | | $p = 0.01$ | |
|---|---|---|---|---|---|---|---|
| | | | | $F_S$ | Level of significance | $F_S$ | Level of significance |
| $T_1$ | 5 | 204,3 | 387,4 | 2,2 | S | 3,0 | S |
| $B$ | 1 | 0,97 | 1,8 | 3,9 | NS | 6,7 | NS |
| $\tau$ | 2 | 124,0 | 235,1 | 3,0 | S | 4,6 | S |
| $T_1B$ | 5 | 10,8 | 20,4 | 2,2 | S | 3,0 | S |
| $T_1\tau$ | 10 | 6,4 | 12,2 | 1,8 | S | 2,3 | S |
| $B\tau$ | 2 | 9,9 | 18,8 | 3,0 | S | 4,6 | S |
| $T_1B\tau$ | 10 | 2,9 | 5,6 | 1,8 | S | 2,3 | S |

N o t e s : $T_1$ – type of SAS; $B$ – species of alga; $\tau$ – duration of SAS action on algae; $T_1B$, $T_1\tau$, $B\tau$ and $T_1B\tau$ – combinations of various factors [Parshikova et al., 1990].

**Table 11.3.** A three-factorial dispersive analysis of the effect of various factors on the velocity of movement by different species of *Dunaliella*.

| Factor | Degrees of freedom | Variance, $\sigma^2$ | Criterion of significance, $F_\Phi$ | $p = 0.05$ | | $p = 0.01$ | |
|---|---|---|---|---|---|---|---|
| | | | | $F_S$ | Level of significance | $F_S$ | Level of significance |
| $C$ | 5 | 295,5 | 244,2 | 2,2 | S | 3,0 | S |
| $T_2$ | 3 | 612,3 | 506,1 | 2,6 | S | 3,8 | S |
| $B$ | 1 | 3,8 | 3,1 | 3,9 | NS | 6,7 | NS |
| $CT_2$ | 15 | 31,0 | 25,6 | 1,7 | S | 2,0 | S |
| $CB$ | 5 | 5,1 | 4,2 | 2,2 | S | 3,0 | S |
| $T_2B$ | 3 | 29,4 | 24,3 | 2,6 | S | 3,8 | S |
| $C\,T_2B$ | 15 | 0,5 | 0,4 | 1,7 | NS | 2,0 | NS |

Notes: $C$ – concentration of SAS; $T_2$ – type of SAS; $B$ – species of algae; $CT_2$, $CB$, $T_2B$ and $CT_2B$ – combinations of various factors [Parshikova et al., 1990].

As shown in Fig. 11.1 *a*, CSAS and ASAS at 1 mg/l result in a stimulating effect on phototopotaxis. At higher concentrations (up to 20 mg/l), the substances suppressed phototopotaxis in both species of *Dunaliella* through immobilization of the cells and changes in the direction of their movement. NSAS and PSAS between 1-10 mg/l stimulated phototopotaxis though at 40 mg/l phototopotaxis decreased more than twofold by 4 hours after inoculation (Fig. 11.1 *b*).

The inhibitory action of CSAS and ASAS at up to 20 mg/l on linear velocity is illustrated in Fig. 11.2 *a*. The inhibitory effect of NSAS and PSAS was less pronounced and was accompanied by a decrease in the velocity of movement to 50 % of the initial level.

The dependence of photomovement parameters on the duration of action of various types of surface-active substances and their combinations (at 10 mg/l) demonstrated a toxic effect on motility ($N_m/N_0$) with both species decreasing in the following sequence: CSAS > CSAS + ASAS > ASAS > CSAS + NSAS > ASAS + NSAS > CSAS + ASAS + NSAS > NSAS > NSAS + PSAS > PSAS. CSAS resulted in complete immobilization of the cells 1 hour after exposure while NSAS and PSAS led to only a 30 % decreases in motility after 3 hours exposure.

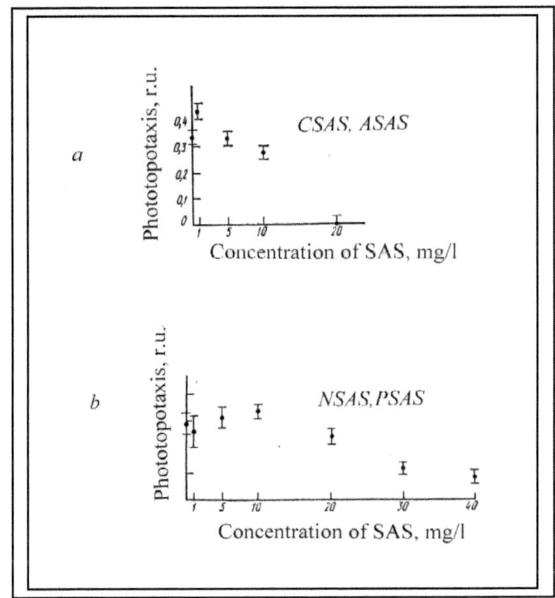


**Fig. 11.1.** Dependence of linear velocity of two species of *Dunaliella* on the concentration of: *a* − CSAS and ASAS; and *b* − NSAS and PSAS, during 4 hours of contact [Parshikova et al., 1990].

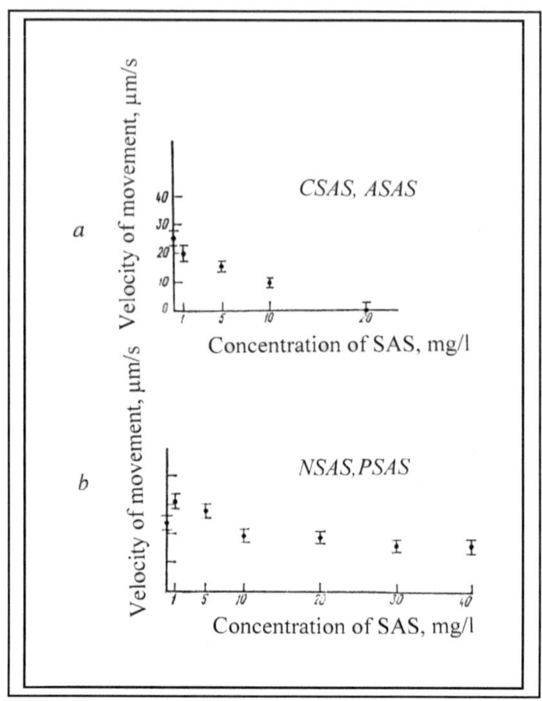


**Fig. 11.2.** Dependence of phototopotaxis in two species of *Dunaliella* on the concentration: *a* − CSAS and ASAS; *b* − NSAS and PSAS, during 4 hours of contact [Parshikova et al., 1990].

The type of surface-active substance affects the motility, velocity, and direction of movement of the cells in both species. The response depends on the chemical composition of compound, its concentration, and its duration of action on the cells. CSAS that possess positive ions, decreased or completely stopped photomovement at 1-20 mg/l. ASAS, which has negative ions in contrast to NSAS and PSAS that do not, induced smaller changes in the velocity and direction of movement of the cells. The higher sensitivity to positive ions indicates the possible influence of surface-active substances on the cell's $\zeta$-potential and motile reactions that are related to it. To date, however, there is not enough experimental data to adequately understand the mechanisms involved in these interactions.

## 11.4. Investigation of the Effect of Heavy Metals on Photomovement in *Dunaliella* Using Laser Doppler Spectroscopy

Laser Doppler spectroscopy is addressed in detail in Section 10.3.2. Application of the method for assessing the effect of heavy metals in an aquatic medium on photomovement parameters such as motility and linear velocity is critiqued. A Doppler correlation spectrometer consists of a laser, thermostatically controlled measuring cuvette, photodetector, correlator, and computer (Fig. 11.3).

*D. salina* and *D. viridis* control and experimental (with toxicant) suspensions were irradiated and fluctuations in light scattering by the motile cells monitored. The toxic effects of substances were estimated using changes in energetic expenses $W$ − a parameter determined as follows [Begma et al., 1989; Vlasenko, 2004]:

$$W = \frac{\dfrac{m\langle \upsilon_{on}\rangle^2}{2}}{\dfrac{m\langle \upsilon_{c}\rangle^2}{2}}\frac{N_m}{N_{im}} = \frac{\langle \upsilon_{on}\rangle^2 \cdot N_m}{\langle \upsilon_{c}\rangle^2 \cdot N_{im}},$$
(11.1)

where $\dfrac{m\langle \upsilon_{exp}\rangle^2}{2}$ and $\dfrac{m\langle \upsilon_{c}\rangle^2}{2}$ are the average kinetic energies of the cells in experimental and control samples, respectively; $\langle \upsilon_{exp}\rangle$ and $\langle \upsilon_{c}\rangle$ are the average velocities of the cells in experimental and control samples, respectively; $N_m$ and $N_{im}$ are the number of mobile and immobile cells.

The dependence of the energetic expense of *Dunaliella* cells on the duration of action of the toxicant ($Cu^{2+}$) at 10 mg/l is presented in Fig. 11.4. Parameter $W$ that characterizes the energy expense changed sufficiently in comparison with its initial value showing that it can be used as a quantitative criterion for evaluating the toxicant.

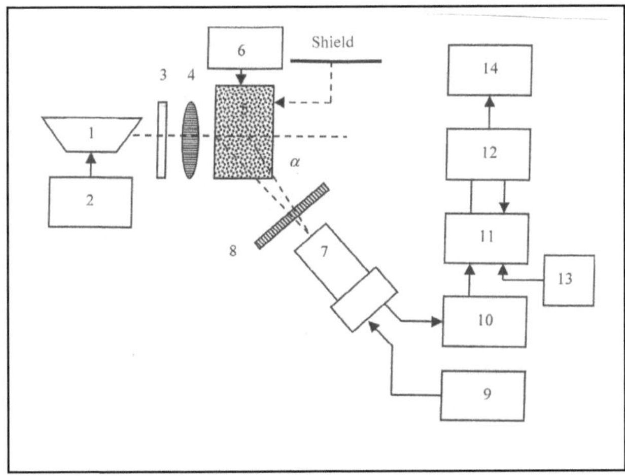


**Fig. 11.3.** Doppler correlation spectrometer: 1 – laser; 2 – power source; 3 – diaphragm; 4 – lens; 5 – cuvette with the sample; 6 – thermostat; 7 – photodetector; 8 – diaphragm; 9 – power source; 10 – amplifier; 11 – correlator; 12 – computer; 13 – timer; 14 – readout system [Begma et al., 1989].

The dependence of parameter $W$ on the concentration of toxicants ($Cu^{2+}$ and triton X-100) is presented in Fig. 11.5.

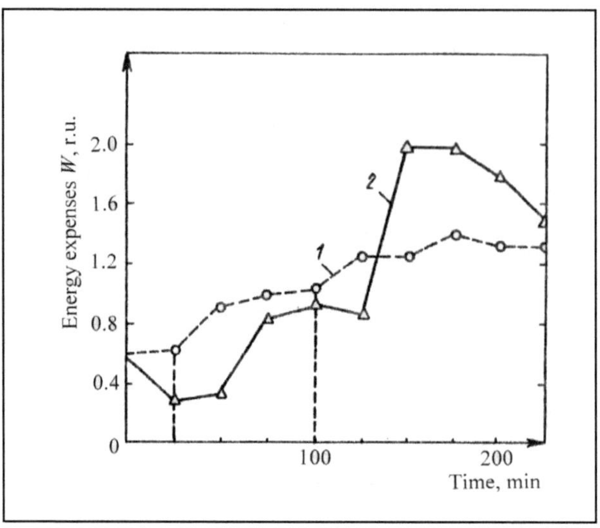


**Fig. 11.4.** Dependence of energy expense $W$ in the cells of *Dunaliella* on the duration exposure to the toxicant ($Cu^{2+}$) at a concentration of 10 mg/l [Begma et al., 1989].

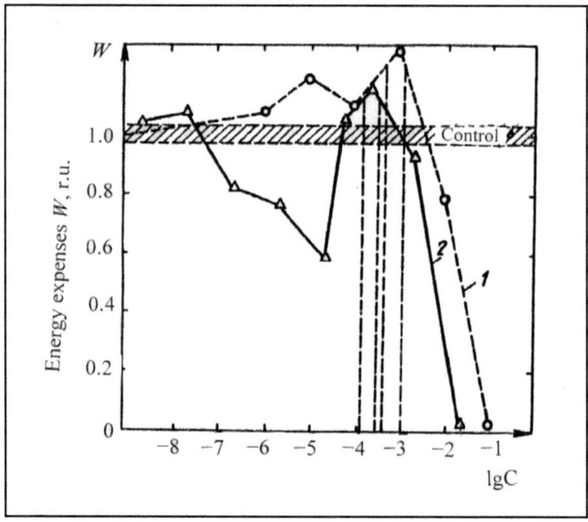


**Fig.11.5.** The dependence of parameter $W$ on the concentration of two toxicants ($Cu^{2+}$ and triton X-100) [Begma et al., 1989].

136

The distinctions in parameter $W$ between experimental and control samples are readily evident. Laser Doppler spectroscopy, based on recording the frequency changes of scattered optical radiation by a moving object, provides a rapid and concise means for evaluating the toxic effect of chemical substances on motile algae in an aquatic medium.

## 11.5. Vector Method of Biomonitoring

The vector method for quantitatively estimating the toxic effects of pollutants on photo-movement parameters of motile microorganisms was initially proposed by our research group [Posudin, 1996$a,b,c$]. Test-objects are placed into treatment (with toxicant) and control cuvettes in a videomicrography system that simultaneously records several photomovement parameters (e.g., linear velocity $v$, rotational velocity $n$, phototopotaxis $F$, and the number of immobile cells $N_{im}$ in relation to the total number $N_0$). The vector method is based on determining the value $r$ and direction $\theta$ of the vector $\vec{R}$ that has the following projections on the axis of a $N$-dimensional system of coordinates: $Xi/X_{1c}$, $X_2/X_{2c}$,..., $X_N/X_{Nc}$ (where $X_1$, $X_2$,..., $X_N$ and $Xi_c$, $X_{2c}$,..., $X_{Nc}$ are the photomovement parameters for the microorganisms in the pollutant and control samples, respectively).

In a two-dimensional system of coordinates ($N=2$), the value $r$ and direction $\theta$ are defined as follows (Fig. 11.6):

$$r = \sqrt{(X_1/X_{1c})^2 + (X_2/X_{2c})^2} \; ; \qquad (11.2)$$

$$\theta = \text{arctg} \; [(X_1/X_{1c})/(X_2/X_{2c})], \qquad (11.3)$$

where $X_i$ and $X_{ic}$ are parameters of movement of the test-objects in experimental and control samples, respectively ($i = 1,2$).

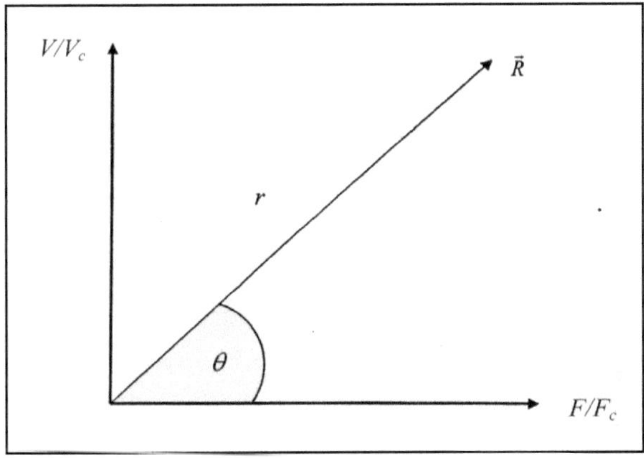


**Fig. 11.6.** Value $r$ and direction $\theta$ of vector $\vec{R}$ in a two-dimensional system of coordinates ($v/v_c$,$F/F_c$). "C" indicates the control sample [Posudin et al., 1996].

In a three-dimensional system of coordinates ($N = 3$), the value $r$ and directions ($\theta_1$ and $\theta_2$) are determined as follows (Fig. 11.7):

$$r = \sqrt{(X_1/X_{2c})^2 + (X_2/X_{2c})^2 + (X_3/X_{3c})^2}\ ; \tag{11.4}$$

$$\theta_1 = \arccos[(X_1/X_{1c})/r], \tag{11.5}$$

$$\theta_2 = \operatorname{arctg}[(X_3/X_{3c})/(X_1/X_{1c})]. \tag{11.6}$$

The value $r$ and the direction ($\theta_1$, $\theta_2$ and $\theta_3$) of the vector $\vec{R}$ in the four-dimensional system of coordinates ($N=4$) are determined from the following equations:

$$r = \sqrt{(X_1/X_{2c})^2 + (X_2/X_{2c})^2 + (X_3/X_{3c})^2 + (X_4/X_{4c})^2}\ , \tag{11.7}$$

$$X_1/X_{1c} = r\cos\theta_1, \tag{11.8}$$

$$X_2/X_{2c} = r\sin\theta_1\cos\theta_2, \tag{11.9}$$

$$X_3/X_{3c} = r\sin\theta_1\sin\theta_2\cos\theta_3, \tag{11.10}$$

$$X_4/X_{4c} = r\sin\theta_1\sin\theta_2\sin\theta_3. \tag{11.11}$$

A $N$-dimensional system of coordinates can also be used, but in such situations (e.g., $N>4$) a graphical description of the vector $\vec{R}$ is impossible; its value and direction can only be tabulated.

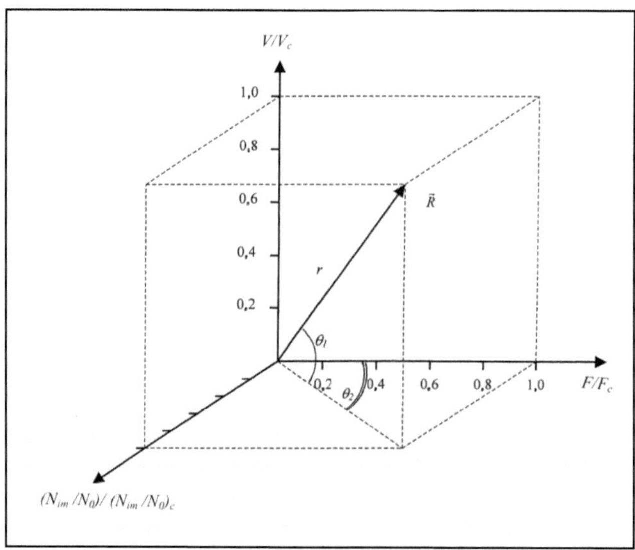


**Fig. 11.7.** Value $r$ and direction ($\theta_1$ and $\theta_2$) of vector $\vec{R}$ in a three-dimensional system of coordinates ($v/v_c$; $F/F_c$; $(N_{im}/N_0)/(N_{im}/N_0)_c$) [Posudin et al., 1996].

## 11.5.1. Dependence of Vector $\vec{R}$ on the Type and Concentration of Surface-Active Substances

The effects of surface-active substances on photomovement in *D. viridis* are presented in Table 11.3. The dependence the value $r$ and direction $\theta$ of the vector $\vec{R}$ on the type and concentration of the surface-active substances in a two-dimensional system of coordinates ($v/v_c$ and $P/P_c$, where $v$ is linear velocity and $P$ – phototopotaxis of the cells) is presented in Fig. 11.8. Increasing the concentration of the surface-active substance causes a decrease in $r$ and turns the vector $\vec{R}$ clockwise.

Table 11.4. Effect of type and concentration of SAS on photomovement parameters of *Dunaliella viridis* [Posudin et al., 1996]

| Type of SAS | Velocity $v$, phototopotaxis $P$, value r and direction $\theta$ of vector $\vec{R}$ | Concentration of SAS (mg/1) | | | | | | |
|---|---|---|---|---|---|---|---|---|
| | | Control | 1 | 5 | 10 | 20 | 30 | 40 |
| ASAS | $v$ | 25.4±4.9 | 24.0±7.6 | 20.8±8.1 | 14.1±3.9 | 2.6±0.8. | 0 | 0 |
| | $P$ | 0.29±0.04 | 0.31±0.02 | 0.31±12 | 0.32±0.03 | 0.16±0.16 | 0 | 0 |
| | $r$ | 1.41 | 1.42 | 1.35 | 1.22 | 0.56 | – | – |
| | $\theta$ | 45 | 41.3 | 37.5 | 26.6 | 10.3 | | |
| CSAS | $v$ | 25.4±4.9 | 19.8±5.5 | 15.2±3.6 | 11.0±4.1 | 0 | 0 | 0 |
| | $P$ | 0.29±0.04 | 0.30±0.05 | 0.31±0.03 | 0.31±0.05 | 0 | 0 | 0 |
| | $r$ | 1.41 | 1.29 | 1.22 | 1.15 | - | - | - |
| | $\theta$ | 45 | 37.1 | 29.3 | 21.9 | - | - | - |
| NSAS | $v$ | 25.4±4.9 | 24.0±5.0 | 23.9±16.4 | 21.0±6.2 | 16.4±4.1 | 16.2±5.5 | 14.9±2.0 |
| | $P$ | 0.29±0.04 | 0.31±.03 | 0.34±0.03 | 0.35±0.07 | 0.32±.09 | 0.30 ±0.10 | 0.30±0.03 |
| | $r$ | 1.41 | 1.39 | 1.37 | 1.30 | 1.19 | 1.18 | 1.16 |
| | $\theta$ | 45 | 43.8 | 43.2 | 39.7 | 32.6 | 31.8 | 30.5 |
| PSAS | $v$ | 25.4±4.9 | 25.1±4.4 | 24.8±5.9 | 25.2±5.2 | 21.4±6.0 | 16.5±2.4 | 12.9±2.5 |
| | $P$ | 0.29±0.04 | 0.32+0.06 | 0.30±.01 | 0.32±0.03 | 0.23+0.04 | 0.25±02 | 0.25±0. |
| | $r$ | 1.41 | 1.41 | 1.40 | 1.39 | 1.30 | 1.19 | 1.12 |
| | $\theta$ | 45 | 44.7 | 44,4 | 44.1 | 40.0 | 33.0 | 27.0 |

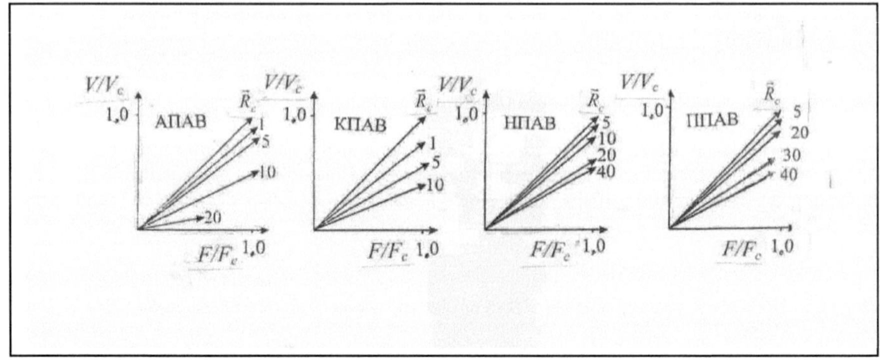


**Fig. 11.8.** The dependence of value  $r$  and the direction  $\theta$  of vector  $\vec{R}$ , in a two-dimensial system of coordinates ($v/v_c$; $F/F_c$), on the type and concentration of SAS. [CSAS – cation-active SAS (catamine); ASAS – anion-active SAS – sodium salt of dodecyl sulfoacid; NSAS – nonion-active SAS – hydropol; PSAS – a natural compound of polysaccharide nature extracted from cyanobacteria; SAS concentrations were: 1; 5; 10; 20; 30; 40 mg/l.  $\vec{R}_c$  (Fig.11.9-11.10,11.14-11.16) indicates the control sample [Posudin et al., 1996$a$,$b$].

Graphical depiction of their dependence is presented in Fig. 11.9. The data was used for the construction of the vector  $\vec{R}$  in two-, three-, and four-dimensional systems of coordinates.

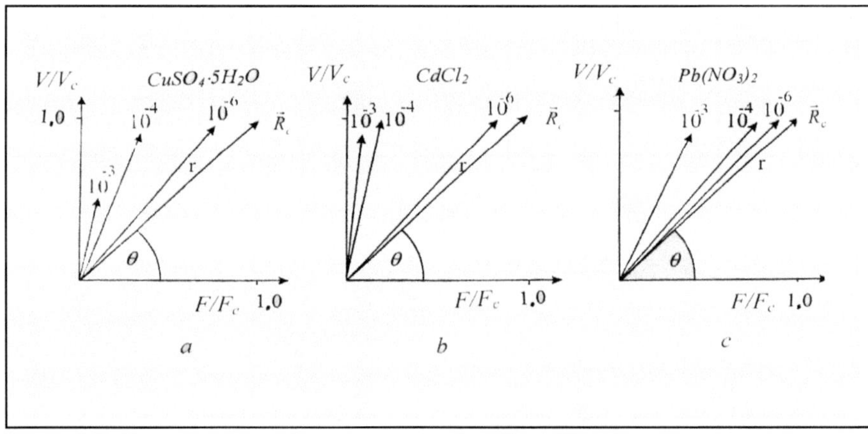


**Fig. 11.9** Dependence of the value $r$ and direction  $\theta$  of vector  $\vec{R}$  for *Dunaliella viridis* in two-dimensional system of coordinates ($v/v_c$; $F/F_c$) on the type and concentration of heavy metal salts: $a$ – CuSO$_4$·5H$_2$O; $b$ – CdCl$_2$; $c$ – Pb(NO$_3$)$_2$ at concentrations of 10$^{-3}$; 10$^{-4}$; 10$^{-5}$; 10$^{-6}$ M [Posudin et al., 1996$a$, $b$].

## 11.5.2. Dependence of vector $\vec{R}$ on the type and concentration of heavy metals

The effect of various heavy metals and concentrations thereof on photomovement parameters in *D. viridis* (e.g., velocity $v$, phototopotaxis $F$, relative number $N_{im}/N_0$ of immobile cells, velocity $n$ of rotation of the cells) is presented in Table 11.5 with averages and error values given.

**Table 11.5.** Effect of the type and concentration of heavy metal salts on photomovement parameters in *Dunaliella viridis* [Posudin et al., 1996*a, b*].

| Salt | Photomovement parameters | Concentration of salt, M | | | | | | |
|---|---|---|---|---|---|---|---|---|
| | | 0 (control) | $10^{-7}$ | $10^{-6}$ | $10^{-5}$ | $10^{-4}$ | $10^{-3}$ | $10^{-2}$ |
| $CuSO_4 5H_2O$ | $V$ ($\mu$m/s) | 25,1±0,6 | 24,7±1,3 | 24,4±0,1 | 23,3±0,6 | 23,1±0,2 | 13,0±0,9 | 0 |
| | $F$(r.u.) | 0,28±0,05 | 0,26±0,04 | 0,21 ±0,07 | 0,20±0,02 | 0,09±0,04 | 0,02±0,2 | 0 |
| | $N_{im}/N_o$ (r.u.) | 0,10±0,005 | 0,11 ±0,02 | 0,13±0,02 | 0,18±0,04 | 0,27±0,08 | 0,54±0,06 | 0 |
| | $n$ (1/s) | 1,86±0,05 | 1,89±0,03 | 1,92±0,04 | 1,94±0,02 | 2,02±0,06 | 2,69±0,12 | 0 |
| $Pb(NO_3)_2$ | $V$ ($\mu$m/s) | 24,5±0.8 | 24,2±0,8 | 23,8±0,2 | 23,6±0.5 | 23,4±0,8 | 22,5±0.7 | 18,6±1,1 |
| | $F$(r.u.) | 0,27±0,05 | 0,25±0.05 | 0,24±0,02 | 0,20±0,005 | 0,20±0,03 | 0,10±0,08 | 0,05±0,07 |
| | $N_{im}/N_o$ (r.u.) | 0,07±0,01 | 0,08±0,01 | 0,09±0,02 | 0,18±0,01 | 0,28±0,01 | 0,38±0,04 | 0.48±0,03 |
| | $n$ (1/s) | 1,88±0,06 | 1,94±0,04 | 1,95±0,03 | 1,98±0,04 | 1,99±0,06 | 2,14±0,02 | 2,39±0,04 |
| $CdCl_2$ | $V$ ($\mu$m/s) | 23,1±0,9 | 22,9±0,8 | 22,7±0.1 | 22,2±0,5 | 21,4±0,2 | 19,7±0,5 | 0 |
| | $F$(r.u.) | 0,28±0,04 | 0,26±0,02 | 0,24±0,05 | 0,14±0,11 | 0,05±0,04 | 0,03 ±0,66 | 0 |
| | $N_{im}/N_o$ (r.u.) | 0,11±0,04 | 0,11±0,03 | 0,11 ±0,01 | 0,12±0,01 | 0,21±0,02 | 0,36±0,04 | 0 |
| | $n$ (1/s) | 2,01±0,05 | 2,03±0,04 | 2,08±0,1 | 2,10±0,02 | 2,13±0,02 | 2,16±0,02 | 0 |

The dependence of value $r$ and direction $\theta$ of vector $\vec{R}$ on the type and concentration of heavy metal salts at a $10^{-4}$ M is presented in Table 11.6. All possible pairs of photomovement parameters responded to the action of copper.

**Table 11.6.** Dependence of the value $r$ and direction $\theta$ of vector $\vec{R}$ in a three-dimensional system of coordinates on the type and concentration of heavy metal salts[Posudin et al., 1996a, b].

| Salt | Photomovement parameters | $r$ and $\theta$ | Concentration of salt, M | | | | | | |
|---|---|---|---|---|---|---|---|---|---|
| | | | 0 (control) | $10^{-7}$ | $10^{-6}$ | $10^{-5}$ | $10^{-4}$ | $10^{-3}$ | $10^{-2}$ |
| $CuSO_4 \cdot 5H_2O$ | $v/v_c$ and $F/F_c$ | $r$ | 1.41 | 1.35 | 1.22 | 1.17 | 0.97 | 0.52 | 0 |
| | | $\theta$ | 45.0 | 46.5 | 52.3 | 52.6 | 70.8 | 82.3 | - |
| $CuSO_4 \cdot 5H_2O$ | $v/v_c$ and $(N_{im}/N_o)/(N_{im}/N_o)_c$ | $r$ | 1.41 | 1.48 | 1.52 | 2.03 | 2.85 | 5.42 | - |
| | | $\theta$ | 45.0 | 41.7 | 37.8 | 27.3 | 18.8 | 5.5 | - |
| $CuSO_4 \cdot 5H_2O$ | $n/n_k$ and $(N_{im}/N_o)/(N_{im}/N_o)_c$ | $r$ | 1.41 | 1.50 | 1.58 | 2.08 | 2.91 | 5.59 | 0 |
| | | $\theta$ | 45.0 | 47.4 | 49.3 | 60.0 | 68.0 | 75.0 | - |
| $Pb(NO_3)_2$ | $v/v_c$ and $F/F_c$ | $r$ | 1.41 | 1.35 | 1.31 | 1.26 | 1.20 | 0.99 | 0.78 |
| | | $\theta$ | 45.0 | 47.1 | 47.8 | 49.8 | 52.1 | 68.1 | 76.7 |
| $CdCl_2$ | $v/v_c$ and $F/F_c$ | $r$ | 1.41 | 1.36 | 1.30 | 1.19 | 1.02 | 0.91 | 0 |
| | | $\theta$ | 45.0 | 46.8 | 48.7 | 62.5 | 79.0 | 82.6 | - |

Notes. Here and further in Tables 11.6-11.7: $v$ – linear velocity; $F$ – phototopotaxis; $N_{im}/N_o$ – relative number of immobile cells; $n$ – velocity of rotation of the cells in the experimental sample; $v_c$ – linear velocity; $F_c$ – phototopotaxis; $(N_{im}/N_o)_c$ – relative number of immobile cells; $n_c$ – velocity of rotation of the cells in the control sample; defies means the uncertainty of the angle $\theta$ under zero value of $r$.

Fig. 11.10 displays the value $r$ and direction $\theta$ of vector $\vec{R}$ in response to various heavy metal salts at a $10^{-4}$ M concentration. The errors of measurements are also indicated. The behavior of the vector $\vec{R}$ depends on the level of toxicity to the photomovement parameter in question.

The principal tendencies for alterations in the value $r$ and direction $\theta$ of the vector $\vec{R}$ are the following (Fig. 11.11): both parameters are decreasing but the first parameter $(X_1/X_{1c})$ decreases at a greater rate than the second $(X_2/X_{2c})$ (see. Fig. 11.11 $a$); the first parameter is increasing and the second is decreasing (see Fig. 11.11 $b$); the first parameter is decreasing and the second is increasing (see Fig. 11.11 $c$); both parameters are decreasing, but the first parameter is decreasing with slower rate than the second (see Fig. 11.11 $d$); and both parameters are increasing but the first parameter is increasing with a greater rate than the second (see Fig. 11.11 $e$).

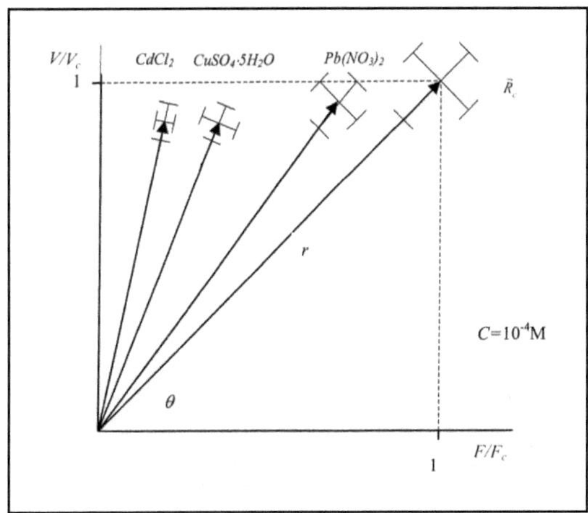


**Fig.11.10.** Dependence of the value $r$ and direction $\theta$ of vector $\vec{R}$, that is constructed in two-dimensional system of coordinates, on the type of the salts of heavy metals [$CuSO_4\cdot 5H_2O$, $CdCl_2$ and $Pb(NO_3)_2$] at the same ($10^{-4}$M) concentration [Posudin et al., 1996, *a,b*].

A two-dimensional system of coordinates constructed using the simultaneous measurement of velocity of movement and phototopotaxis is presented in Fig. 11.11.

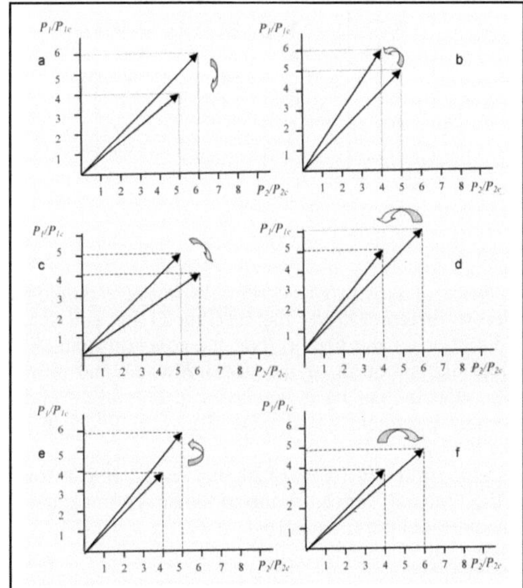


**Fig. 11.11.** The primary tendencies for changes in the value $r$ and direction $\theta$ of the vector $\vec{R}$ in a two-dimensional system of coordinates ($P_1/P_c$; $P_2/P_c$) [Posudin et al., 1996].

Construction of the vector $\vec{R}$ in a three-dimensional system of coordinates using the simultaneous measurement of three photomovement parameters $[v/v_c, F/F_c$ and $(N_{im}/N_0)/(N_{im}/N_0)_c]$ is presented in Table 11.7.

**Table 11.7.** Dependence of the value $r$ and direction ($\theta_1$ and $\theta_2$) of vector $\vec{R}$ in a three-dimensional system of coordinates on the type and concentration of heavy metal salts[Posudin et al., 1996a,b].

| Heavy metal | Parameters of photomovement | Value $r$ and direction $\theta_1$ and $\theta_2$ of vector $\vec{R}$ | Concentration (M) | | | | | | |
|---|---|---|---|---|---|---|---|---|---|
| | | | Control | $10^{-7}$ | $10^{-6}$ | $10^{-5}$ | $10^{-4}$ | $10^{-3}$ | $10^{-2}$ |
| Cuprum | $v, F, N/N_0$ | $r$ | 1.73 | 1.74 | 1.79 | 2.14 | 2.87 | 5.42 | |
| | | $\theta_1$ | 54.69 | 55.72 | 57.20 | 64.24 | 71.72 | 84.49 | |
| | | $\theta_2$ | 44.90 | 49.68 | 60.00 | 68.37 | 83.26 | 89.25 | |
| Lead | $v, F, N/N_0$ | $r$ | 1.73 | 1.77 | 1.83 | 2.86 | 4.20 | 5.52 | 6.90 |
| | | $\theta_1$ | 54.69 | 56.01 | 57.99 | 70.39 | 76.94 | 86.73 | 89.0 |
| | | $\theta_2$ | 44.90 | 51.16 | 60.25 | 72.50 | 82.27 | 88.54 | 89.7 |
| Cadmium | $v, F, N/N_0$ | $r$ | 1.73 | 1.68 | 1.64 | 1.54 | 2.13 | 3.38 | |
| | | $\theta_1$ | 54.69 | 54.02 | 53.30 | 51.32 | 64.09 | 75.44 | |
| | | $\theta_2$ | 44.90 | 43.88 | 49.16 | 65.42 | 84.61 | 88.07 | |

The diagrams constructed for a three-dimensional system of coordinates in correspondence with Table. 11.6 allow comparing the action of different heavy metals (Fig. 11.12) and determining the dependence of the value and direction of the vector $\vec{R}$ on the concentration (Fig. 11.13). The effect of changes in the value and direction of the vector $\vec{R}$ when one photomovement parameter is decreasing and the other increasing is displayed in Fig. 11.14 and a simultaneous increase in both photomovement parameters is displayed in Fig. 11.15 in response to an increasing concentration of the heavy metal.

The dependence of the value $r$ and direction ($\theta_1$, $\theta_2$, $\theta_3$) of the vector $\vec{R}$ in a four-dimensional system of coordinates due to the type and concentration of the heavy metal using simultaneous measurements of four photomovement parameters $[v/v_c, F/F_c, (N_{im}/N_0)/(N_{im}/N_0)_c$ and $n/n_c]$ in *D. viridis* are presented in Table 11.8.

144

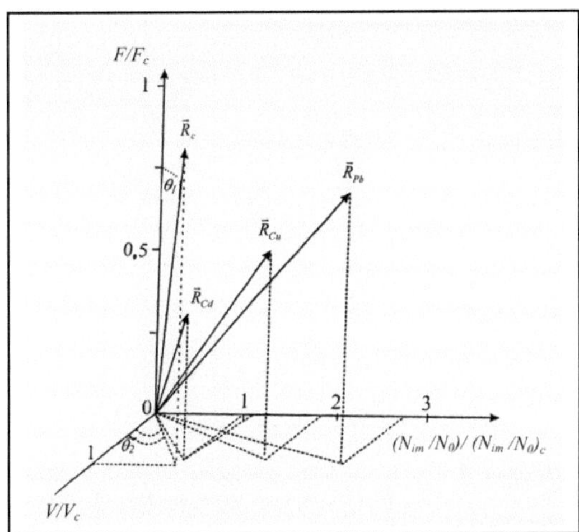

**Fig. 11.12.** The dependence of the value *r* and direction ($\theta_1$ and $\theta_2$) of the vector $\vec{R}$ in a three-dimensional system of coordinates ($v/v_c$; $F/F_c$; $(N_{im}/N_0)/(N_{im}/N_0)_c$) [Posudin et al., 1996].

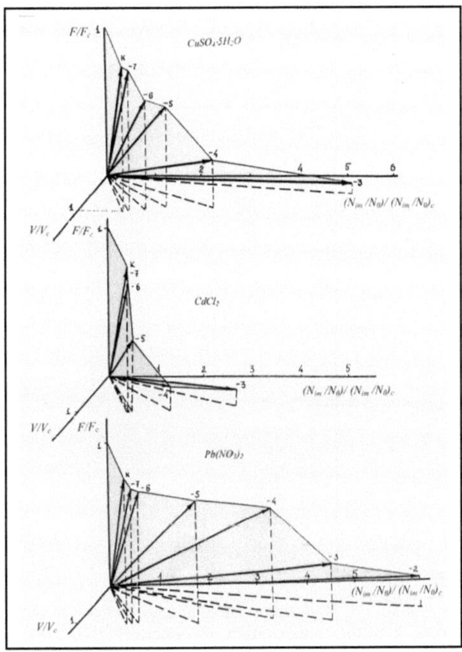

**Fig. 11.13.** The dependence of the value *r* and direction ($\theta_1$ and $\theta_2$) of a vector $\vec{R}$ on the concentration of heavy metal salts in a three-dimensional system of coordinates ($v/v_c$; $F/F_c$; $(N_{im}/N_0)/(N_{im}/N_0)_c$). Figures −7, −6, −5, … correspond to the concentrations $10^{-7}$, $10^{-6}$, $10^{-5}$,…$10^{-1}$ M [Posudin et al., 1996].

145

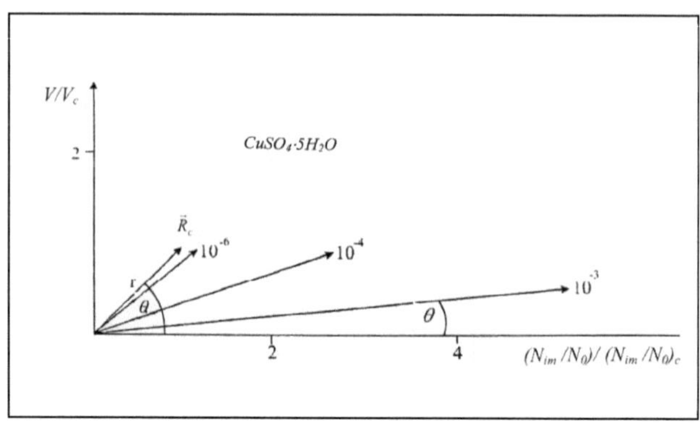


**Fig. 11.14.** Changes of the value $r$ and direction $\theta$ of vector $\vec{R}$ when one parameter $(v/v_c)$ is decreasing and there is a simultaneous increase the second $((N_{im}/N_0)/N_{im}/N_0)_k)$ in response to an increase of concentration of copper $(10^{-6}; 10^{-4}; 10^{-3}$ M $CuSO_4 \cdot 5H_2O)$ [Posudin et al., 1996].

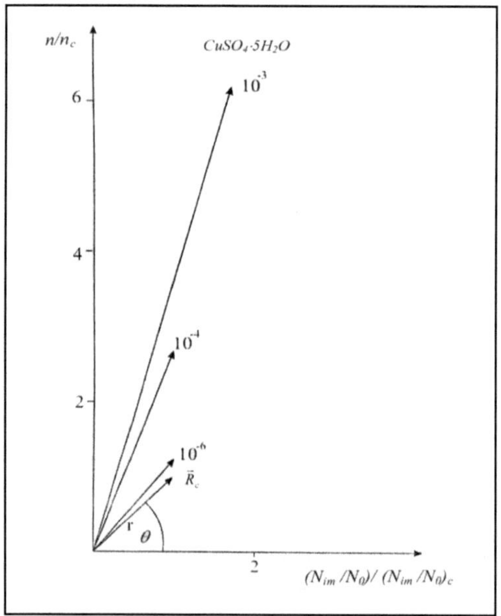


**Fig. 11.15.** Changes of the value $r$ and direction $\theta$ of vector $\vec{R}$ during a simultaneous increase in parameters $(v/v_c)$ and $((N_{im}/N_0)/N_{im}/N_0)_k)$ in response to an increase in the concentration of copper $(10^{-6}; 10^{-4}; 10^{-3}$ M $CuSO_4 \cdot 5H_2O)$ [Posudin et al., 1996].

146

**Table 11.8.** The dependence of value $r$ and the direction ($\theta_1$, $\theta_2$, $\theta_3$) of vector $\vec{R}$ in a four-dimensional system of coordinates on the type and concentration of heavy metals using the simultaneous measurement of four photo-movement parameters in *Dunaliella viridis* [Posudin et al., 1996].

| Salt | Photomovement parameter | $r$, $\theta_1$, $\theta_2$ | Concentration of salt, M | | | | | | |
|---|---|---|---|---|---|---|---|---|---|
| | | | 0 (control) | $10^{-7}$ | $10^{-6}$ | $10^{-5}$ | $10^{-4}$ | $10^{-3}$ | $10^{-2}$ |
| $CuSO_4 \cdot 5H_2O$ | $v/v_c$, $F/F_c$, $(N_{im}/N_o)/(N_{im}/N_o)_c$ and $n/n_c$ | $r$ | 2.00 | 2.01 | 2.06 | 2.38 | 3.07 | 5.61 | - |
| | | $\theta_1$ | 60.00 | 60.90 | 61.90 | 67.05 | 72.55 | 84.68 | - |
| | | $\theta_2$ | 54.73 | 56.86 | 56.86 | 71.10 | 71.10 | 83.73 | - |
| | | $\theta_3$ | 45.24 | 41.58 | 41.58 | 29.77 | 29.77 | 21.97 | - |
| $Pb(NO_3)_2$ | $v/v_c$, $F/F_c$, $(N_{im}/N_o)/(N_{im}/N_o)_c$ and $n/n_c$ | $r$ | 2.00 | 2.04 | 2.11 | 3.05 | 4.31 | 5.63 | 7.01 |
| | | $\theta_1$ | 60.00 | 61.04 | 62.59 | 71.63 | 77.26 | 80.60 | 83.78 |
| | | $\theta_2$ | 54.73 | 58.97 | 61.98 | 73.75 | 79.86 | 96.18 | 88.52 |
| | | $\theta_3$ | 45.24 | 41.82 | 39.27 | 22.35 | 22.35 | 11.52 | 10.04 |
| $CdCl_2$ | $v/v_c$, $F/F_c$, $(N_{im}/N_o)/(N_{im}/N_o)_c$ and $n/n_c$ | $r$ | 2.00 | 1.96 | 1.93 | 1.86 | 2.38 | 3.55 | - |
| | | $\theta_1$ | 60.00 | 59.74 | 59.65 | 58.88 | 66.97 | 76.14 | - |
| | | $\theta_2$ | 54.73 | 56.68 | 58.91 | 71.70 | 85.28 | 88.17 | - |
| | | $\theta_3$ | 45.24 | 45.01 | 45.48 | 43.86 | 28.96 | 18.35 | - |

It is possible to present the minimum tendencies in behavior of vector $\vec{R}$ in respect to the concentration of a heavy metal. If both parameters $X_1/X_{1c}$ and $X_2/X_{2c}$ are decreasing with an increasing heavy metal concentration but the first parameter decreases at a greater rate than the second, the value $r$ is decreasing and the direction $\theta$ changes in such a way that vector $\vec{R}$ deviates from the axis of the first parameter (see Fig. 11.11*a*). If one parameter is increasing with increasing heavy metal concentration and the second is decreasing, and the change of both parameters occurs at the same rate, the value $r$ is increasing and the direction $\theta$ changes in such a way that vector $\vec{R}$ moves nearer the axis of the first parameter (see Fig. 11.11*b,c*).

If both parameters are decreasing but the first parameter decreases more slowly, the value $r$ is decreasing and the direction $\theta$ changes in such a way that vector $\vec{R}$ deviates from the axis of the second parameter (see Fig. 11.11*d*). When there is a simultaneous increase of both parameters but the first increases more rapidly than the second with increasing heavy metal concentration, the increase in the value $r$ and the direction $\theta$ change in such a way that vector $\vec{R}$ deviates from the axis of the second parameter (see Fig. 11.11*e*). The simultaneous increase of both parameters but the first increases more slowly than the second with increasing heavy metal concentration, the increase in the value $r$ and the direction $\theta$ change in such a way that the vector $\vec{R}$ deviates from the axis of the first parameter (see. Fig. 11.11*f*).

The level of accuracy in determining the effect of heavy metals can be increased by increasing the number of parameters that are measured. The application of the vector method in our experiments allowed estimating the toxicity of three heavy metals (Pb > Cu > Cd) and determining the most toxic concentration for their salts (i.e., $10^{-2}$ M for $CuSO_4 \cdot 5H_2O$ and $CdCl_2$, $10^{-1}$ M for $Pb(NO_3)_2$).

The concentrations ($10^{-4}$–$10^{-1}$ M) of the salts of the heavy metals [$CuSO_4 \cdot 5H_2O$; $CdCl_2$; $Pb(NO_3)_2$] tested are comparable to those found in municipal sewage. It is important to note that the metals themselves do not have a direct toxic effect, however, a number of chemicals act as antagonists or synergists. Likewise, there is the possible formation of complexes between the metals and other compounds that lead to a loss in toxicity.

### 11.5.3. Dependence of Vector $\vec{R}$ on the Type and Concentration of Pesticides

Effect of different types of pesticides and concentrations thereof on photomovement parameters in *D. viridis* are presented in Table 11.9.

The cells exhibited a differential in sensitivity depending upon the type of pesticide. An effect on photomovement parameters due to certain pesticides exhibited a threshold at $10^{-4}$ M [e.g., DPC (diphenylcarbazide), harmoni], while the others [e.g., arylon, eradicane (2,2-dichloro-N,N-di-2-propen-1-yl-acetamide), acetazine, tecto] exibited a threshold at $10^{-9}$ M. The inhibitory action of some pesticides [e.g., alachlor, arylon] on phototopotaxis in *D. viridis* was at concentrations that did not affect the velocity of movement of the cells. As a rule, with increasing pesticide concentration the response of the vector $\vec{R}$ is characterized by a decreasing the value $r$ and rotation of the vector $\vec{R}$ counter-clockwise (see Fig. 11.16).

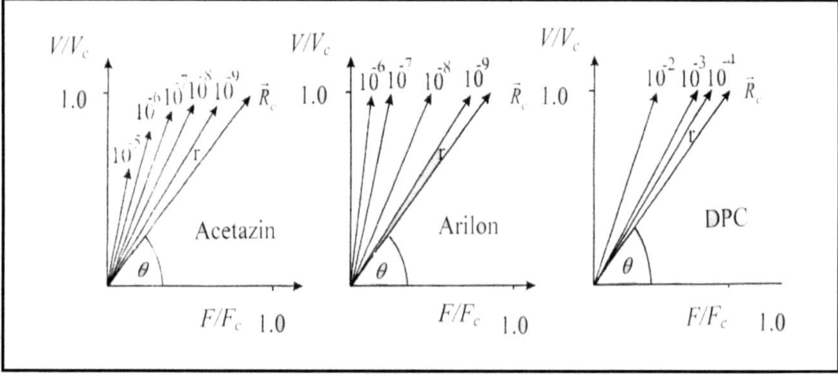


**Fig. 11.16.** The dependence of the value $r$ and direction $\theta$ of vector $\vec{R}$ in a two-dimensional system of coordinates using simultaneous monitoring of two parameters $(v/v_c)$ and $(F/F_c)$ on the type and concentration ($10^{-9}$; $10^{-8}$; $10^{-7}$; $10^{-6}$; $10^{-5}$; $10^{-4}$; $10^{-3}$; $10^{-2}$ M) of pesticides [Posudin et al., 1996].

**Table 11.9** .Effect of type and concentyration of pesticieds on photomovement parameters of *Dunaliella viridis* [Posudin et al., 1996, *a,b*]

| Type of pesti-cides | Parameters of photo-movement and vector $\vec{R}$ | Concentration of pesticide (M) | | | | | | | | |
|---|---|---|---|---|---|---|---|---|---|---|
| | | Control | $10^{-9}$ | $10^{-8}$ | $10^{-7}$ | $10^{-6}$ | $10^{-5}$ | $10^{-4}$ | $10^{-3}$ | |
| Arylon | $\upsilon$ | 43±5 | 43+4 | 43±5 | 43±4 | 42±4 | 38±5 | 37±6 | | |
| | $P$ | 0.35±0.03 | 0.30+0.04 | 0.20±0.05 | 0.10±0.04 | 0.05±0.04 | 0 0.88 | 0 | | |
| | $r$ | 1.41 | 1.31 | 1.15 | 1.07 | 1.02 | 90 | 0.86 | | |
| | $\theta$ | 45 | 49.3 | 60.3 | 74.3 | 81.9 | | 90 | | |
| Furare | $\upsilon$ | 49±3 | 48±4 | 47±4 | 46±4 | 42±4 | 28±5 | 0 | | |
| | $P$ | 0.31+0.03 | 0.30±0.04 | 0.25±0.04 | 0.19±0.05 | 0.15±0.04 | 0.15±0.04 | 0 | | |
| | $r$ | 1.41 | 1.38 | 1.26 | 1.16 | 1.05 | 0.55 | - | | |
| | $\theta$ | 45 | 45.3 | 49.8 | 54.1 | 54.6 | 49.9 | - | | |
| Eradi-can | $\upsilon$ | 48±5 | 46+5 | 45±4 | 44±5 | 43±5 | 42±3 | 40±5 | | |
| | $P$ | 0.32±0.04 | 0.26±0.04 | 0.22±0.05 | 0.14±0.04 | 0.11±0.05 | 0.05±0.04 | 0 | | |
| | $r$ | 1.41 | 1.26 | 1.17 | 1.02 64.4 | 0.95 | 0.88 | 0.83 | | |
| | $\theta$ | 45 | 49.8 | 53.7 | | 69.1 | 79.6 | 90 | | |
| Aceta-zin | $\upsilon$ | 48±5 | 47+4 | 46±4 | 45±4 | 41±6 | 31±3 | 0 | | |
| | $P$ | 0.42±0.04 | 0.33±0.05 | 0.25±0.04 | 0.18±0.04 | 0.12±0.07 | 0.05±0.04 | 0 | | |
| | $r$ | 1.41 | 1.25 | 1.13 58.4 | 1.03 | 0.89 | 0.65 | | | |
| | $\theta$ | 45 | 51.5 | | 65.4 | 71.8 | 79.4 | | | |
| Tecto | $\upsilon$ | 44±5 | 44+5 | 43±4 | 42±7 | 41±6 | 40±3 | 35±5 | | |
| | $P$ | 0.44I0.05 | 0.32±0.05 | 0.27±0.04 | 0.17±0.06 | 0.10±0.05 | 0 0.91 | 0 0.79 | | |
| | $r$ | 1.41 | 1.23 | 1.15 | 1.03 | 0.96 | | | | |
| | $\theta$ | 45 | 53.9 | 58.1 | 67.7 | 76.1 | | | | |
| Acetal | $\upsilon$ | 44±5 | 44+3 | 44±4 | 43±5 | 42±3 | 40±5 | 0 | | |
| | $P$ | 0.33±0.05 | 0.33±0.04 | 0.32±0.05 | 0.27±0.03 | 0.19±0.04 | 0.08±0.05 | 0 | | |
| | $r$ | 1.41 | 1.41 | 1.39 | 1.28 | 1.23 | 0.94 | | | |
| | $\theta$ | 45 | 45 | 45.9 | 50.1 | 59.0 | 75.2 | | | |
| Alach-lor | $\upsilon$ | 46+5 | 46±3 | 46±4 | 45±5 | 43±4 | 40±5 | 0 | | |
| | $P$ | 0.12±0.04 | 0.12±0.04 | 0.12±0.05 | 0.05±0.04 | 0.03±0.04 | 0 | 0 | | |
| | $r$ | 1.41 | 1.41 | 1.41 | 1.07 | 0.96 | 0.87 | - | | |
| | $\theta$ | 45 | 45 | 45 | 66.8 | 74.9 | 90 | - | | |
| Ladok | $\upsilon$ | 42±3 | 42±5 | 42±4 | 40±4 | 35±5 | 30±4 | 27±2 | | |
| | $P$ | 0.26±0.05 | 0.24+0.04 | 0.22+0.04 | 0.17±0.04 | 0.10±0.04 | 0.05±0.04 | 0 | | |
| | $r$ | 1.41 | 1.36 | 1.31 | 1.15 | 0.83 | 0.73 | 0.64 | | |
| | $\theta$ | 45 | 47.4 | 49.6 | 55.6 | 65.4 | 75.0 | 90 | | |
| Basta | $\upsilon$ | 48±4 | 48±4 | 48±4 | 48+5 | 48±4 | 48±5 | 44±4 | 0 | |
| | $P$ | 0.37±0.04 | 0.37±0.04 | 0.37±0.04 | 0.32±0.05 | 0.30±0.05 | 0.20±0.07 | 0.10±0. | 0 | |
| | $r$ | 1.41 | 1.41 | 1.41 | 1.05 | 1.04 | 1.00 | 07 | - | |
| | $\theta$ | 45 | 45 | 55 | 49.3 | 50.9 | 87.1 | 0.92 | - | |
| | | | | | | | | 88.1 | | |

$\upsilon$ = Velocity, P= photopotaxis, r= value, $\theta$=direction of vector $\vec{R}$

| Dual | $v$ | 48+4 | 48±4 | 47±3 | 46±4 | 41±4 | 3.3±3 | | | |
|---|---|---|---|---|---|---|---|---|---|---|
| | $P$ | 0.4±0.05 | 0.38±0.04 | 0.35±0.04 | 0.33±0.04 | 0.30±0.05 | 0.24±0.06 | | | |
| | $r$ | 1.41 45 | 1.07 | 1.04 | 1.02 | 0.90 | 0.73 | 0 | 0 | 0 |
| | $\theta$ | | 47.1 | 49.1 | 50.5 | 49.3 | 49.9 | - | - | - |
| DPC | $v$ | 46±5 | 46±5 | 46±5 | 46±5 | 46±5 | 46±5 | 46±6 | 46±5 | 46±7 |
| | $P$ | 0.38±0.04 | 0.38±0.04 | 0.38±0.04 | 0.38±0.04 | 0.38±0.04 | 0.38±0.05 | 0.33±0.04 | 0.30±0.05 | 0.18±0.06 |
| | $r$ | 1.41 | 1.41 | 1.41 | 1.41 | 1.41 | 1.41 | 1.32 | 1.27 | 1.10 |
| | $\theta$ | 45 | 45 | 45 | 45 | 45 | 45 | | 51.7 | 64.8 |
| Har-moni | $v$ | 45±5 | 45±5 | 45±5 | 45±5 | 45±5 | 45±5 | 45±4 | 45±5 | 45±5 |
| | $P$ | 0.40±0.05 | 0.40±0.05 | 0.40±0.05 | 0.40±0.05 | 0.40±0.05 | 0.40+0.04 | 0.35±0.04 | 0.10±0.04 | 0 |
| | $r$ | 1.41 | 1.41 | 1.41 | 1.41 | 1.41 | 1.41 | 1.32 | 1.03 | 1.00 |
| | $\theta$ | 45 | 45 | 45 | 45 | 45 | 45 | 48.9 | 75.9 | 90 |

## 11.5.4. Advantages of the Vector Method for Biomonitoring

By increasing the number of photomovement parameters and assessing them simultaneously using the vector method, it is possible to more precisely elucidate differences in response of the test-objects due to various toxicants. Quantitative evaluation of a specific concentration of toxicant can be determined by comparing the data with a calibration curve for the effect of the toxicant at a range of concentrations. The proposed method makes it possible to fairly accurately qualitatively estimate the toxicity and its identity. Further progress in this area of research involves the use of the vector method to estimate the effect at various levels of pollutants in aquatic environments (freshwater or seawater).

## 11.6. Summary

The high sensitivity of the two species of *Dunaliella* to environmental factors enhances their potential for use as test-objects in biomonitoring. The great advantage of these organisms is their microscopic size, ability to reproduce at high temperatures, active movement, photokinetic and photovector reactions, salt-tolerance and euryhalinity.

The sensitivity of various photomovement parameters of *Dunaliella salina* and *D. viridis* to the presence of surface-active substances in an aquatic medium [e.g., cation-active catamine (CSAS), anion-active sodium salt of dodecyl sulphoacid (ASAS), non-ionogenic hydropol (NSAS) and natural surface-active substances of polysaccharide origin (PSAS) that were extracted from a cyanobacteria], salts of heavy metals [e.g., copper
($CuSO_4 \cdot 5H_2O$), cadmium ($CdCl_2$) and lead ($Pb(NO_3)_2$) at concentration from $10^{-7}$ to $10^{-2}$ M], and pesticides [e.g., acetal (55 %), acetazine (50 %), alachlor (45 %), arylon (75 %), basta (20 %), dual (96 %), DPC (20 %), harmoni (75 %) and tecto (45 %) at $10^{-7}$ to $10^{-2}$ M] are described. The data indicates the possibility of using the linear and rotational velocities of the cell, frequency of flagella beating, and phototopotaxis values in *Dunaliella* species as test-functions during biomonitoring of aquatic environments.

The use of simultaneous measurement of several photomovement parameters is proposed in that it allows increasing the sensitivity of biomonitoring. The vector method for biomonitoring is recommended for estimating the effect of toxicant concentration in aquatic environments using simultaneous measurement of two or more parameters of movement. This method facilitates processing of data from large-scale measurements, allows the quantitative estimation of the effect of toxicant concentration, and can also facilitate toxicant identification.

# Chapter 12

# *Dunaliella* Biotechnology

*Dunaliella salina* Teod. and *Dunaliella viridis* Teod. are very interesting from scientific and practical points of view in that they represent models for studying mechanisms of tolerance to extreme conditions of salinity, temperature, and pH. Identifying the genes that encode proteins responsible for the species remarkable tolerance to extreme conditions may facilitate increasing resistance to these conditions in other plants via transgenic means. The species can also be cultivated in large volumes and may therefore be a viable source for the industrial production of β-carotene (provitamin A), ascorbic and dehydroascorbic acids, glycerol, and forage for piscine industry and other uses [Massjuk, 1973].

D. *salina*, an exceptionally rich source of β-carotene [Drokova, 1961; Ben-Amotz et al., 1982*a*], could be used for the prevention and treatment of cardiovascular and ophthalmic diseases, avitaminosis, arthrosis, cancer (skin, liver, stomach and leukemia), macular degeneration, and asthma. β-Carotene containing preparations from this species are known to promote an increase in appetite and decrease sleeplessness. It has also been proposed using D. *salina* as hepatopathy inhibitors. Powders produced from this alga are claimed to act as inhibitors that are useful in the treatment of hepatitis, liver cirrhosis, and fatty liver [Mizoguchi, 2006].

The cultivation of *Dunaliella* for biomass in reservoirs for various uses  has been described in a number of publications [Massjuk, 1966, 1967, 1973; Massjuk and Abdulla, 1969; Nosova et al., 1979;  Avron and Ben-Amotz, 1992; Ben-Amotz and Avron, 1982, 1989, 1990; Borowitzka et al., 1984, 1986; Moulton et al., 1987; Rashkova and Vlakhov, 1988; Borowitzka and Borowitzka, 1988, 1990; Mohn and Contreras, 1990; Ben-Amotz et al., 1991; Markovits et al., 1993; Zhou, Q., 1995; Ventosa and Nieto, 1995;  Ben-Amotz, 1995, 1996; Krol et al., 1997; Hong et al., 1998; Orset and Young, 1999; Jin and Melis, 2003; Mohammad R.H. and  Mansour, S., 2003; Borowitzka, 2005; Pisal and Lele, 2005; Raja et al., 2007; Del Campo et al., 2007;  Prasanna et al., 2007].

## 12.1. Carotenoids, β-carotene Biosynthesis and Stereoisomers

Carotenoids are yellow, orange, red or brown colored pigments found widely in nature. Their aliphatic or alicyclic structure consists of isoprene subunits that absorb strongly in the violet-blue portion of the spectrum. Carotenoids are long polyisoprene chains that contain conjugated double bonds. The majority of the carotenoids are comprised of a 40-carbon polyene chain. Carotenoids are divided into two classes: the carotenes − the hydrocarbon carotenoids and the xanthophylls − oxygenated derivatives of these hydrocarbons. The nomenclature of the carotenoids is based on the 9 carbon end groups of which there are 7 primary types that can be arranged in various combinations on the methylated straight chain portion of the molecule: for example, α-carotene is β,ε-carotene while  β-carotene is β,β-carotene. β-Carotene, also called provitamin A, is the most widespread and important of the diverse structures in that it can be readily converted into vitamin A in the liver of animals.

The isolation, identification, stereochemistry, properties, functions and distribution of carotenoids have been detailed in a wealth of scientific publications [Bensasson, 1975; Goodwin, 1980, 1988; Bauernfeind, 1981; Britton, 1988; Hong et al., 1998; Del Campo et al., 2007;

Prasanna et al., 2007]. The direct chemical synthesis of β-carotene was first reported in 1956. The molecular formula is $C_{40}H_{56}$ with a mass of 536.9. It is violet-red in color in the crystalline state. The cost of synthetic β-carotene is approximately \$500/kg which has in part been responsible for an increasing demand for the chemical from natural sources [Borowitzka and Borowitzka, 1990].

The presence of stereoisometric isomers is a well-known peculiarity of β-carotene. Each double bond in the aliphatic chain of β-carotene can exist in two configurations. As result, 272 *cis/trans*-isomers of β-carotene can theoretically be formed, 12 of which have been identified in nature [Ben-Amotz and Shaish, 1992, in: Avron and Ben-Amotz, 1992]. Light absorption by the carotenoids is altered by isomerisation. For example, *cis*-stereoisomers are characterized by a shift in the absorption maximum with the peak in ultraviolet portion of the spectrum. In nature the *trans*-forms predominate. Natural β-carotene, extracted from various sources, contains significant mono- bi- and poly-*cis*-forms [Avron and Ben-Amotz, 1992].

The content of β-carotene in plants ranges from 0.01 to 10 mg/100 g with green leaves (e.g., parsley, spinach, broccoli), yellow-orange fruits (e.g., mandarin, mango, peach), and certain vegetables (e.g., carrots, sweetpotato, pumpkin) being rich sources. Some species of microorganisms, such as the fungus *Phycomyces blakesleanus* (Bgtt). Arch. and the yeast *Rhodotorula*, accumulate large quantities of β-carotene (5 and 0.5 mg/g dry weight, respectively). These natural sources generally contain a mixture of carotenoids, carotenoid esters, carotenoid isomers together with varying amounts of β-carotene.

*D. salina* can accumulate a great quantity of β-carotene and is considered the most dense natural source of provitamin A known [Drokova, 1961; Milko, 1963; Aansen et al., 1969; Ben-Amotz et al., 1982a; Loeblich, 1982; Orset and Young, 1999; Jin and Melis, 2003; Pisal and Lele, 2005]. A number of methods for extracting carotenoids and β-carotene, in particular, have been reported [Yamaoka, 1994; Garcia Gonzalez et al., 2003; Chen et al., 2008].
β-Carotene that accumulates in *D. salina*, consists mainly of two stereoisomers, the ratio of which depends on quantity of light absorbed during a cell cycle [Ben-Amotz et al., 1982*a*, 1987, 1988; Tsukida et al., 1982].

The biosynthesis of β-carotene in *D. salina* proceeds in four stages [Ben-Amotz and Shaish, 1992, in: Avron and Ben-Amotz, 1992]: 1) formation of geranylgeranyl diphosphate (GGDP) from mevalonic acid; 2) condensation with the formation of phytoene; 3) desaturation of phytoene to lycopene and 4) cyclization of lycopene with formation of β-carotene. The intermediates in β-carotene synthesis such as phytoene, phytofluene, ξ-carotene, neurosporene, β-zeacarotene, lycopene, γ-carotene were formed between prephytoene and β-carotene [Ben-Amotz et al., 1987; Ben-Amotz and Shaish, 1992, in: Avron and Ben-Amotz, 1992].

The cells of *D. salina* have a green coloration under conditions suitable for growth and reproduction and have 0.3 % β-carotene on a dry weight basis, similar to the content in plant leaves and the cells of carotene-containing algae. β-carotene accumulates, under conditions that delay growth and reproduction of the cells, within the orange oily globules located in the interthylakoid space of the chloroplast.

The important parameters modulating the growth, reproduction and formation of carotene are light intensity and duration, salt concentration, temperature, and nutrient availability [Milko, 1963; Massjuk, 1966, 1973; Massjuk and Abdula, 1969; Semenenko and Abdulaev, 1980; Ben-Amotz et al., 1982*a*; Loeblich, 1982]. High light intensity and slow growth in *D. salina* result in a higher rate of carotenogenesis. Elevated salt concentrations (i.e., >4 M NaCl) that osmotically alter the growth medium, extreme temperatures, and nutrient deficiencies (nitrogen, in particular) reduce the accumulation of β-carotene in the cells.

The possibility of cultivating *D. salina* in commercial salt water ponds used for NaCl production (Societe Scherifienne des sels, Larache, Morocco) was assessed [Riyahi et al., 2006]. *D. salina* was the only algal species surviving at salinity levels up to 25 % (w/v). The biosynthesis of β-carotene in *D. salina* can easily be regulated. It is probable that the production costs for β-carotene derived from algae grown in open, non-sterile conditions will decrease with new technological improvements [Borowitzka and Borowitzka, 1990; Borowitzka, 1990, 2005]. At the same time, mass cultivation in closed industrial production systems with rigidly controlled conditions for β-carotene by the pharmaceutical industry is also a viable option.


## 12.2. Use of *Dunaliella salina* for the Commercial Production of β-carotene

Between 1958 and 1960, a group of scientists at the Institute of Botany and the Institute of Biochemistry of National Academy of Science in the Ukraine established that *D. salina* has the highest β-carotene content of all plant sources (i.e., up to 1100 mg % for air-dried algae) [Drokova, 1960, 1961; Massjuk, 1961*a,b*; Vendt, 1963; Geleskul, 1968].

*D. salina* grows naturally in incredible quantities in salt reservoirs found in the Ukraine and causes the red "flowering" of *rapa* in estuaries, salt lakes, and artificial reservoirs. The fresh weight of natural populations of the alga in Cremia reservoirs was approximately 40 tons between July and August in 1960 [Massjuk, 1961*b*]. Morphological peculiarities of the alga (absence of a cell wall) facilitates extraction of carotene from the cells. This led to the realization that the commercial production of carotene in natural reserves maybe an economically viable enterprise [Drokova, 1961; Massjuk, 1961*a, b*]. It was evident that *D. salina*, due to its unique biochemical, physiological, ecological, and morphological traits was an excellent potential source of β-carotene [Drokova,1960, 1961; Massjuk, 1961 *a,b*; Vendt, 1963; Gelescul, 1968; Cifuentes, 1996; Gomez, 1999; Haouazine et al., 1999]. A number of investigations have shown that *D. salina* is an ideal candidate for commercial cultivation for the production of β-carotene (see review [Massjuk, 1973]). In addition, the euryhalinity, eurythermity, geliophylness, shade-tolerance, and resistance to variation in the chemical composition of the nutrient medium, the content of main biogenic elements, and the hydroxyl ions concentration are the principal reasons *D. salina* is a superior organism for biotechnology [Heidari et al., 2000].

A typical halobiont, *D. salina* growth is seldom impeded due to competition from other organisms. As a consequence, monocultures readily develop in natural reservoirs [Масюк, 1961*a,b*]. This greatly simplifies and facilitates its cultivation under open air conditions.

The concentration, temperature, and light optima for the growth and reproduction of the cells are known to differ from the optimum conditions for the biosynthesis of carotene (Fig. 12.1) [Milko, 1963; Yurkova, 1965; Massjuk 1965*b*, *c*, 1966, 1967; Massjuk and Abdula, 1969].

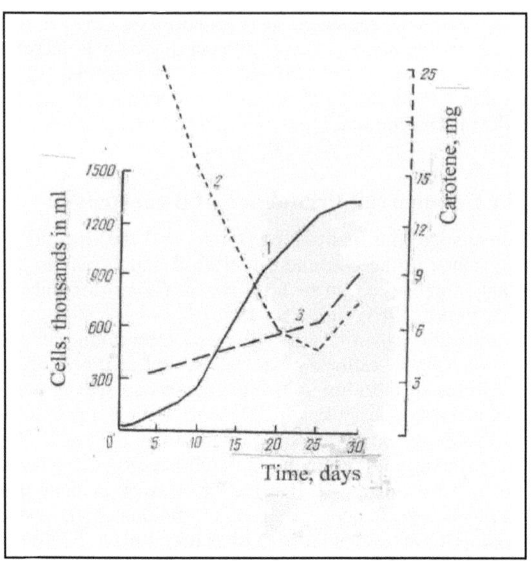


**Fig. 12.1.** Accumulation of algae and carotene in biomass of *Dunaliella salina* when produced in mass culture: 1 – density of the cells; 2 – content of carotene in the cells; 3 – carotene content/1 L of suspension [Massjuk, 1973].

Since the main biological processes that help to maximize the yield of carotene (reproduction of the cells and biosynthesis of β-carotene) require different conditions, a two-stage method of cultivation was proposed for carotene production. The first stage facilitates the accumulation of biomass by using a 2M NaCl medium with appropriate biogenic elements at 25-30 °C and a light intensity of 5,000-6,000 lx. The second stage of growth is in open reservoirs 4-5 M NaCl, 35-40 °C and illuminance around 100,000 lx without the addition of biogenic elements [Massjuk, 1965a,b,c, 1966, 1967; Massjuk and Abdula, 1969, 1971]. We tested the method near Kiev in 1963-1965 and obtained greater than 30 kg/ha of carotene per vegetation season (5 months) or an average carotene production of 24 mg m$^{-2}$ day$^{-1}$ [Massjuk, 1966]. This semi-industrial method for the mass cultivation of *D. salina* for β-carotene was tested using an experimental area of 0.5 ha at the Saksky chemical plant in Crimea using an inexpensive chlorine-magnesium brine and supplemental fertilizer [e.g., superphosphate, ammoniac saltpeter (nitric acid potassium salt), potassium salt]. The station was equipped with 15 plastic 200 l trays, 4 concrete pools of 4 m$^3$ and 4 pools of 5 m$^3$ where the algae was bred (Photo 12.1) and the carotene accumulated.

Experiments in 1965-1968 demonstrated the potential for production in the southern part of the Ukraine, with yields of up to 120 kg of carotene per hectare during a growing season of 7 months [Massjuk and Abdula, 1969; Massjuk et al., 1970; Massjuk, 1973]. The technology for the repeated utilization of the brackish water produced during cyclic production of the algae has also been studied [Massjuk, 1973].

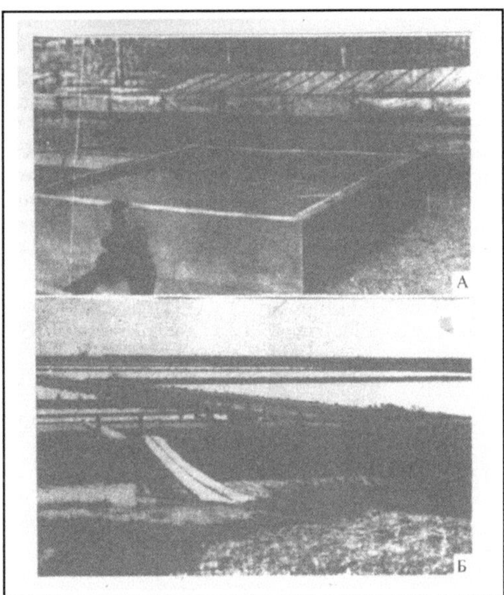

**Photograph 12.1.** Experimental carotene production station at the Saksky chemical plant, Crimea, 1965-1969: A – trays and concrete pools where the first stage of *Dunaliella salina* cultivation occurred; B – industrial pools with a ground floor [Massjuk, 1973].

The technology for the extraction of the carotene using plant oils was also developed. It involves four stages: 1) mechanical fragmentation of the cells; 2) flotation of the cell residue that includes carotene; 3) co-sedimentation of the residues using ferrum oxide hydrates and the removal of the sediment from the rapa; and 4) extraction of the pigment using organic solvents [Vendt et al., 1965; Geleskul, 1966, cited by: Massjuk, 1973]. While the methods and economics appeared to be attractive, industrial production of β-carotene in Ukraine has not yet occurred. While research on the production of carotene-containing algae in Ukraine has been terminated, there continues to be considerable interest elsewhere. For example, this area of research continued in the 1970s in Israel, USA, Australia, China and other countries where the mechanisms of salt tolerance, osmoregulation, and the biosynthesis of glycerol, and β-carotene were studied [Ben-Amotz et al., 1982a,b; see reviews: Borowitzka and Borowitzka, 1990; Avron and Ben-Amotz, 1992].

At the end of the 1970s, several commercial companies were interested in the possibility of production of *D. salina* for β-carotene which accelerated this area of research. At the beginning of the 1980s, a carotene experiment station was established in a sea lagoon in Western Australia near Perth. The first experimental ponds were 10, 100, 250, and 600 m² in size. Another five industrial ponds of much greater size (i.e., 5×) were built in 1986 and the carotene manufacturing plant officially opened. The number of the ponds was doubled in 1988. At the same time, biotechnological and engineering research improved productivity [Borowitzka and Borowitzka, 1990].

At present, the cultivation of *Dunaliella* has expanded in volume and geographical location. For example, production has been undertaken in hypersaline lakes, ponds and lagoons under a range of climatic conditions in various countries (e.g., Larache, Maroc [Haouazine et al., 1999]; the north of Chile [Gomez et al., 1999]; East Coast of Thailand [Bhumibhamon et al., 2003]; southern Spain [Garcia-Gonzalez, 2003]; Venezuela (Araya, Coche, Peonia, Cumaraguas. and Boca Chica) [Guevara et al., 2005]; Urmia hypersaline lake, northwest of Iran [Fazeli et al., 2006] and central region of Iran [Tafreshi and Shariati, 2006]).

Separation of the algae from its aqueous environment is the most difficult, labor-consuming and expensive stage in the technology of carotene production. Likewise, methods for the separation of the β-carotene such as filtration, centrifugation, flotation and flocculation, sedimentation, and the concentrating of the cells in a salinity gradient have been tested [Vendt et al.,1965; Geleskul, 1966, cited by: Massjuk, 1973; Borowitzka and Borowitzka, 1990; Mohn and Contreras, 1990].

*D. salina* is typically processed into three general products: 1) a capsulated solution of 1.6-4 % β-carotene in plant oil for use as a dietary supplement; 2) a 30 % suspension of crystalline β-carotene in plant oil that is used as a dye in the food industry; and 3) a dried product that contains 2-3 % β-carotene for use as a additive in animal feeds [Borowitzka and Borowitzka, 1990].

Similar carotene factories were created in South Australia and the USA (California). There were also smaller scale companies producing β-carotene from *D. salina* in Israel (see photographs 12.3, 12.4)[2], Chile, and Great Britain [Borowitzka, 1990; Borowitzka and Borowitzka, 1990; Mohn and Contreras, 1990].

**Photograph 12.2.** A general view of the reactor used for biomass production of *Dunaliella* near the city Eilat (courtesy of Prof. A. Ben-Amotz).

---

[2] The photographs courtesy of Prof. A. Ben-Amotz, National Oceanographic Institute, Israel.

**Photograph 12.3.** Details of the reactor used for biomass production of *Dunaliella* near the city Eilat (courtesy of Prof. A. Ben-Amotz).

Worldwide the country producing the greatest quantity of β-carotene is Australia. Glycerol, which can reach 30 % of the algal dried biomass, high-quality protein which remains after extraction of β-carotene, ascorbic and dehydroascorbic acids and other valuable biologically active organic compounds are by-products of β-carotene production [Massjuk, 1973; Ben-Amotz et al., 1982a, b; Borowitzka and Borowitzka, 1990; Avron and Ben-Amotz, 1992].

The current state of β-carotene production from *Dunaliella*, based on either open-pond systems or closed photobioreactors, is discussed in the following articles: [Yamaoka et al., 1994; Ben-Amotz, 1995; Yamaoka et al., 1996, 1997; Chen and Wang, 2003; Leon et al., 2003; Hejazi et al., 2003; Chai et al., 2004; Hejazi and Wijffels, 2004]. Likewise, the effect of various abiotic and chemical factors on *Dunaliella* cell growth and carotene content is reviewed in a number of papers. This includes the effect of high light intensity, pH, NaCl concentration, temperature, and trace nutrients [Yamaoka et al., 1992]; high irradiance and areal densities [Grobbelaar, 1995]; differences in the chemical composition of the media [Cifuentes et al., 1996]; various stress conditions (nutrient deficiency or high salt concentration) [Haouazine et al., 1999]; temperature, pH and illumination [Markovits et al., 1993]; high light intensity, high salt concentration, and nitrate deficiency [Bhumibhamon et al., 2003]; mixing, flow rate, culture depth, cell density, and dilution cycles [Garcia-Gonzalez, 2003]; salinity [Mohammad and Mansour, 2003]; temperature and irradiance [Gomez and Gonzalez, 2005]; different salt concentrations [Fazeli et al., 2006]; and high light intensity, high salinity, temperature and availability of nutrients [Raja et al., 2007]. Laser mutagenesis of the organism and its effect on β-carotene production was investigated by Zhao et al. (1992).

Utilization of *Dunaliella* by the food industry includes the introduction of β-carotene into foods as a source of vitamin A and as an antioxidant, the production of β-carotene tablets, and as a component of mixed carotenoid products are described by Honda [1997]. *Dunaliella* spp. have significant amounts of lipid, protein, chlorophyll, carotenoids, vitamins, minerals, and unique pigments and are used as nutrient-dense foods and sources of fine chemicals [Kay, 1991; Honda, 1997; Hatanaka, 2002]. Certain strains of *Dunaliella* are harvested to produce dried algal meals for the extraction of polyunsaturated fatty acids and oils for the health food industry and for coloring agents to be used in the food and cosmetic industries [Jin and Melis, 2003]. Due to their health benefits, algal pigments have great commercial value as natural colorants in food industry [Prasanna et al., 2007].

Interest in β-carotene has increased significantly due to the discovery of its anti-tumor activity and its role as an antioxidant in human nutrition. At the present time *D. salina* is cultivated as a source of β-carotene on an industrial scale in Australia, USA, Japan, Taiwan, China, and Indonesia [Borowitzka, 2005]. β-Carotene, astaxanthin, fucoxanthin, halocynthaxanthin and peridinin are believed to inhibit lung, breast, buccal pouch, and nerve cell cancers [Hong et al., 1998].

The problems of production, functions, and quality standards of *D. salina* powder and soft capsules are discussed by Guo et al. (2003). The preliminary experiments with laboratory animals have shown that the capsules containing the powder inhibit radiation damage and tumor development and enhance immunity. Certain carotenoids in *Dunaliella* (e.g., lutein) are thought to be effective agents in the prevention and treatment of a variety of degenerative diseases [Campo et al., 2007].

Preparations of β-carotene are exported from Australia to Japan, USA, Korea, Taiwan and other countries with total exports to Japan and USA in 1990 of 2 million Australia dollars [Borowitzka, 1990]. The cost of β-carotene from *D. salina* in the 1990s was estimated by Australian manufacturers as 50-100 Australia dollars per kilogram while the price of natural β-carotene on the world market at the same time varied between 500-1000 dollars per kg. It is estimated that the worldwide market value of carotenoids will exceed US $1,000 million by the end of the decade [Campo et al., 2007].

It is worth mentioning the potential application of *D. salina* as an educational model for biotechnology students [Bosma and Rouke, 2003]. Students can develop strategies for optimization of the growth rate, and how and when to stress the alga to initiate β-carotene production. The students should be responsible for their own results and creating a competition among the students can stimulate their interest. The approach also provides considerable practical experience for the students.


## 12.3. Summary

Several species of *Dunaliella* can be cultivated as a natural source of β-carotene, ascorbic and dehydroascorbic acids, glycerol, and other valuable organic compounds. In addition, the genes that encode for proteins that confer exceptional tolerance to extreme environmental conditions are viable candidates for facilitating the resilience of other organisms to environmental stress using recombinant DNA technology.

Mass cultures of *Dunaliella* can be used to provide a nutritionally excellent forage for the piscine industry, especially for the cultivation of valuable breeds of sturgeon. *D. salina* is the richest source of β-carotene in the plant kingdom. Mass cultivation of selected strains is utilized in a number of countries for the industrial production of β-carotene preparations that are used in food and pharmaceutical industries and in medicine for the prevention and treatment of tumors, cardiovascular, ophthalmic diseases, avitaminosis, arthrosis, and other diseases. Further study of photomovement regularities and peculiarities in *Dunaliella* species can facilitate solving certain technological problems associated with carotene production.

# Chapter 13

# General Results and Perspectives of Further Investigations

The senior author of the present work started her studies of the genus *Dunaliella* at the N.G. Kholodny Institute of Botany, the National Academy of Science of Ukraine in the late 1950s in the context of projects on surveying algae as prospective sources of β-carotene (provitamin A). In the 1960s-1970s, the composition and ecology of the genus in watersheds of the Ukraine, the Trans-Caucasian region, Central Asia, and the Far East were investigated in detail along with issues of phylogeny and taxonomy. A collection of *Dunaliella* strains was created and peculiarities in their morphology, physiology, and biochemistry were investigated. In addition, methods for their laboratory and semi-industrial commercial cultivation [Massjuk 1973] were developed. In the 1980's, several species of *Dunaliella* that had attracted the attention of a biophysicist were selected as possible models for studying photomovement using a series of complex optical, spectroscopic, and laser methods [Posudin et al., 1992]. An experimental laboratory was established and the instrumentation needed for studying the photomovement of cells of *Dunaliella* developed in the Department of Biophysics at the National Agricultural University in Kiev (now National University of Life and Environmetal Sciences of Ukraine).

The main objective of the present work is to summarize our research on the photomovement of *Dunaliella* and to compare our results with data of other scientists that have studied similar phenomena in other organisms, identifying both common and specific peculiarities in photomovement in representatives of various taxa and assess differences relative to phylogenetic relationships. Experimental studies were preceded by a detailed analysis of previous results published by the photobiology community on the photomovement of flagellates. Particular attention has been focused on discrepancies in terminology covering the various aspects of photomovement in microorganisms.

## 13.1. Problems of Terminology

Critical considerations in the terminology and classification of different types of light-induced behavior of freely motile organisms have shown that the existing classification systems create considerable terminological confusion. As a consequence, we developed a parametrical classification system for the light-dependent behavour of either individual motile cells (individual effect or micro-effect, Table 2.1) or their aggregations (group effect or macro-effect, Table 2.2). To this end, we retained the original meaning of the term *phototaxis* as any light-induced movement of freely motile organisms in space. Light-dependent *reactions* of motile organisms (*photoresponse, photoreaction*) are considered any immediate motion responses by the organism to any change in the light stimulus.

*Motility* of biological objects is a special case in the general physical phenomenon of movement (mobility). Therefore, motility of organisms can be described by such well known parameters as speed or velocity ($v$), direction ($r$), and trajectory ($l$) of movement. The light stimulus, in turn, can be characterized by such parameters as intensity ($I$), direction ($s$), spectral composition ($\lambda$), and polarization ($P$) of light and by the duration, frequency, and the shape of light pulses. Considering the parametrical characteristics of both factors (light and movement), we believe that any classification of the dependence of movement (phototaxis) of microorganisms on light should be based on a parametrical principle (Tables 2.1, 2.2). We therefore propose defining *photokinesis* as any dependence of speed of individual organisms or their groups on any parameters of a light stimulus. *Phototopotaxis* is any dependence of the direction of movement of individual organisms or their groups on any light parameter.

The proposed classification can be further developed in greater detail by accounting for additional parameters of movement and light (e.g., the rhythm of a light flux), their possible interactions (e.g., wavelength and intensity of light, velocity and direction of movement), or specific features of certain parameters (e.g., velocity of movement can be linear or angular, light intensity can be characterized by its absolute value ($I$) or its gradient in space ($dI/dx$) and time ($dI/dt$)). The suggested principles not only promote an improvement in the existing terminology but also facilitate planning further research.


## 13.2. Phenomenology of Photomovement

Species of *Dunaliella*, as well as other flagellate algae, move freely in an aquatic environment under the influence of light, a response termed *photomovement* (*phototaxis*). Photomovement involves a translational movement of a cell that is accompanied by rotation of the cell around its longitudinal axis, sidewise turns from the main direction, and oscillatory movements ("staying in one place"). Sometimes the cell is attached by the distal ends of the flagella to the substratum, convulsively twitching around the attachment site. When the cell becomes detached from the substratum, it continues free "navigation".

In contrast to the pattern observed in *Chlamydomonas reinhardtii* P.A. Dang. and *Euglena gracilis* G.A. Klebs, only the ciliary type of flagella movement is observed in *Dunaliella* during photomovement. Only in unusual cases, when mechanical obstacles are present and/or when a *Dunaliella salina* Teod. cell becomes constrained in a very thin preparation between the cover glass and the slide, does it switch to a more ancient undulating type of flagella movement.

The trajectory of locomotion of a cell resembles a sine wave and its plane projection looks like a non-uniform zigzag. It is believed that, similar to *Chlamydomonas*, *Dunaliella* performs rotary cell movements around its longitudinal axis that are mediated by the beating of its flagella in the three-dimensional space. The sinusoid movement is a result of an almost synchronous but unequal number of beatings between the cell's two flagella. In contrast to the pattern observed in *Chlamydomonas*, when a change occurs in the movement direction, one flagellum of the *Dunaliella* cell ceases beating, while the second continues to beat, causing the cell to turn. Subsequently, the first flagellum renews its beating and the cell moves in the new direction.


## 13.3. Photoreactions

Species *of Dunaliella* are capable of photokinetic and photovector reactions. In contrast, the presence of photokinetic reactions in *Chlamydomonas* has been challenged in the literature. We did not observe photophobic reactions in *Dunaliella*, though they are present in *Chlamydomonas* and *Euglena*. There are, however, several publications indicating the possibility of such reactions in *Dunaliella*.


## 13.4. Photokinesis

The locomotion velocity of *Dunaliella* cells (*photokinesis*) depends on the intensity of the light stimulus and the environmental conditions, such as temperature, strength of electric and electromagnetic fields, intensity and duration of ionizing radiation, and the presence and concentration of calcium channel blockers (such as isoptin and cinnarizine), sodium aside, surfactants, salts of heavy metals, and pesticides. Likewise, a combination of factors can modulate velocity. We did not find the locomotion velocity of individual *Dunaliella* cells depended on the wavelength of the light, dosage of preliminary UV-irradiation, concentration of cobalt and calcium ions, the presence of ionophores that increase the permeability of the cellular membrane to calcium ions, and the presence of ionotropic preparations that stimulate $Na^+$-$K^+$-ATPase.

The average cell velocity of translational movement of hyperhalobic species of *Dunaliella* was $36 \pm 2$ μm/s for *Dunaliella viridis* and $48 \pm 2$ μm/s for *D. salina*. The average velocity of movement for these two closely related species among experiments varied over a wide range, indicating that distinctions between the species based on velocity probably are not valid. The modal value for the average velocity of movement of the marine species *Dunaliella bioculata* Butcher was $105 \pm 5$ μm/s. The values for the *Dunaliella* species exceed by 1-3 orders of magnitude those of microorganisms not possessing a flagellar apparatus and are within the limits known for others flagellates, both prokaryotic and eukaryotic. However, the average locomotion velocity of hyperhalobic species of *Dunaliella* is lower (sometimes by an order of magnitude) than in the marine species *D. bioculata* and the freshwater species *C. reinhardtii* and *E. gracilis*. These differences may be caused by variation in the viscosity of the media.

The average cell velocities for rotary movement in hyperhalobic species of *Dunaliella* were nearly identical [i.e., $0.52 \pm 0.04$ rotations (revolutions) per second in *D. salina* and $0.54 \pm 0.04$ rotations per second in *D. viridis*]. These values, however, were lower than in the freshwater algae *C. reinhardtii*. The maximum average velocity values for forward and rotary cell photomovement in *Dunaliella* were observed under the following conditions: white-light intensity of 0.22–0.81 $W/m^2$, illuminance of 150–550 lx, temperatures of 20–30 °C, and pH 8.

The rhythmic regulation of cell movement in *Haematococcus pluvialis* Flotow is caused by rhythmic pulses, presumably related to the functioning of contractile vacuoles. However, in hyperhalobic species *of Dunaliella*, the contractile (pulsating) vacuoles are absent and the beating rhythm of flagella is apparently related to other oscillators. The flagellar beating frequency in hyperhalobic species of *Dunaliella* is 25–50 Hz while in freshwater *C. reinhardtii* it is as much as 64 Hz.

## 13.5. Phototopotaxis

The direction of cell movement of *Dunaliella* in relation to the light source (*phototopotaxis*) depends on the parameters of the light signal (e.g., light intensity, spectral composition, gradients of intensity in space and time, polarization) and environmental conditions (e.g., temperature, intensity of electric fields, level of ionizing and UV radiation, concentration of $Ca^{2+}$, $Co^{2+}$, cinnarizine, isoptin, sodium aside, surfactants, salts of heavy metals and pesticides), and combinations of these factors. Phototopotaxis of *Dunaliella* species, however, is not modulated by ionotropic preparations that stimulate a membrane $Na^+$-$K^+$-ATPase.

Phototopotaxis of both hyperhalobic species of *Dunaliella* in laboratory cultures was observed at an illuminance of 500 lx (positive) and 40,000 lx (negative). Transition from positive to negative phototopotaxis was observed at 1,500 lx. These parameters are within the limits known for other algae. However, sensitivity thresholds to weak and strong illuminance, and transitions from positive to negative phototopotaxis differ substantially among algal species. They allow assessing shade-tolerance, sun-tolerance, and resistance to high-level light exposure of these species.

*D. viridis* is more sensitive to weak light (30 lx) than *D. salina*, which corresponds to behavioral peculiarities between the two species in nature. In laboratory culture, both species are more sensitive to high illuminance than *C. reinhardtii*; in the latter, the transition to negative phototopotaxis occurred at 100,000 lx.

The transition of *Dunaliella* species from positive to negative phototopotaxis differs from that in chlamydomonads. In contrast to chlamydomonads, the change in flagellar beating from a ciliary to an undulate mode was not observed in the *Dunaliella* species. The beating of only one flagellum was observed, which caused turning of the cell followed by its subsequent movement in the opposite direction to the light source.

Under the conditions found in Crimean hyperhaline watersheds with high illuminance (i.e., above 100,000 lx), natural populations of *D. salina* are found in the "red form" and display a complete absence of negative phototopotaxis. Thus, hyperhalobic species *D. salina* and *D. viridis* differ in their degree of sensitivity to both high- and low-light intensity which explains differences in the ecological niches occupied by the species in nature.

The maximum values for positive and negative phototopotaxis were found at 20–30°C and the maximum value for positive phototopotaxis was at pH 7.35.

Electric fields suppress phototopotaxis in *Dunaliella* as well as in other algae, which indicates the participation of bioelectric potentials in this process. Increasing the temperature from 18 °C up to 30 °C removes the inhibitory influence of the electric field density and stimulates phototopotaxis in both hyperhalobic species of *Dunaliella*.

Ionizing and UV irradiation inhibit phototopotaxis in *D. salina* and *D. viridis*. The inhibiting effect depends on the irradiation dosage; in the case of UV-irradiation, the response also depends on the wavelength. We described for the first time in *Dunaliella* the transformation from positive phototopotaxis to negative due to the influence of UV-irradiation; at high levels of irradiation, phototopotaxis can be completely blocked.

The maximum values for phototopotaxis in hyperhalobic species of *Dunaliella* were observed when $CaCl_2 \cdot 6H_2O$ was in the medium at concentrations between $10^{-5}$ and $10^{-3}$ M. Increasing the calcium chloride concentration up to $10^{-2}$ M suppressed phototopotaxis by 10–20 %. The addition of the ionophore A23187 to the medium increased the permeability of the cellular membrane to calcium ions causing a complete inhibition of phototopotaxis both in *D. salina* and *D. viridis*. The addition of $CoCl_2$, that blocks membrane calcium channels at concentrations from $10^{-6}$ to $10^{-3}$ M, suppressed phototopotaxis in both species of *Dunaliella*. Other calcium channel blockers, such as cinnarizine, isoptin, and sodium aside, elicit a similar effect on phototopotaxis in *Dunaliella*. Ouabain, which stimulates $Na^+$-$K^+$-ATPase, does not influence phototopotaxis in *Dunaliella*.

The phototopotaxis action spectrum for the two hyperhalobic species of *Dunaliella* was identical. It occurred between 400 and 520 nanometers (nm) and has maxima at 410–415 nm and 465–475 nm. The action spectrum for phototopotaxis in *Dunaliella* differs somewhat from those in *Chlamydomonas reinhardtii* and *Haematococcus pluvialis* which display a wide band in the 400–600 nm range, with the maximum at 500 nm. In contrast to the situation observed in representatives of the *Chlorophyceae*, *Tetraselmis viridis* Rouch. (*Chlorodendrophyceae*) exhibited phototopotaxis in the UV region of the electromagnetic spectrum. In *E. gracilis* (*Euglenophyta*), phototopotaxis occurs in the 300–550 nm range with two basic maxima at 385 nm and 460 nm and two smaller maxima at 410 nm and 490 nm. Thus, representatives of different genera, classes, and divisions differ distinctly in their action spectra for phototopotaxis indicating differences in their photoreceptor systems and the composition of photoreceptor pigments.

## 13.6. Motility

One parameter of photomovement is the motility of cells or the relative number of motile cells ($N_m/N_0$), where $N_m$ is the number of motile cells, and $N_0$ – the total number of motile and non-motile cells. In populations of *Dunaliella* this varies from 0 to 100 % and displays the same dependence on the characteristics of the light stimulus and environmental conditions as phototopotaxis. The number of motile cells differs with the presence or absence and degree of dependence due to certain factors (e.g., wavelength of light, dose of preliminary ionizing and UV irradiation, concentration of compounds opening or blocking calcium channels) on the velocity of individual cells (photokinesis).


## 13.7. Photoreceptor System

The photoreceptor system of *Dunaliella* species, as well as that of other green algae, consists of a photoreceptor, presumably located in plasmalemma and membranes of the chloroplast (in the area near the stigma), and a stigma that consists in different species of one to two layers of lipid globules located in the peripheral zone of the plastid. It has been shown that, in contrast to algae such as *E. gracilis*, species of *Dunaliella* do not possess a photoreceptor with a dichroic structure.


## 13.8. Mechanisms of Photoreception

Photoreception in *Dunaliella* and probably in certain other motile microorganisms with flagella, is based on the interaction of several mechanisms: modulation, diffraction, and interference. The modulation mechanism is due to the rotary movement of the cell around its longitudinal axis during which the stigma modulates the light signal affecting the photoreceptor. With the presence of more than one layer of pigmented globules in the stigma, an interference mechanism in photoreception is possible (similar to that in chlamydomonads). We believe the photoreception diffraction mechanism is universal for all flagellates having a globular stigma structure. Thus, during photoreception in these flagellates (including *Dunaliella*), there is a cooperative effect during the simultaneous functioning of several mechanisms, increasing the efficiency level of the light signal.

Our data provide evidence that the composition of photoreceptor pigments in *Dunaliella* is neither flavins (as in *E. gracilis*) nor rhodopsin (as in *E. gracilis*, *C. reinhardtii* and *H. pluvialis*), but is other carotenoids or carotenoproteins, though some authors [Wayne et al., 1991] believe there is participation of rhodopsin in the photoreception of *D. salina*.

With regard to photoreception mechanisms, we accept Nultsch's hypothesis (1983) that the absorption of a quantum of light by a photoreceptor molecule is accompanied by its excitation and conformational changes in the photoreceptor protein(s). We proposed as further development of this hypothesis a photoregulation model for the movement of flagellate algae that is based on conformational changes in protein molecules that are components of either the photoreceptor system or the flagellar apparatus. The absorption of light by photoreceptor molecules is accompanied by the excitation of exciton or soliton conditions in $\alpha$-spiral sites of membrane proteins. Reorganization of the configuration of the membrane protein results in the formation of ion channels through which calcium ions freely diffuse inside the cell thus stimulating locomotory activity. The discovery of the contractile protein *centrin* in the fibrous structures of the flagellar apparatus in *Dunaliella* indicates its possible participation in photoorientation of basal bodies in these algae, as occurs in other flagellates.

Our experiments using imposing electric fields provide evidence that in the processes of photoreception the *Dunaliella* species, as well as in other green algae, light-induced changes of membrane potentials play a role.

## 13.9. Sensory Transduction of the Light Signal

As it has been demonstrated with ions of calcium and cobalt, and substances blocking or stimulating ion channels, sensory transduction of an absorbed quantum of light into a locomotory reaction in species of *Dunaliella*, as well as in other green algae, is most likely of ionic nature, thus confirming the critical role of $Ca^{2+}$ in these processes. At the same time, $Na^+$-$K^+$-ATPase does not participate in the photoregulation of locomotion in *Dunaliella* cells nor does ouabain have an effect on photomovement parameters. It is assumed that the intake of calcium ions into the cells of these algae is controlled not by a $Na^+$-$K^+$ pump, as is the case in *E. gracilis*, but probably occurs directly, through light-induced membrane channels, as in *Chlamydomonas*.

In contrast to *C. reinhardtii*, in many cases the *non-specific* reaction (autotomy of flagella) of *Dunaliella* cells to substances stimulating or blocking ionic channels does not result in the loss of flagella. Therefore, their locomotory reactions can be regarded as a *specific* response to the blocking or stimulating of ionic processes.

We demonstrated for the first time that different photomovement parameters in *Dunaliella* (phototopotaxis and the relative number of motile cells, on the one hand, and photokinesis, on the other hand) are controlled by different mechanisms. This is confirmed by the presence/absence or different degrees of dependence of these parameters on such environmental factors as the wavelength of incident light, dose of preliminary ionizing and UV-irradiation, concentration of ions of calcium or cobalt, and compounds stimulating or blocking calcium channels, etc.


## 13.10. Importance of Data on Algal Photomovement for Related Fields of Science

Results of studies of photomovement processes in microorganisms are of interest not only for photobiology but also for related fields of science, such as evolutionary biology, phylogenetics, systematics, ecology, and applied aspects of biology. Comparative studies of photomovement in two closely related species of *Dunaliella* (i.e., *D. salina* and *D. viridis*) have shown that they generally do not differ from each other in key parameters. They also further substantiate the structure of their photoreceptor systems and photoregulation mechanisms for cell locomotion, that have developed as a result of a long process of joint evolution. At the same time, the species differ in their sensitivity to white light at low and high intensities and also to the threshold for the transition from positive to negative phototopotaxis. The two species also react differently to the interaction of some factors correlated with light (temperature, electric field, and intensity of light). In contrast to *D. viridis*, *D. salina* under conditions of natural hyperhaline watersheds does not display negative phototopotaxis at an illuminance above 100,000 lx due to the protective function of β-carotene that accumulates in its cell. The differential in sensitivity between the two species of *Dunaliella* to light is a result of their adaptation to the differing ecological niches for light. Such adaptation provides an opportunity for the two species to coexistence in the same reservoirs but in different niches: a) on the brightly illuminated surface of the salt solution (*D. salina*) and b) in more shaded benthic layers (*D. viridis*).

The sensitivity of *D. salina* and *D. viridis* to ionizing radiation also differs, which is most likely caused by differences in the size of their photoreceptor systems as targets responsible for effecting locomotory reactions. Thus, intrageneric differences do not involve structural features of the photoreceptor systems and mechanisms of photoreception and sensory transformation of a light signal into locomotory reactions. The differences are the result of ecological adaptations or are caused by dimensional characteristics of cells and their organelles. The ability of one species to accumulate large quantities of β-carotene plays an important protective function.

Critical differences in photobehaviour are found among representatives of various genera belonging to different orders of green algae in the class *Chlorophyceae*: species of *Dunaliella* (*Dunaliellales*) on the one hand, and *C. reinhardtii* and *H. pluvialis* (*Chlamydomonadales*), on the other hand. Differences in the action spectra and maxima for phototopotaxis leads to the assumption that there are different sets of photoreceptor pigments: carotenoids and carotenoproteins in *Dunaliella* and rhodopsin as the basic photoreceptor pigment in *C. reinhardtii* and *H. pluvialis*.

Taking into account differences in the stigma structure between the species of *Dunaliella* and *C. reinhardtii*, it is probable that photoreception in the former, in addition to the modulation mechanism, light diffraction plays an essential role, while in the latter, the critical mechanism involves interference in light flux.

Introduction of chemical compounds into the medium that stimulate or block membrane ionic channels causes in the *Dunaliella* species specific locomotory reactions, while in *C. reinhardtii* the reactions are nonspecific (autotomy of flagella). Likewise, in the *Dunaliella* species photophobic reactions like those in *Chlamydomonas* have not been observed, while the presence of photokinetic reactions, so well expressed in *Dunaliella,* are questionable in *Chlamydomonas*.

Photomovement in *Chlamydomonas* and *Dunaliella* differ in the beating mode of their flagella. Forward movement of a *Chlamydomonas* cell is caused by ciliary beating of the flagella, while backward movement utilizes undulate beating. Since the cells of *Dunaliella* are not capable of photophobic reactions, the undulate mode of flagella beating has not been observed in the genus. *Chlamydomonas* cells change direction by using unequal beating frequencies between the *cis-* and *trans-*flagella. In contrast, *Dunaliella* cells temporarily stop the beating of one flagellum.

Thus, species of *Dunaliella* differ from *C. reinhardtii* in a complex array of fundamental features that include the composition of their photoreceptor pigments, mechanisms of photoreception, certain details in the sensory transduction of the light signal into a locomotory reaction, the presence or absence of photophobic and photokinetic reactions, functioning of the flagellar apparatus, etc. The results obtained correlate with data of molecular cladistics. The data of molecular phylogeny demonstrate that *C. reinhardtii,* as a representative of the heterogeneous genus *Chlamydomonas,* belongs to the group of chlamydomonads (which is closely related to colonial *Volvocales)* that is very distant from another group that is closer to *Dunaliella* (e.g., *Chlamydomonas applanata* Pringsh.). The evolutionary distance between these two groups is comparable to the distance between soybeans and cycads.

Interesting data are available on photomovement of *Tetraselmis viridis* (Roukhiyajnen) Norris et al., a representative of a separate class of green algae, the *Chlorodendrophyceae* [Massjuk, 2006]. Photomovement in the ultraviolet region of the spectrum and the presence of a phototopotaxis maximum in that area indicates the possible presence of flavins and/or pterins as part of the composition of photoreceptor pigments of this species. Thus, evolution of photoreceptor systems, in particular sets of photoreceptor pigments, may occur in different ways and in various branches (clades) of the phylogenetic tree of green plants.

Even greater differences in the structure of the photoreceptor were found between *Dunaliella* spp. and *E. gracilis*, a representative of the division *Euglenophyta*. According to some authors, *Euglena* and *Dunaliella* belong to different kingdoms: *Euglenozoa* Cavalier-Smith, 1981, or *Euglenobionta* Kussakin et Drozdov, and *Plantae* Leedale, 1974, or *Viridiplantae* Cavalier-Smith, 1981, respectively. Other authors place them in different super-kingdoms of the organic world: *Discicristata* (Mirabdullaev) Leontiev and Akulov (2002) ex Zmitrovich, 2003 and *Lamellicristata* (Taylor) Starobogatov (1986) emend. Zmitrovich (2003).

*Dunaliella* differs from *Euglena* in its low speed of cell movement, higher sensitivity to low and high light intensities, and the lower threshold for the transition from positive to negative phototopotaxis, which is probably partly explained by the differing ecological requirements at the species level. Until now the transition from positive to negative phototopotaxis with the subsequent complete suppression of phototopotaxis in response to UV irradiation was observed only in species of *Dunaliella*. Differences in the phototopotaxis action spectra also indicate differences in photoreceptor pigments. There are also distinct differences in the structure of the photoreceptor: crystal and dichroic in *Euglena* and non-crystal and non-dichroic in *Dunaliella*.

Taking into account the peculiarities of the photoreceptor system of euglenoid algae, in particular, their stigma structure, it is possible to believe that they possess only the most ancient, basic modulation mechanism for photoreception inherited from prokaryotes or common ancestors of these groups. In contrast, in green algae three mechanisms (modulation, diffraction, and interference) can function simultaneously, increasing the level of the light signal absorbed by a photoreceptor. Though processes of sensory transduction of the light signal into a locomotory reaction are most likely of an ion nature in all flagellate algae, in *E. gracilis*, in contrast to green algae, a $Na^+$-$K^+$-pump participates in the control of these processes. *Euglena* also differs from species of *Dunaliella* and *C. reinhardtii* in the unique functioning of its flagellar apparatus. During beating, the flagellum of *E. gracilis* looks like a "broken (interrupted) spiral" where the spiral parts of the flagellum are interrupted by its straight parts. Neither typical ciliary nor undulate type flagella beating are observed in euglenids.

Thus, the greater the phylogenetic distance between taxa, the greater number and degree of distinctions such as differences in their photobehaviour, structure and functioning mechanisms in the photoreceptor systems, sensory transduction of the light signal into the different types of locomotory reaction and the functioning of the flagellar apparatus. These differences make it possible to use variation in the photomovement traits as additional diagnostic criteria in the *evolutionary biology, phylogenetics, systematics,* and *taxonomy* of algae.

Along with the differences in photomovement processes and its photoregulation, representatives of different taxa of flagellates have common features. These include the structure of the photoreceptor system (which, with a few exceptions, consists of the photoreceptor and stigma), the primarily modulation mechanism in photoreception (which in representatives of different taxa, depending on the stigma structure, can be associated with additional mechanisms, such as diffraction and interference), the ionic nature of photoreception and sensory transduction of the light signal into a locomotory reaction coupled with the dominant role of calcium ions in these processes, the possible participation in these processes in light-induced changes in membrane electric potentials, conformational changes in protein molecules as the basis of photoreception and sensory transduction, and the participation of centrin in photoorientation of basal bodies.

At the same time, it is evident that the current experimental data on algal photomovement are far from complete and are based on only a small number of model organisms. Too often the studies are not carried out according to a coherent plan at an appropriate methical level. Therefore, the results obtained in studies using different model organisms and by representatives of different photobiology schools are not always comparable. As a consequence, the conclusions made here must continue to be considered preliminary and represent a critique of the initial results of ongoing studies. It is anticipated that future studies will add to, modify, or even negate one or more of these conclusions.

Studies of peculiarities in photomovement are of interest with respect to the *ecology* and *geography* of algae, in particular, for *autecology*. Such studies allow specifying characteristics of selected species focusing on their reaction to parameters of light, determining optimum, maximum and minimum values of these parameters among species, promoting their subdividing the species into groups of shade-preferring, shade-resistant, photophilous, and light-resistant organisms, and enhancing our understanding of laws governing their distribution on the planet Earth.

## 13.11. Applied Importance of Data on the Photomovement of Algae

The applied importance of studying photomovement parameters in species of *Dunaliella* is determined mainly by the position of the genus in the kingdom of green plants *(Viridiplantae)*. Plants dominate our planet and provide the everyday needs of mankind. Hyperhalobic species of *Dunaliella* are classical models for studying the mechanisms of salt tolerance, osmotic regulation, permeability of membranes, and processes governing the biosynthesis of carotene in plants and photomovement. Due to their copious synthesis of β-carotene, osmotically active compounds of strategic importance, and high content of additional physiologically active substances, species of *Dunaliella* are valuable organisms for *biotechnology*. Their high sensitivity to environmental factors provides an opportunity for their use as *biomonitoring* test organisms. The advantages of these organisms include their microscopic sizes, high rate of reproduction, active motility, and photokinetic and photovector reactions.

In our experiments, the sensitivity of various parameters of photomovement in *D. salina* and *D. viridis* were studied in the presence of surface-active substances in the environment (surfactants: cation-active catamin, anion-active sodium dodecylsulfonate, non-ionactive hydropol, a natural polysaccharide compound isolated from blue-green algae, agents "water bloom" in the Dnieper reservoirs, and also their combinations in the concentration range from 1 mg/L to 40 mg/L), salts of heavy metals [$CuSO_4 \cdot 5H_2O$, $CdCl_2$ , and $Pb(NO_3)_2$ in the $10^{-7}$-$10^{-2}$ M concentration range] and pesticides (Acetal 55 %, Acetazine 50 %, Alachlor 45 %, Arylon 75 %, Basta 20 %, Dual 96 %, Harmoni 75 %, Tecto 45 % in concentrations from $10^{-7}$ to $10^{-2}$ M). The data obtained indicate the possibility of using the velocity of forward and rotary movement of the cells, the frequency of their flagella beating, and phototopotaxis values in the species of *Dunaliella* as biological monitors to assess the health of aquatic environments. The simultaneous assessment of several photomovement parameters has the potential to significantly increase the sensitivity of the method. The vector method of biotesting is proposed for assessing the concentration of various toxicants in aquatic environments and allows simultaneously monitoring of two or more movement parameters. The method facilitates processing the data from large-scale measurements, allows obtaining a quantitative estimation of toxicant concentration, and provides the potential of identifying the toxicant.

Species of *Dunaliella* are excellent sources of β-carotene, ascorbic and dehydroascorbic acids, glycerol and other valuable organic compounds and are therefore candidates for biotechnological manipulation. Strains of *Dunaliella* are grown in many countries on an industrial scale for the production of β-carotene for use in the food and pharmaceutical industries and in medicine for the prevention and treatment of tumors, cardiovascular and ophthalmic diseases, avitaminosis, arthrosis, and other pathologies. Studying photomovement regularities and peculiarities in species of *Dunaliella* can help solve certain technological problems currently confronting the production of carotene from these algae.

The main tendencies and perspectives for further photomovement investigations in flagellates are discussed.

# References


Aansen, A.J., Eimhjellen, K.E., and Liaaen-Jensen, S. 1969. An extreme source of β-carotene. *Acta Chem. Scand.* 23: 2544.

Abdel-Raouf, N. and Ibraheem, I.B.M. 2001. Efficiency of *Dunaliella* sp. and *Aphanocapsa elachista* in removing of copper and nickel from culture media. *Al-Azhar J. Microbiology* 54: 192–200.

Aguirre-von-Wobeser, E., Figueroa, F.L., and Cabello-Pasini, A. 2000. Effect of UV radiation on photoinhibition of marine macrophytes in culture systems. *J. Applied Phycology* 12(2): 159–168.

Ahmed, A.M. and Zidan, M.A. 1987. Glycerol production by *Dunaliella bioculata. J. Basic Microbiology* 27(8): 419–25.

Ahmed, A.M., Zidan, M.A., and Adam, M.S. 1988. Effects of high boron levels on growth and some metabolic activities of the halotolerant *Dunaliella tertiolecta. Biol. Plantarum (Praha)* 30(5): 357–361.

Angelicini, F., Ascoli, C., Frediani, C., and Petracchi, D. 1986. Transient photoresponses of a phototactic microorganism, *Haematococcus pluvialis*, revealed by light scattering. *Biophys. J.* 50: 929.

Aristotle (384–322 B.C.) *On the Motion of Animals* (or *De Motu Animalium).* On the Motion of Animals By Aristotle. Translated by A.S.L. Farquharson. 5 January 2010. At: http://classics.mit.edu/Aristotle/motion_animals.html

Arnoldi, V.M. 1908. *Introduction to Studying Lower Organisms.* Kharkov (In Russian)

Arnott, H.J. and Brown, R.M., Jr. 1967. Ultrastructure of the eyespot and its possible significance in phototaxis of *Tetracystis excentrica. J. Protozool.* 14: 529–539.

Artari, A. 1903. *To the Problem of Effect of Medium on Growth and Development of Algae.* Moscow. (In Russian)

Ascoli, C. 1975. New techniques in photomotion methodology. pp.109–120, In: *Biophysics of Photoreceptors and Photobehaviour of Microorganisms.* G. Colombetti (ed.). Lito Felici, Pisa.

Ascoli, C., Barbi, M., Frediani, C., and Mure, A. 1978. Measurement of *Euglena* motion parameters by laser light scattering. *Biophys. J.* 24: 585–599.

Ascoli, C. and Frediani, C. 1980. Quasi-elastic light scattering in the measurement of the motion of flagellated algae. pp.183–198. In: *Light Scattering in Liquids and Macromolecular Solutions.* V. Degiorgij, M. Corti, and M. Giglio. (eds.). Plenum Press, New York.

Ascoli, C. and Petracchi, D. 1991. Light scattering techniques in studying photoresponses. pp. 111–123. In: *Biophysics of Photoreceptors and Photomovement in Microorganisms.* F. Lenci, F. Ghetti, G. Colombetti, D.P. Häder, and P.S. Song (eds.). Plenum Press, New York.

Avron M. and Ben-Amotz A. (Eds.). 1992. *Dunaliella: Physiology, Biochemistry and Biotechnology.* CRC Press, Boca Raton, Abb Arbor, London, Tokyo.

Baas-Becking, L.G.M. 1931. Salt effects on swarmers of *Dunaliella viridis* Teod. *The Journal of General Physiology* 14:765–779. Obtenido de "http://es.wikipedia.org/wiki/Dunaliella_viridis"

Baas-Becking, L.G.M. 1930. Observations on *Dunaliella viridis* Teod. pp.102–114. In: *Contributions to Marine Biology.* Stanford Univ. Press.

Bachofen, R. 1980. Use of solar energy biological systems – possibilities with microorganisms. *Experientia* 36(12): 1429–33.

Baldi, C., Gambassi, F., Giordano, M., Grilli, A., Pietrini, R., Sbrilli, G., Bucci, M., and Armani, G. 1993. Integration of toxicity tests (algal and Microtox tests) and physicochemical analyses in studies of industrial wastewater. *Rivista Italiana d'Igiene* 53(1-2): 63–74.

Balnokin, G.M., Medvedev, A.V., and Bodnar, I.V. 1983. Systems of potassium transport in the cells of halophylic algae *Dunaliella. Plant Physiology* 30(5): 955–963. (In Russian)

Balayan, A.E. and Stom, D.I. 1988. pp.21–23. Method of biotesting immobilization of the cells of alga *Dunaliella.* In: *Methods of Biotesting Waters.* Chernogolovka (In Russian)

Barenboim, G.M. and Malenkov, A.G. 1986. pp.363. *Biologically Active Substances. New Principles of Searching.* Nauka, Moscow. (In Russian)

Barghigiani, C., Colombetti, G., Lenci, F., Banchetti, R., and Bizzaro, M.P. 1979. Photosensory transduction in *Euglena gracilis*: Effect of some metabolic drugs on the photophobic response. *Arch. Microbiol.* 120: 239–245.

Barghigiani, C., Ferrara, R., Ravera, O., and Serritti, A. 1981. p. 44. Biological effects under laboratory conditions.. In: *NATO-ASI Workshop on Trace Element Specification in Surface Waters and Its Ecological Implications.* Genova, 2–4 May.

Barghigiani, C., Maserti, E., Serritti, A., and Ferrara, R. 1983. pp. 634–637.Accumulation and release of Cu and Pb by *Dunaliella salina.* In: Proc. Intern. Conf. Heavy Metals in Environment, Heidelberg.

Barlow, S.B. and Cattolico, R.A. 1980. Fine structure of the scalecovered green flagellate *Mantoniella squamata* (Manton and Parcke) Desikachary. *Br. Phycol. J.* 15: 321–333.

Barltrop, J., Martin, B.B. and Martin, D.F. 1983. *Ptychodiscus brevis* as a model system for photodynamic action. *Microbios.* 37: 95–103.

Barsanti, L., Evangelista, V., Gualtieri, P., and Passarelli, V. 2004. 121/1–121/14. Photoreception in microalgae. In: *CRC Handbook of Organic Photochemistry and Photobiology* (2nd Edition), CRC Press LLC, Boca Raton, Fla.

Barsanti, L., Passarelli, V., Lenci, F., and Gualtieri, P. 1992. Elimination of photoreceptor (paraflagellar swelling) and photoreception in *Euglena gracilis* by means of the carotenoid biosynthesis inhibitor nicotine. *J. Photochem. Photobiol.* 13: 135–144.

Barsanti, L., Pasarelli, V., Lenci, F., Walne, P.L., Dunlap, J.R., and Gualtieri, P. 1993. Effects of hydroxylamine, digitonin and triton X-100 on photoreceptor (paraflagellar swelling) and photoreception in *Euglena gracilis. Vision Res.* 33: 2043–2050.

Barsanti, L., Pasarelli, V., Walne, P.L., and Gualtieri, P. 1997. *In vivo* photocycle of the *Euglena gracilis* photoreceptor. *Biophys. J.* 72: 545–553.

Batra, P.P. and Tollin, G. 1964. Phototaxis in *Euglena.* I. Isolation of the eye-spot granules and identification of the eye-spot pigments. *Biochimica et Biophysica Acta,* Specialized Section on Biophysical Subjects 79(2): 371–8.

Batschelet, E. *Circular Statistics in Biology.* 1981. Acad. Press, New York.

Bauernfeind, J.C. 1981. *Carotenoids, Colorants and Vitamin A Precursors, Technological and Nutritional Applications.* Acad. Press, New York.

Beardall, I. and Entwisle, L. 1984. Internal pH of the obligate acidophile *Cyanidium caldarium* Geitler (Rhodophyta). *Phycologia* 23 (3): 397–399.

Beardall, I., Heraud, Ph., Roberts, S., Shelly, K., and Stojkovic, S. 2002. Effects of UV-B radiation on inorganic carbon acquisition by the marine microalga *Dunaliella tertiolecta* (Chlorophyceae). *Phycologia* 41(3): 268–272.

170

Beckmann, M. and Hegemann, P. 1991. *In vitro* identification of rhodopsin in the green alga *Chlamydomonas. Biochemistry* 30: 3692–3697.

Begma, A.A., Vlasenko, V.V., Matskivskii, V.I., Pen'kov, F.M., Posudin, Y.I., and Frolov, G.V. 1989. Method of assessing toxic effect of chemicals contained in aqueous medium. URXXAF SU 1482887 A119890530.

Ben-Amotz, A. 1995. New mode of Dunaliella biotechnology: two-phase growth for β-carotene production. *Journal of Applied Phycology* 7(1): 65–8.

Ben-Amotz, A. 1996. Effect of low temperature on the stereoisomer composition of β-carotene in the halotolerant alga *Dunaliella bardawil* (Chlorophyta). *Journal of Phycology* 32(2): 272–275.

Ben-Amotz, A. and Avron, M. 1982. pp. 207–214. The potential use of *Dunaliella* for the production of glycerol, β-carotene and high protein feed. In: *Biosaline Research: A Look to the Future.* San-Pietro (Ed.). Plenum Publ. Corp., New York.

Ben-Amotz, A. and Avron, M. 1989. Chapter 4. The biotechnology of mass culturing *Dunaliella* for products of commercial interest, In: *Algal and Cyanobacterial Biotechnology.* R.C. Cresswell, T.A. Rees, and N. Shah (Eds.) London: Longman Sci. Tech. Press.

Ben-Amotz, A. and Avron, M. 1990. The beta-carotene accumulating halotolerant alga *Dunaliella*: a new venture in biotechnology. *Biotech Forum Europe* 7(1): 48–50, 52-3.

Ben-Amotz, A. and Avron, M. 1990. The biotechnology of cultivating the halotolerant alga *Dunaliella. Trends in Biotechnology* 8(5): 121–6.

Ben-Amotz, A., Gressel, J., and Avron, M. 1987. Massive accumulation of phytoene induced by norflurazon in *Dunaliella bardawil (Chlorophyceae)* prevents recovery from photoinhibition. *J. Phycol.* 23: 176.

Ben-Amotz, A., Gressel, J., and Avron, M. 1988. Stereoisomers of β-carotene and phytoene in the alga *Dunaliella bardawil. Plant Physiol.* 86: 1286.

Ben-Amotz, A., Katz, A., and Avron, M. 1982*a*. Accumulation of β-carotene in halotolerant algae: purification and characterisation of β-carotene-rich globules from *Dunaliella bardawil (Chlorophyceae). J. Phycol.* 18: 529.

Ben-Amotz, A., Polle J.,and Rao S. (Eds.). 2009. *The alga Dunaliella: Biodiversity, Physiology, Genomics and Biotechnology.* Science Publishers, Enfield, Jersey, Plymouth..

Ben-Amotz, A., Shaish, A., and Avron, M. 1991. The biotechnology of cultivating *Dunaliella* for production of β-carotene rich algae. Bioresource Technology 38(2–3): 233–5.

Ben-Amotz, A., Sussman, I., and Avron, M. 1982*b*. Glycerol production by *Dunaliella. Experientia.* 38 (1): 49–52.

Bendix, S.W. 1960. Phototaxis. *The Botanical Review* 26:145–208.

Benedetti, P.A. and Checcucci, A. 1975. Paraflagellar body (PFB) pigments studied by fluorescence microscopy in *Euglena gracilis. Plant Sci. Lett.* 26: 315–318.

Bensasson, R.V. 1975. pp. 146–163. Spectroscopic and biological properties of carotenoids. In: *Biophysics of Photoreceptors and Photobehaviour of Microorganisms.* Proc.of the Intern. School of the CNR of Italy, Badia Fiesolana (Firenze), 1–5 September, 1975.

Berg, H.C. 1985. Physics of bacterial chemotaxis. pp.19–30. In: *Sensory Perception and Transduction in Aneural Organisms.* Colombetti, G., Lenci, F., and Song, P.S. (Eds.) Plenum Press, New York.

Bernstein, N.A. (1896–1966) Nikolai Bernstein. 25 September 2009. At: <http://en.wikipedia.org/wiki/Nikolai_Bernstein>.

Bhattacharya, D., Steinkotter, J., and Melkonian, M. 1993. Molecular cloning and evolutionary analysis of the calcium-modulated contractile protein, centrin, in green algae and land plants. *Plant Mol. Biol.* 23 (6): 1243–1254.

Bhumibhamon, O., Sittiphuprasert, U., Boontaveeyuwat, N., and Praiboon, J. 2003. The optimum use of salinity, nitrate and pond depth for β-carotene production of *Dunaliella salina. Kasetsart Journal: Natural Sciences* 37(1): 84–89.

*Biomechanics*: Galileo, Harvey, Borelli, Young and Helmholtz. 5 January 2010. At: "http://yunus.hacettepe.edu.tr/~saritan/biom_hist.htm"

Birkbeck, T.E., Stewart, K.D., and Mattox, K.R. 1974. The cytology and classification of *Schizomeris leibleinii* (Chlorophyceae) II. The structure of quadriflagellate zoospores. *Phycologia* 13: 71–79.

Bischof, K., Janknegt, P.J., Buma, A.G.J., Rijstenbil, J.W.; Peralta, G., and Breeman, A.M. 2003. Oxidative stress and enzymatic scavenging of superoxide radicals induced by solar UV-B radiation in *Ulva* canopies from southern Spain. *Scientia Marina* 67(3): 353–359.

Bischof, K., Kraebs, G., Wiencke, C., and Hanelt, D. 2002. Solar ultraviolet radiation affects the activity of ribulose-1,5-bisphosphate carboxylase-oxygenase and the composition of photosynthetic and xanthophyll cycle pigments in the intertidal green alga *Ulva lactuca* L. *Planta* 215(3): 502–509.

Bischof, K., Peralta, G., Krabs, G., Van de Poll, W.H., Perez-Llorens, J.L., and Breeman, A. M. 2002. Effects of solar UV-B radiation on canopy structure of *Ulva* communities from southern Spain. *Journal of Experimental Botany* 53(379): 2411-2421.

Bjorn, L.O., Sundqvist, C., and Oquist, G. 2007. A tribute to Per Halldal (1922–1986), a Norwegian photobiologist in Sweden. *Photosynthesis research* 92(1): 7–11.

Blakefield, M.K. and Calkins, J. 1992. Inhibition of phototaxis in *Volvox aureus* by natural and simulated solar ultraviolet light. *Photochem. Photobiol.* 55(6): 867–72.

Bobkova, O.N. 1997. The formation of lipid hydroperoxides as a test indicator of the sea water toxicity. *Gidrobiologicheskii Zhurnal* 33(5): 87–92 (In Russian).

Born, M. and Wolf, E. *Principles of Optics*, Nauka, Moscow, 1973. (In Russian).

Borovkov A.B. Green microalga *Dunaliella salina* Teod. 2005. *Sea Ecology* 67: 5–17 (In Russian).

Borowitzka, M.A. 1990. The mass culture of *Dunaliella salina*. In: *Regional seafarming development and demonstration project ras/90/002,* 27–31 august 1990, Cebu city, Philippines.

Borowitzka, L.J. 1990. Status of the Australian Algal Biotechnology Industry in 1990. *Austr. J. Biotechnol.* 4(4): 239–240, 250.

Borowitzka, M.A. 2005. Cultivation of microalgal from village level to industrial scale. *Phycologia* 44 (4):11–12.

Borowitzka, M.A. and Borowitzka, L.J. 1988*a. Dunaliella. Microalgal biotechnology.* Borowitzka, M.A. and Borowitzka, L.J. (Eds.) Cambridge, pp. 27–58.

Borowitzka, M.A. and Borowitzka, L.J. 1988*b.* Limits to growth and carotenogenesis in laboratory and large-scale outdoor cultures of D. salina. pp. 139–150. In: *Algal Biotechnology,* Stadler, T., Mollion, J., Berdus, M.C., Karamanos, Y., Morvan, H., and Christiane, D. (Eds.) Elsevier Applied Science, Barking.

Borowitzka, L.J. and Borowitzka, M.A. 1989*a.* β-carotene (Provitamin A) production with algae. pp. 15–26. In: *Biotechnology of Vitamins, Pigments and Growth Factors*, Vandamme, E.J. (Ed.) Elsevier Applied Science, London.

Borowitzka, L.J. and Borowitzka, M.S. 1989*b*. Industrial production: methods and economics. pp. 294–316. In: *Algal and Cvanobacterial Biotechnology*, Cresswell, R.C., Rees, T.A.V., and Shah, N. (Eds.) Longman Scientific, London.

Borowitzka, L.J. and Borowitzka, M.A. 1990. Commercial production of β-carotene by *Dunaliella salina* in open ponds. *Bull. Mar. Sci.* 47(1): 244–252.

Borowitzka, L.J., Borowitzka, M.A., and Moulton, T.P. 1984. The mass culture of *Dunaliella salina* for fine chemicals: from laboratory to pilot plant. *Hydrobiologia* 116/117: 115–134.

Borowitzka, L.J., Moulton, T.P., and Borowitzka, M.A. 1986. Salinity and the commercial productivity of beta-carotene from *Dunaliella salina. Nova Hedw.* 83: 224–229.

Bosma, R. and Wijffels, R.H. 2003. Marine biotechnology in education: a competitive approach. *Biomolecular Engineering* 20(4-6): 125–131.

Bound, K.E. and Tollin G. 1967. Phototactic response of *Euglena gracilis. Nature* 216: 111–114.

Bovee, E.C. 1968. Movement and locomotion of *Euglena.* pp.143–168. In: *The Biology of Euglena,* Vol. III Physiology, D.E. Buetow (Ed.), Acad. Press, New York.

Bovee, E.C. and Jahn, T.L. 1972. Theory of piezoelectric activity and ion-movements in the relation of flagellar structures and their movements to the phototaxis of *Euglena. Journal of Theoretical Biology* 35(2): 259–76.

Braginsky, L.P. 1986. Some regularities and mechanisms of response of freshwater ecosystem on action of pesticides and surface-active substances. *Experimental aquatic toxicology* 2: 12–18. (In Russian)

Branton, D., Bullivant, S., Gilula, N.B., Karnovsky, M.J., Mühlethaler, K., Northcote, D.H., Packer, L., Satir, B., Satir, P., Speth, V., Staeheilin, L.A., Steere, R.L., and Weinstein, R.S. 1975. Freeze-etching nomenculature. *Science* 190: 54–56.

Braun, F.J. and Hegemann, P. 1999. Direct measurement of cytosolic calcium and pH in living *Chlamydomonas reinhardtii* cells. *Eur. J. Cell Biol.* 78: 199–208.

Braune, W., Häder, D.P., and Hagen, C. 1994. Copper toxicity in the green alga *Haematococcus lacustris*: Flagellates become blind by copper (II) ions. *Cytobios* 77(308): 29–39.

Bray, D.F., Nakamura, K., Costerton, J.W., and Naganaar, E.B. 1974. Ultrastructure of *Chlamydomonas eugametos* as revealed by freeze-etching: cell wall plasmalemma and chloroplast membrane. *J. Ultrastr. Res.* 47: 125–141.

Britton, G. 1986. *Biochemistry of natural pigments.* Moscow. Mir (In Russian)

Britton, G. 1988. *Biosynthesis of carotenoids. Plant Pigments.* T.W. Goodwin (Ed.), Acad. Press, New York, Chapter 3, pp.133–180.

Brodhun, B. and Häder, D.P. 1990. Photoreceptor proteins and pigments in the paraflagellar body of the flagellate *Euglena gracilis. Photochemistry and Photobiology* 52(4): 865–71.

Brodhun, B. and Häder, D.P. 1993. UV-induced damage of photoreceptor proteins in the paraflagellar body of *Euglena gracilis. Photochemistry and Photobiology* 58(2): 270–4.

Brodhun, B. and Häder, D.P. 1995. A novel procedure to isolate the chromoproteins in the paraflagellar body of the flagellate *Euglena gracilis. J. Photochem.Photobiol.,* B: Biology 28(1): 39–45.

Brodhun, B., Neumann, R., Hertel, R., and Häder, D.P. 1994. Riboflavin-binding sites in the flagella of *Euglena gracilis* and *Astasia longa. J.Photochem.Photobiol.,* B: Biology 23(2–3): 135–9.

Brokaw, C.J., Luck, D.J.L., and Huang, B. 1982. Analysis of the movement of *Chlamydomonas* flagella: the function of the radial-spoke system is revealed by comparison of wild-type and mutant flagella. *J. Cell Biol.* 92(3): 722–32.

Buchheim, M.A., Buchheim, J.A., and Chapman, R.L. 1997. Phylogeny of *Chloromonas* (*Chlorophyceae*): a study of 18S ribosomal RNA gene sequences. *J. Phycol.* 33: 286–293.

Buchheim, M.A., Turmel, M., Zimmer, E.A., and Chapman, R.L. 1990. Phylogeny of *Chlamydomonas* (*Chlorophyta*) based on cladistic analysis of nuclear 18S rRNA sequence data. *J. Phycol.* 26: 689–696.

Buder, J. 1917. Zur Kenntnis der phototaktischen Richtungsbewegungen. *Jhb. Wiss. Bot.* 58: 105-220.

Burkholder, P.R.1933. Movement in the Cyanophyceae: effect of pH upon movement of *Oscillatoria*. *J. Gen. Physiol.* 16: 875-881.

Burr, A.H. 1984. Photomovement behaviour in simple invertebrates. pp. 179–215. In: *Photorecept. and Vision Invertebr.* Proc. NATO Adv. Study Inst., Lennoville New York, London.

Butterworth F.M., Corcum L.D., Guzman-Rincon J. 1995. *Biomonitors and Biomarkers as Indicators of Environmental Change. A Handbook.* Plenum Press, New York and London.

Cai, A., Li, W., and Zheng, A. 1999. Biological effects of organophosphorus pesticide on marine microalgae. *Xiamen Daxue Xuebao, Ziran Kexueban* 38(4): 589–593.

Cai, H., Tang, X., Zhang, P., Dong, D., and Qu, L. 2005. Effects of UV-B radiation on the growth interaction of *Ulva pertusa* and *Alexandrium tamarense*. *J.Env. Sci.* (Beijing, China) 17(4): 605–610.

Campo, J.A., Garcia-Gonzalez, M., and Guerrero, M.G. 2007. Outdoor cultivation of microalgae for carotenoid production: current state and perspectives. *Applied Microbiology and Biotechnology* 74(6): 1163–1174.

Cantatore, G., Ascoli, C., Colombetti, G., and Frediani, C. 1989. Doppler velocimetry measurements of phototactic response in flagellated algae. *Bioscience reports* 9(4): 475–80.

Castiello, A., Mikolajczyk, E., and Passarelli, V. 1980. pp. 243–5. Photosensory transduction in *Euglena gracilis*: ineffectiveness of some protonophores. In: *Dev. Biophys. Res.*, Proc. Congr. Meeting Borsellino, A., Omodeo, P., and Strom, R. (Eds.) 1979.

Cavalier-Smith T. 1981. Eukaryote Kingdoms: seven or nine? *BioSystems* 14: 461–481.

Cecconi, C., Ascoli, C., and Petracchi, D. 1996. Step-up photophobic responses of the unicellular alga *Haematococcus pluvialis* and their interpretation in terms of photoreceptive apparatus characteristics. *J. Photochem.Photobiol.* B: Biology. 33(3): 201–209

Celekli, A. and Doenmez, G. 2006. Effect of pH, light intensity, salt and nitrogen concentrations on growth and β-carotene accumulation by a new isolate of *Dunaliella* sp. *World Journal of Microbiology & Biotechnology*, 22(2):183–189.

Chai, Y., Wang, T., and Xue, L., 2004. Recent progress in a new bioreactor: *Dunaliella salina*. *Zhongguo Shengwu Gongcheng Zazhi* 24(2): 30–33.

Chamberlain, A.H.L. 1976. Algal settlement and secretion of adhesive materials. pp. 424–426. Proc. 3rd Intern. Biodegrad. Symp. Kingston, 1975, London.

Chardard, R. 1987. L'infrastructure du plasmalemme de *Dunaliella bioculata* (Algue verte). Mise en évidence d'un cell-coat; essai de localisation des charges négatives. *Crypt. Algol.* 8 (3): 173–189.

Chardard, R. 1990. Nouvelles observations sur la structure et la composition du cell-coat de *Dunaliella bioculata* (Algue verte). *Cryptogam. Algol.* 11(2): 137–152.

Checucci, A., Colombetti, G., Ferrara, R., and Lenci, F. 1976. Action spectra for photoaccumulation of green and colorless *Euglena*: Evidence for identification of receptor pigments. *Photochem. Photobiol.* 23: 51–54.

Checcucci, G., Sgarbossa, A., and Lenci, F. 2004. Photomovements of microorganisms: An introduction. pp.120/1–120/10. In: *CRC Handbook of Organic Photochemistry and Photobiology* (2nd Edition) Istituto di Biofisica CNR, Fr., Horspool, W. and Lenci, F. (Eds.).

Chen, L. and Wang, G. 2003. Application of photobioreactors to cultivation of microalgae. *Journal of Harbin Institute of Technology* (English Edition) 10(2): 221–224.

Chen, F. and Jiang, Y. (Eds.). 2001. *Algae and Their Biotechnological Potential.* Proceedings of the 4th Asia-Pacific Conference on Algal Biotechnology held 3–6 July 2000 in Hong Kong., Dordrecht ; Boston : Kluwer Academic Publishers. 305 pp.

Chen, Z., Yang, M., Dang, J., Wu, Y., Hu, H., and Liu, J. 2008. Method for extracting carotenoid and edible glycerol from *Dunaliella salina. Faming Zhuanli Shenqing Gongkai Shuomingshu*, 7 pp.

Chung, Y.H., Park, Y.M., Moon, Y.J., Lee, E.M., and Choi, J.S. 2004. Photokinesis of cyanobacterium *Synechocystis* sp. PCC 6803. *Journal of Photoscience* 11(3): 89–94.

Cifuentes, A.S., Gonzalez, M.A., and Parra, O.O. 1996. The effect of salinity on the growth and carotenogenesis in two Chilean strains of *Dunaliella salina* Teodoresco. *Biological research* 29(2): 227–36.

Clayton, R.K. 1964. Phototaxis in microorganisms. 2: 51–77. In: *Photophysiology*, A.C. Giese (Ed.). Acad. Press, New York, London.

Clegg, M.R. 2005. Behavioural responses of phytoplanktonic flagellates to the quantity and quality of light and their ecological implications. *Phycologia.* 44(4): 22.

Cohn, F. 1865*a*. Ueber die Gesetze der Bewegung Mikroskopischer Tiere und Pflanzen unter Einfluss des Lichtes. *Jber. Schles. Ges. Vaterl. Kult.* 42: 35.

Cohn, F. 1865*b*. *Chlamydomonas marina* Cohn. *Hedwigia* 4: 97.

Colombetti, G., Ghetti, F., Lenci, F., Polacco, E., Posudin, Y., and Campani, E. 1980. Microspettrometria laser di fotopigmenti *"in vivo".* pp. 223–227. II Congr. Naz. *"Elettronica quantistica e plasmi"*, 20–22 Maggio 1980. Palermo.

Colombetti, G., Ghetti, F., Lenci, F., Polacco, E., Posudin, Y., and Campani, E. 1981. Laser microspectrofluorometry *in vivo* of photopigments. *Quantum Electronics* 8(12): 2680–2683. (In Russian)

Colombetti, G., Häder, D.P., Lenci, F., and Quaglia, M. 1982. Phototaxis in *Euglena gracilis*: effect of sodium azide and triphenyl-methyl phosphonium ions on the photosensory transduction chain. *Curr. Microbiol.* 7: 281–284.

Colombetti, G. and Lenci, F. 1980. Identification and spectroscopic characterisation of photoreceptor pigments. pp. 173–188. In: *Photoreception and Sensory Transduction in Aneural Organisms.* F. Lenci and G. Colombetti (Eds.).

Colombetti, G. and Lenci, F. 1982. Photoreception and sensory responses in microorganisms. *Medicine, Biologie, Environment.* 10: 319 325.

Colombetti, G., Lenci, F., and Diehn, B. 1982. Responses to photic, chemical and mechanical stimuli. 3: 169–195. In: *The Biology of Euglena.* Buetow (Ed.).Acad. Press, New York.

Colombetti, G. and Marangoni, R. 1991. Mechanisms and strategies of photomovement in Flagellates. pp. 53–71. In: *Biophysics of Photoreceptors and Photomovement in Microorganisms.* F. Lenci, F. Ghetti, G. Colombetti, D.P. Häder, and P.S. Song (Eds.), Plenum Press, New York.

Colombetti, G. and Petracchi, D. 1989. Photoresponse mechanisms in flagellated algae. *Crit. Rev. Plant. Sci.* 8: 309–355.

Cooper, D.G. and Zajic J.E. 1980. Surfaceactive compounds from microorganisms. *Adv. Appl. Microbiol.* 26: 240–244.

Cordi, B., Donkin, M.E., Peloquin, J., Price, D.N., and Depledge, M.H. 2001. The influence of UV-B radiation on the reproductive cells of the intertidal macroalga, *Enteromorpha intestinalis. Aquatic Toxicology* p 56(1): 1–11.

Cosens, D.J. and Nultsch, W. 1975. Phototaxis and photokinesis. pp. 29–90. In: *Primitive Sensory and Communication Systems: The Taxes and Tropisms of Microorganisms and Cells.* M.J. Carlile (Ed.). Acad. Press, New York, San Francisco.

Crawley, J.C.W. 1966. Some observations on the fine structure of the gametes and zygotes of *Acetabularia. Planta* 69: 365–376.

Crawley, J.C.W. 1970. The fine structure of the gametes and zygotes of *Acetabularia.* pp. 73–83. In: *Biology of Acetabularia.* J. Brachet and S. Bonotto (Eds.). Acad. Press, New York.

Creutz, C. and Diehn, B. 1976. Motor responses to polarized light and gravity sensing in *Euglena gracilis. J. Protozool.* 23: 552-556.

Cubeddu, R., Ramponi, R., and Taroni, P. 1991. Biological applications of time-gated fluorescence spectroscopy. *NATO ASI Series, Series B: Physics, 252* (Laser Syst. Photobiol. Photomed.), pp. 217–27.

Darwin, Charles (1872), *On The Origins of Species,* Sixth Edition. On the Origin of Species. 2 January 2010. At: <http://www.darwiniana.org/eyes.htm>; <http://en.wikipedia.org/wiki/Origin_of_species>.

Davydov, S.A. and Suprun, A.D. 1974. Configurational changes and optical properties of α-spiral protein molecules. *Uspekhi phys. nauk* 19: 44–50. (In Russian).

Deason, T.R.and Floyd, G.L. 1987. Comparative ultrastructure of three species of *Chlorosarcina* (*Chlorosarcinaceae, Chlorophyta*). *J. Phycol.* 23: 187–195.

De Busk, T.A. and Ryther, J.H. 1984. Effects of seawater exchange, pH and carbon supply on the growth of *Gracilaria tikvahiae* (*Rhodophyceae*) in largescale cultures. *Bot. Mar.* 27(8): 357–362.

Del Campo, J.A, Garcia-Gonzalez, M., and Guerrero, M.G 2007. Outdoor cultivation of microalgae for carotenoid production: current state and perspectives. *Applied microbiology and biotechnology* 74(6): 1163–74.

DeLorenzo, M.E. and Serrano, L. 2003. Individual and mixture toxicity of three pesticides; atrazine, chlorpyrifos, and chlorothalonil to the marine phytoplankton species *Dunaliella tertiolecta. Journal of environmental science and health* Part. B. Pesticides, food contaminants, and agricultural wastes 38(5): 529–38.

DeLorenzo, M.E. and Serrano, L. 2006. Mixture toxicity of the antifouling compound irgarol to the marine phytoplankton species *Dunaliella tertiolecta. Journal of environmental science and health* Part. B. Pesticides, food contaminants, and agricultural wastes 41(8): 1349–60.

DeLorenzo, M.E. and Fleming, J. 2008. Individual and Mixture Effects of Selected Pharmaceuticals and Personal Care Products on the Marine Phytoplankton Species *Dunaliella tertiolecta. Archives of environmental contamination and toxicology* 54(2): 203–10.

Derby, S.B. and Ruber, E. 1970. Primary production: depression of oxygen evolution in algal cultures by organophosphorus insecticides. *Bulletin of Environmental Contamination and Toxicology* 5(6): 553–8.

Di Pasquale, M.A., Bizzaro, M.P., and Ciliberti, M. 1980. Photomotile responses in *Ochromonas danica*. pp. 237–41. In: *Dev. Biophys. Res.* [Proc. Congr.], Borsellino, A., Omodeo, P., and Strom, R. (Eds.), Meeting Date 1979.

Diao, S. 2007. Research of comparison of poison in algae acute toxicity monitor. *Huanjing Kexue Yu Guanli* 32(9): 139–141.

Diehn, B. 1969. Action spectra of the phototactic responses in *Euglena*. *Biochimica et Biophysica Acta,* General Subjects 177(1): 136–43.

Diehn, B. 1970. Mechanism and computer stimulation of the phototactic accumulation of *Euglena* in a beam of light. *Photochem. Pholobiol.* 11: 407–413.

Diehn, B. 1973. Phototaxis and sensory transduction in *Euglena*. *Science* 181: 1009–1015.

Diehn B. 1979. Photic responses and sensory transduction in Protists. pp. 23–66. In: *Handbook of Sensory Physiology, Comparative Physiology and Evolution of Vision in Invertebrates.* VII/6A, H. Autrum (Ed.), Springer Verlag, Berlin.

Diehn, B. 1980. Photomovement and photosensory transduction in microorganisms. *Photochemistry and Photobiology* 31(6): 641–4.

Diehn, B., Feinleib, M., Haupt, W., Hildebrand, E., Lenci, F., and Nultsch, W. 1977. Terminology of behavioural responses of motile microorganisms. *Photochem. Photobiology.* 26: 559–560.

Diehn, B. and Kint, B. 1970. Flavin nature of the photoreceptor molecule for phototaxis in *Euglena*. *Physiological Chemistry and Physics* 2(5): 483–8.

Diehn, B. and Tollin, G. 1967. Phototaxis in *Euglena*. IV. Effect of inhibitors of oxidative and photophosphorylation on the rate of phototaxis. *Archives of Biochemistry and Biophysics* 121(1): 169–77.

Dmitriev, Y.V., Klimova E.V., and Boyko V.V. 2005. Application of *Dunaliella viridis* Teod. as test-system during estimation of pathological states of organism. pp.47–48. In: *Actual problems of modern algology*. Materials of III Int. Conf., Kharkov, 20–23 April.

Dodge, J.D. 1973. *The fine structure of algal cells*. Acad. Press, London.

Dodge, J.D. and Crawford, R.M. 1969. Observations on the fine structure of the eye-spots and associated organelles in the dinoflagellate *Glenodinium foliaceum*. *J. Cell Sci.* 5: 479–493.

Doehler, G., Hagmeier, E., and David, C. 1995. Effects of solar and artificial UV irradiation on pigments and assimilation of 15N ammonium and 15N nitrate by macroalgae. *Journal of Photochemistry and Photobiology*, B: Biology 30(2-3): 179–87.

Dolle, R. and Nultsch, W. 1988. Effects of calcium ions and of calcium channel blockers on galvanotaxis of *Chlamydomonas reinhardtii*. *Botanica Acta* 101(1): 18–23.

Dolle, R., Pfau, J., and Nultsch, W. 1987. Role of calcium ions in motility and phototaxis of *Chlamydomonas reinhardtii*. *J. Plant Physiol.* 126: 467–473.

Donk E., Hessen D.O. 1996. Loss of flagella in the green alga *Chlamydomonas reinhardtii* due to *in situ* UV-exposure. 60 (Suppl. 1): 107–112. In: *Underwater Light and Algal Photobiology*. F.L. Figueroa, C. Jimenez, J.L. Perez-Llorens, and F.X. Niell (Eds.), Malaga: Scientia Marina.

Doughty, M.J. 1991. Mechanism and strategies of photomovement in protozoa. 73–101. In: *NATO ASI Series*, Series A: Life Sciences, 211 (Biophys. Photorecept. Photomovements Microorg.).

Doughty, M.J. and Diehn, B. 1979. Photosensory transduction in the flagellated alga, *Euglena gracilis*: I. Action of divalent cations, $Ca^{2+}$ antagonists and $Ca^{2+}$ ionophore on motility and photobehaviour. *Biochem. Biophys. Acta.* 588: 148–168.

Doughty, M.J. and Diehn, B. 1980. Flavins as photoreceptor pigments for behavioral responses in motile microorganisms, especially in the flagellated alga, *Euglena* sp. *Structure and Bonding* 41: 45–70.

Doughty, M.J. and Diehn B. 1982. Photosensory transduction in the flagellated alga, *Euglena gracilis*: III. Induction of $Ca^{2+}$-dependent responses by monovalent cation ionophores. *Biochem. Biophys. Acta.* 682: 32–43.

Doughty, M.J., Diehn, B., and Grieser, R. 1980. Photosensory transduction in the flagellated alga, *Euglena gracilis*: II. Evidence that blue light effects alteration in $Na^+/K^+$ permeability of the photoreceptor membrane. *Biochem. Biophys. Acta.* 602: 10–23.

Drews, G. 2005. Contributions of Theodor Wilhelm Engelmann on phototaxis, chemotaxis, and photosynthesis. *Photosynthesis research* 83(1): 25–34.

Drokova, I.G. 1961. Alga *Dunaliella salina* Teod. as the source of β-carotene. *Ukr. Bot. J.* 18(4): 110–112 (In Ukrainian).

Drokova, I.G. 1969. Investigation of algae β-carotene in algae. *Ukr. Bot. J.* 17(2): 39–42 (In Ukrainian).

Druez, D., Marano, F., Calvayrac, R., Volochine, B., and Soufir, J.C. 1989. Effect of gossypol on the morphology, motility and metabolism of a flagellated protist, *Dunaliella bioculata. Journal of Submicroscopic Cytology and Pathology* 21(2): 367–74.

Dujardin, F. 1841. *Histoire Naturelle des Zoophytes (Infusoires).* Paris.

Dunal, F. 1838. Extrait d'un mémoire sur les algues qui colorent en rouge certains eaux des marais salants méditerranéens. *Ann. Sc. Nat. Bot. 2 Ser.* 9: 172.

Duvnjak, Z., Cooper, D.G., and Kosaric, N. 1982. Production of surfactant by *Arthrobacter paraffineus* ATCC 19558. *Biotechnol. and Bioeng.* 24(24): 170–174.

Ebrey, T.G. 2002. A new type of photoreceptor in algae. Proceedings of the National Academy of Sciences of the United States of America, 99(13): 8463–8464.

Ehrenberg C.G. Christian Gottfried Ehrenberg. 22 December 2009. At: "http://en.wikipedia.org/wiki/Christian_Gottfried_Ehrenberg"

Ehrenberg, C.G. 1838. *Die Infusoriensthierchen als vollkommen Organismen.* Leipzig.

Ekelund, N.G.A. 1993. The effect of UV-B radiation and humic substances on growth and motility of the flagellate, *Euglena gracilis. Journal of Plankton Research* 15(6): 715–22.

Ekelund, N.G.A. 1996. Effects of protein synthesis inhibitors on photoinhibition by UV-B (280-320 nm) radiation in the flagellate *Euglena gracilis.* 60 (Suppl.1): 95–100. In: *Underwater Light and Algal Photobiology.* F.L. Figueroa, C. Jimenez, J.L. Perez-Llorens and F.X. Niell (Eds.) Malaga: Scientia Marina.

Ekelund, N.G.A. and Bjorn, L.O. 1990. Ultraviolet radiation stress in dinoflagellates in relation to targets, sensitivity and radiation climate. Proceedings of Workshop, Scripps Institution of Oceanography, University of California, San Diego, La Jolla, 1990.

Ekelund, N.G.A., Dolle, R., and Nultsh, W. 1988. Phototactic responses of *Chlamydomonas reinhardtii* in electric field. *Physiol. Plant.* 73: 265–270.

Elenkin, A.A. 1925. *Biology of lower plants.* Leningrad (In Russian).

Emerson, P.L. 1980. Fast Fourier transform fundamentals and applications. *Creative Computing* 6: 58–63.

Engelmann, T.W. (1843–1909) Engelmann, T.W. (1843-1909) About: Theodor Wilhelm Engelmann. 3 June 2006. At: <http://en.wikipedia.org/wiki/Thomas_Engelmann]; 5 January 2010. http://dbpedia.org/page/Theodor_Wilhelm_Engelmann.

Engelmann T.W. 1882*a.* Über die Licht und Fasbenperzeption niederster Organismen. *Pflugers Arch. Ges. Physiol.* 29: 387–400.

Engelmann, T.W. 1882*b*. Über Sauerstoffausscheidung von Pflanzellen im Mikrospectrum. *Bot. Ztg.* 26: 419–426.

Enhuber, G. and Gimmler, H. 1980. The glycerol permeability of the plasmalemma of the halotolerant green alga *Dunaliella parva (Volvocales). J. Phycol.* 16(4): 524–532.

En-Zanfeily, H.T. and Nawar, S.S. 1980. Degradation of anionic surfactants and the effect on bacterial indices of pollution. *Zbl. Bacteriol. Parasitenk., Infektionskrankh. und Hyg. 2-Naturwiss. Abt.* 135(8): 486–488.

Ermilova, E.V., Zalutskaya, Z.M., Lapina, T.V., Nikitin, M.M., and Gromov, B.V. 2000. Flagellum functioning in the control of Chlamydomonas reinhardtii tactic movements. *Russian Journal of Plant Physiology* (Translation of Fiziologiya Rastenii - Moscow) 47(5): 660–663.

Esquivel, D.M.S. and de Barros, H.G.P.L. 1986. Motion of magnetotactic microorganisms. *Exp. Biol.* 121: 153.

Eswaran, K., Ganesan, M., Periyasamy, C., Rao, and P.V. Subba. 2002. Effect of ultraviolet-B radiation on biochemical composition of three *Ulva* species (Chlorophyta) from southeast coast of India. *Indian Journal of Marine Sciences* 31(4): 334–336.

Ettl, H. 1980. *Grundriß der allgemein Algologie.* Jena: Fischer, 549 S.

Ettl, H. 1981. Die neue Klasse *Chlamydophyceae*, eine natürliche Grouppe der Grünalgen (*Chlorophyta*). *Plant Syst. Evol.* 137: 107–126.

Ettl, H. 1983. *Chlorophyta* I. *Phytomonadina* Süßwasserflora von Mitteleuropa. Bd.9. Jena: Fischer, 807 S.

Ettl, H. and Green, J.C. 1973. *Chlamydomonas reginae* sp. nov. (*Chlorophyceae*), a new marine flagellate with unusual chloroplast differentiation. *J. Mar. Biol. Ass. U.K.* 53: 975–985.

Ettl, H. and Manton, I. 1964. Die feinere Struktur von *Pedimonas minor* Korschikoff. *Nova Hedwigia* 8: 421–451.

Ettl, H. and Moestrup, Ø. 1980. Light and electron microscopical studies on *Hafniomonas* gen. nov. (*Chlorophyceae, Volvocales*), a genus resembling *Pyramimonas* (*Prasinophyceae*) *Pl. Syst. Evol.* 135: 177–210.

Evdokimov, M.V, Priezzhev, A.V., Romanovskii, I.M., and Cherniaeva, E.B. 1982. Study of protoplasm mobility in the cells of *Nitella* by means of optical Doppler spectroscopy. *Biofizika* 27(5): 918–20.

Evtodienko, Y.V. 1985. *Problems of bioenrgetics.* Moscow, Znanie. (In Russian).

Eyden, B.P. 1975. Light and electron microscope study of *Dunaliella primolecta* Butcher (*Volvocida*). *J. Protozool.* 22(3): 336–344.

Fabregas, J., Patino, M., Arredondo-Vega, B.O., Tobar, J.L., and Otero, A. 1995. Renewal rate and nutrient concentration as tools to modify productivity and biochemical composition of cyclostat cultures of the marine microalga *Dunaliella tertiolecta. Applied Microbiology and Biotechnology* 44(3-4): 287–92.

Famintsin, A.S. 1866. Action of light on algae and some other close to them organisms. St.-Petersbourg (In Russian).

Famintzin, A.S. 1867*a*. Die Wirkung des Lichtes auf Algen und einige andere ihnen nahe verwandte Organismen. *Jb. Wiss. Bot.* 6: 1–48.

Famintzin, A.S. 1867*b*. Die Wirkung des Lichtes auf die Bewegung der *Chlamydomonas pulvisculus* Ehr., *Euglena viridis* Ehr. und *Oscillatoria insignis* Tw. *Bull. Acad. Sci.* (St.-Petersbourg.) 10: 534–548.

Famimtsin, A.S. 1887*a*. Action of light and darkness on the location of chlorophyll grains in the leaves of *Mnium. Naturalist* 13-20: 194–197. (In Russian).

Famimtsin, A.S. 1887*b*. *A Text-Book of Plant Physiology*, St.-Petersbourg (In Russian).

Fazeli, M.R., Tofighi, H., Samadi, N., and Jamalifar, H. 2006. Effects of salinity on β-carotene production by *Dunaliella tertiolecta* DCCBC26 isolated from the Urmia salt lake, north of Iran. *Bioresource Technology* 97(18): 2453–2456.

Feinleib, M. 1977. Photomovement in microorganism's strategies of response. *Research in Photobiology*, A. Castellani (Ed.), Plenum Publ. Co., New York, pp. 71–84.

Feinleib, M. 1978. Photomovement of microorganisms. *J. Photochem. Photobiology* 27: 849–854.

Feinleib, M. 1980. Photomotile responses in flagellates. pp. 45–68. In: *Photoreception and sensory transduction in aneural organisms*, F. Lenci and G. Colombetti (Eds.), Plenum Press, New York.

Feinleib, M. 1985. Behavioural studies of free-swimming photoresponsive organisms. pp. 119–146. In: *Sensory perception and transduction in aneural organisms,* G. Colombetti and P.S. Song (Eds.), Plenum Press Corp., New York.

Feinleib, M. and Carry, G.M. 1967. Methods for measuring phototaxis of cell population and individual cells. *Physiol. Plant.* 20: 1083–1095.

Figueroa, F.L., Nygard, C., Ekelund, N., and Gomez, I. 2003. Photobiological characteristics and photosynthetic UV responses in two *Ulva* species (Chlorophyta) from southern Spain. *Journal of Photochemistry and Photobiology*, B: Biology 72(1-3): 35–44.

Filenko, O.F. 1988. *Aquatic toxicology.* pp.156. Chernogolovka. Moscow State Univ. Publ. (In Russian).

Flannery, J.G. and Greenberg, K.P. 2006. Looking within for vision. *Neuron* 50(1): 1–3.

Flores-Moya, A., Hanelt, D., Figueroa, F.L., Altamirano, M., Vinegla, B., and Salles, S. 1999. Involvement of solar UV-B radiation in recovery of inhibited photosynthesis in the brown alga *Dictyota dichotoma* (Hudson) Lamouroux. *J. Photochem. Photobiology* 49: 129–135.

Flores-Moya, A., Posudin, Y.I., Fernandez, J.A., Figueroa, F.L., and Kawai, H. 2002. Photomovement of the swarmers of the brown algae *Scytosiphon lomentaria* and *Petalonia fascia*: effect of photon irradiance, spectral composition and UV dose.

*J. Photochem.Photobiol.*, B: Biology 66(2): 134–140.

Foster, K.W. 2001*a*. Action spectroscopy of photomovement. *Comprehensive Series in Photosciences* 1: (Photomovement) 51–115.

Foster, K.W. 2001*b*. Action spectroscopy of photomovement. pp. 55–115. In: *Photomovement*, D.P. Häder and M. Lebert (Eds.). Elsevier Sci. B.V.

Foster, R.M. and Lüning K. 1996. Photosynthetic response of *Laminaria digitata* to ultraviolet A and B radiation. 60 (Suppl. 1): 65–67. In: *Underwater Light and Algal Photobiology*, F.L. Figueroa, C. Jimenez, J.L. Perez-Llorens, and F.X. Niell, (Eds.), Sci. Marina, Malaga.

Foster, K.W., Saranak, J., Zarilli, G., Okabe, M., Kline, T., and Nakanishi, K. 1984. A rodopsin is the functional photoreceptor for phototaxis in the unicellular eukaryote *Chlamydomonas. Nature* 311: 756–759.

Foster, K.W. and Smyth R.D. 1980. Light antennas in phototactic algae. *Microbiol. Rev.* 44: 572–630.

Fox, G.E., Stackebrandt, E., and Hespell, R.B. et al. 1980.The phylogeny of prokaryotes. *Science* 309: 457–463.

Fraenkel, G.S. and Gunn, D.L. 1961. *The Orientation of Animals. Kineses, Taxes and Compass Reactions.* Dover Publ. Inc., New York.

Francis, D. 1967. On the eyespot of the dinoflagellate, *Nematodium. J. Exp. Biol.* 47: 495–501.

Fredersdorf, J. and Bischof, K. 2007. Irradiance of photosynthetically active radiation determines ultraviolet-susceptibility of photosynthesis in *Ulva lactuca* L. (Chlorophyta). *Phycological Research* 55(4): 295–301.

Freeman, H. 1961. On the ecoding of arbitrary geometric configurations.V. EC-10 *IRE Trans.*, pp. 260–268.

Friedl, Th. 1997. The evolution of the Green Algae. *Origins of Algae and their Plastids.* D. Bhattacharya (Ed.), Springer, Wien, pp. 87-114.

Fuggi, A., Pinto, G., Pollio, A., and Taddei, R. 1988. Effects of sodium chloride, sodium sulfate, sulfuric acid and glucose on growth, photosynthesis, and respiration in the acidophilic alga *Dunaliella acidophila* (Volvocales, Chlorophyta). *Phycologia* 27(3): 334–339.

Fujiu, K., Nakayama, Y., Yanagisawa, A., Sokabe, M., and Yoshimura, K. 2009. *Chlamydomonas* CAV2 Encodes a Voltage-Dependent Calcium Channel Required for the Flagellar Waveform Conversion. *Current Biology* 19: 133–139.

Galland, P., Keiner, P., Dörnemann, D., Senger, H., Brodhum, B., and Häder, D.P. 1990. Pterin- and flavin-like fluorescence associated with isolated flagella of *Euglena gracilis. Photochem. Photobiol.* 51: 675–680.

Garcia-Gonzalez, M., Moreno, J., Manzano, C., Florencio, F.J., and Guerrero, M.G. 2003*a*. Culture media for *Dunaliella salina* rich in β-carotene in closed tubular photobioreactor. (Consejo Superior de Investigaciones Cientificas, Spain, Universidad de Sevilla; Junta de Andalucia). Span. 6 pp.

Garcia-Gonzalez, M., Moreno, J., Canavate, J.P., Anguis, V., Prieto, A., Manzano, C., Florencio, F.J., and Guerrero, M.G. 2003*b*. Conditions for open-air outdoor culture of *Dunaliella salina* in southern Spain. *Journal of Applied Phycology* 15(2-3): 177–184.

Garcia-Pichel, F. 1996. The absorption of ultraviolet radiation by microalgae: simple optics and photobiological implications. 60 (Suppl. 1): 73–79. In: *Underwater Light and Algal Photobiology*, F.L. Figueroa, C. Jimenez, J.L. Perez-Llorens, and F.X. Niell (Eds.). Malaga: Scientia Marina.

Geiss, D., Senger, H., and Galland, P. 1997. Pigments associated with the flagellum of *Euglena gracilis. Journal of Plant Research* 110(1100): 421–433.

Geleskul, Y.F. 1968. Production of carotene from alga *Dunaliella* and preparations for animal husbandry. In: *Utilisation of hydrobiological resources of Black Sea as forage for livestocks.* Odessa (In Russian).

George, D.G., Hunt, L.T., and Dayhoff, M.O. 1983. Sequence evidence for the symbiotic origins of chloroplasts and mitochondria. pp. 845–861. In: *Endocytobiology* II., H.F.A. Schenk and W. Schwemmler (Eds.), De Gruyter, Berlin, New York.

Gerber, S. and Häder, D.P. 1992. UV effects on photosynthesis, proteins and pigmentation in the flagellate *Euglena gracilis*: Biochemical and spectroscopic observations. *Biochem. System and Ecology* 20(6): 485–492.

Gerber, S. and Häder, D.P. 1994. Effects of enhanced UV-B irradiation on the red colored freshwater flagellate *Euglena sanguinea. FEMS Microbiology Ecology* 13(3): 177–84.

Ghazi, A., Rudik, V., and Oswald, W.J. 1983. Effect of pH on *Dunaliella bardawil* biomass and production. *News Quarterly* 33(2): 1–3.

Ghetti, F., Checcucci, G., and Lenci, F. 1992. Photosensitized reactions as primary molecular events in photomovements of microorganisms. *Journal of Photochemistry and Photobiology*, B: Biology 15(3): 185–98.

Ghetti, F., Colombetti, G., Lenci, F., Campani, E., Polacco, E., and Quaglia, M. 1985. Fluorescence of *Euglena gracilis* photoreceptor pigment: an *in vitro* microspectrofluorometric study. *J. Photochem. Photobiology* 42: 29–33.

Giese, A.C. 1971. Photosensitization by natural pigments. 6: 77–129. In: *Photophysioiology,* A.C. Giese, (Ed.) Acad. Press, New York.

Gil, T.A., Balayan, A.E., Loginova, A.N., and Stom, D.I. 1993. Biotesting of industrial wastewater. *Khimiya i Tekhnologiya Vody* 15(4): 308-11 (In Russian).

Gilmour, D.J., Kaaden, R., and Gimmler, H. 1985. Vanadate inhibition of ATPases of *Dunaliella parva in vitro* and *in vivo*. *Journal of Plant Physiology* 118(2): 111–26.

Gimmler, H., Kuehnl, E.M., and Carl, G. 1978. Salinity dependent resistance of *Dunaliella parva* against extreme temperatures. I. Salinity and thermoresistance. *Zeitschrift fuer Pflanzenphysiologie* 90(2):133–53.

Gimmler, H., Kugel, H., Leibfritz, D., and Mayer, A. 1988. Cytoplasmic pH of *Dunaliella parva* and *Dunaliella acidophila* as monitored by *in vivo* phosphorus-31 NMR spectroscopy and the DMO method. *Physiologia Plantarum* 74(3): 521–30.

Gimmler, H., Schieder, M., Kowalski, M., Zimmermann, U., and Pick, U. 1991. *Dunaliella acidophila*: an algae with a positive zeta potential at its optimal pH for growth. *Plant, Cell and Environment* 14(3): 261–9.

Gimmler, H., Treffny, B., Kowalski, and M., Zimmermann, U. 1991. The resistance of *Dunaliella acidophila* against heavy metals: the importance of the zeta potential. *Journal of Plant Physiology* 138(6): 708–16.

Gimmler, H. and Weis U. 1992. *Dunaliella acidophila* – life at pH 1.0. pp. 99–133. In: *Dunaliella: Physiology, Biochemistry and Biotechnology*, M. Avron and A. Ben-Amotz (Eds.) CRC Press, Boca Raton.

Ginzburg M. 1987. *Dunaliella*: a green alga adapted to salt.*Adv. Bot. Res.* 14: 93–183.

Goldman, I.C., Azov, Y., Riley, C.B., and Dennett, M.R. 1982. The effect of pH in intensive microalgal cultures. I. Biomass regulation. *J. Exp. Mar.Biol. and Ecol.* 57(1): 1–13.

Goldmann, I.C., Riley, C.B., and Dennett, M.R. 1982. The effect of pH in intensive microalgal cultures. II. Species competition. *J. Exp. Mar. Biol. and Ecol.* 57(1): 15–24.

Gomez, P.I. and Gonzale, M.A. 2005. The effect of temperature and irradiance on the growth and carotenogenic capacity of seven strains of *Dunaliella salina* (Chlorophyta) cultivated under laboratory conditions. *Biol.research* 38(2–3): 151–62.

Gomez, P.I., Gonzalez, M.A., and Becerra, J. 1999. Quantity and quality of β-carotene produced by two strains of *Dunaliella salina* (Teodoresco 1905) from the north of Chile. *Boletin de la Sociedad Chilena de Quimica* 44(4): 463–468.

Goodwin, T.C. 1988. *Algal Pigments*. Acad. Press, New York.

Goodwin T.W. 1980. *The Biochemistry of Carotenoids*, Vol. 1, 2nd ed. Plants, Chapman and Hall, London.

Gossel, I. 1957. Action spectrum of phototaxis of chlorophyll-free Euglena. *Archiv fuer Mikrobiologie* 27: 288–305.

Govorunova, E.G., Altschuler, I.A., Häder, D.P., and Sineshchekov, O.A. 2000. A novel express bioassay for detecting toxic substances in water by recording rhodopsin-mediated photoelectric responses in *Chlamydomonas* cell suspensions. *J. Photochem. Photobiology* 72(3): 320–326.

Govorunova, E.G., Jung, K., Sineshchekov, O.A., and Spudich, J.L. 2004. *Chlamydomonas* sensory rhodopsins A and B: cellular content and role in photophobic responses. *Biophysical Journal* 86(4): 2342–2349.

Govorunova, E.G., Sineshchekov, O.A., Gärtner, C.A.S., and Hegemann, P. 2001. Photoreceptor current and photoorientation in *Chlamydomonas* mediated by 9-demethylchlamyrhodopsin. *Biophys. J.* 81: 2897–2907.

Govorunova, E.G., Sineshchekov, O.A., and Hegemann, P. 1997. Desensitization and dark recovery of the photoreceptor current in *Chlamydomonas reinhardtii. Plant Physiol.* 115: 633–642.

Goyal, A. and Gimmler, H. 1989. Osmoregulation in *Dunaliella tertiolecta*: effects of salt stress, and the external pH on the internal pH. *Archives of Microbiology* 152(2): 138–42.

Graham L.E. and Wilcox L.W. 2000. *Algae.* Prentice Hall. 650 pp.

Greuet, C. 1987. Complex organelles. pp. 119–142. In: *The Biology of Dinoflagellates.* F.J.R. Taylor (Ed.). Oxford: Blackwell.

Grizeau, D. and Navarro, J.M. 1986. Glycerol production by *Dunaliella tertiolecta* immobilized within calcium alginate beads. *Biotechnology Letters* 8(4): 261–4.

Grobbelaar, J.U. 1995. Influence of areal density on β-carotene production by *Dunaliella salina. J. Appl. Phycol.* 7(1): 69–73.

Grobe, C.W. and Murphy, T.M. 1998. Solar ultraviolet-B radiation effects on growth and pigment composition of the intertidal alga *Ulva expansa* (Setch.) S. & G. (Chlorophyta). *J. Exp. Mar.Biol. and Ecol.* 225(1): 39–51.

Gromov, B.V. 1985. *Structure of bacteria.* Leningrad. Leningrad State Univ. Publ. (In Russian).

Gross, W. 2000. Ecophysiology of algae living in highly acidic environments. *Hydrobiologia* 433: 31–37.

Gualtieri, P. 2001. Rhodopsin-like-proteins: light detection pigments in *Leptolyngbya, Euglena, Ochromonas, Pelvetia.* pp. 283–294. In: *Photomovement.* D. P. Häder and M. Lebert (Eds.). Elsevier Sci. B.V.

Gualtieri, P., Barsanti, L., and Passarelli, V. 1989. Absorption spectrum of a single isolated paraflagellar swelling of *Euglena gracilis. Biochim. Biophys. Acta* 993: 293–296.

Gualtieri, P., Barsanti, L., Passarelli, V., and Verni, F. 1988. Morphological investigations of the *Euglena gracilis* isolated paraflagellar body. *Micron and Microscopica Acta* 19(4): 241–246.

Gualtieri, P., Barsanti, L., and Rosati, G. 1986. Isolation of the photoreceptor (paraflagellar body) of the phototactic flagellate *Euglena gracilis. Arch. Microbiol.* 145: 303–305.

Gualtieri, P., Pelosi, P., Passarelli, V., and Barsanti, L. 1992. Identification of a rhodopsinic photoreceptor in *Euglena gracilis. Biochim. Biophys. Acta* 1117: 55–59.

Guevara, M., Lodeiros, C., Gomez, O., Lemus, N., Nunez, P., Romero, L., Vasquez, A., and Rosales, N. 2005. Carotenogenesis of five strains of the algae *Dunaliella* sp. (Chlorophyceae) isolated from Venezuelan hypersaline lagoons. *Revista de biologia tropical* 53(3–4): 331–7.

Guo, L., Zhang, J., Cui, Z., Wang, F., and Shi, F. 2003. Research and development of *Dunaliella salina* powder and *Dunaliella salina* powder soft capsule. *Haihuyan Yu Huagong* 32(2): 23–25.

Häder, D.P. 1977. Influence of electric fields on photophobic reactions in blue-green algae. *Archives of microbiology* 114(1): 83–6.

Häder, D.P. 1979. Control of locomotion. pp. 268–309. In: *Physiology of movements.* W. Haupt and M. Feinleib (Eds.). Springer-Verlag, Berlin.

Häder, D.P. 1985. Effect of UV-B on motility and photobehavior in the green flagellate, *Euglena gracilis. Arch. Microbiol.* 141: 159–163.

Häder, D.P. 1986*a.* Effects of solar and artificial UV irradiation on motility and phototaxis in the flagellate, *Euglena gracilis. Photochem. Photobiol.* 44(5): 651–656.

Häder, D.P. 1986*b.* New evidence for the mechanisms of phototactic orientation of *Euglena gracilis. Curr. Microbiol.* 14:157–163.

Häder, D.P. 1987*a.* Photomovement in eukaryotic microorganisms. *Photobiochem. Photobiophys. Supp.,* pp. 203–214.

Häder, D.P. 1987*b.* Polarotaxis, gravitaxis and vertical phototaxis in the green flagellate, *Euglena gracilis. Arch. Microbiol.* 147:179–183.

Häder, D.P. 1987*c.* Movement. pp. 101-133. In: *Blue Light Responses: Phenomena and Occurence in Plants and Microorganisms.* H. Senger (Ed.). CRS Press, Boca Raton. Häder, D. P. 1997. Oben oder unten – Schwerkraftperzeption bei dem einzelligen flagellaten *Euglena gracilis. Microkosmos.* 86: 351–356.

Häder, D.P. 1991. Effects of enhanced solar ultraviolet radiation on aquatic ecosystems. pp. 157–172. In: *Biophysics of Photoreceptors and Photomovement in Microrganisms.* F. Lenci et al. (Eds.), Plenum Press, New York.

Häder, D.P. 1996*a.* Mechanisms of photoreception: energy and signal transducers. pp. 185–194. In: *Light and Energy Source and Information Carrier in Plant Physiology.* Plenum Press, New York.

Häder, D.P. 1996*b.* Effects of enhanced solar UV-B radiation on phytoplankton. 60 (Suppl. 1): 59–63. In: *Underwater Light and Algal Photobiology.* F.L. Figueroa, C. Jimenez, J.L. Perez-Llorens, and F.X. Niell (Eds.), Sci. Marina, Malaga.

Häder, D.P. 1994*a.* UV-B effects on aquatic systems. In: NATO ASI Series, Series I: *Global Environmental Change* 18: 155–61.

Häder, D.P. 1994*b.* Real-time tracking of microorganisms. *Binary* 6: 81–86.

Häder, D.P., Colombetti, G., Lenci, F., and Quaglia, M. 1981. Phototaxis in the flagellates, *Euglena gracilis* and *Ochromonas danica. Arch. Microbiol.* 130: 78–82.

Häder, D.P. and Grienbow, K. 1987. Versatile digital image analysis by microcomputer to count microorganisms. *EDV in Med. und Biol.* 18(2-3): 37–42.

Häder, D.P. and Häder, M.A. 1988. Inhibition of motility and phototaxis in the green flagellate, *Euglena gracilis,* by UV-B radiation. *Arch. Microbiol.* 150: 20–25.

Häder, D.P. and Häder, M.A. 1989*a.* Effects of solar radiation on photoorientation, motility and pigmentation in a freshwater *Cryptomonas. Botanica Acta* 102: 236.

Häder, D. P. and Häder, M.A.1989*b.* Effects of solar UV-B irradiation on photomovement and motility in photosynthetic and colorless flagellates. *Environ. Exp. Bot.* 29: 273–282.

Häder, D.P. and Häder, M.A. 1990. Effects of solar radiation on photoorientation, motility and pigmentation in the marine *Cryptomonas maculate. J. Photochem. Photobiol.* 5: 105.

Häder, D.P. and Häder, M.A. 1991. Effects of solar and artificial U.V. radiation on motility and pigmentation in the marine *Cryptomonas maculata. Env. Exp. Botany* 31(1): 33–41.

Häder, D.P., Häder, M.A., Liu, S.M., and Ullrich, W. 1990. Effects of solar radiation on photoorientation, motility and pigmentation in a freshwater *Peridinium. BioSystems* 23: 335.

Häder, D.P. and Lebert M. 1998. The photoreceptor for phototaxis in the photosynthetic flagellate *Euglena gracilis. Photochem. Photobiol.* 68: 260–265.

Häder, D.P. and Lebert, M. 2000. Photoreception and phototaxis in flagellated algae. *Res. Adv. in Photochem. Photobiol.* 1: 201–226.

Häder D.P. and Lebert M. (Eds.) 2001.*Photomovement.* Elsevier Sci. B.V. Amsterdam.

Häder, D.P. and Lipson, E.D. 1986. Fourier analysis of angular distributions of motile microorganisms. *Photochem. Photobiol.* 42: 657–663.

Häder, D.P., Porst, M., Tahedl, H., Richter, P., and Lebert, M. 1997. Gravitactic orientation in the flagellate *Euglena gracilis. Micrograv. Sci. Technol.* 10: 53–57.

Häder, D.P. and Reinecke, E. 1991. Phototactic and polarotactic responses of the photosynthetic flagellate, *Euglena gracilis. Acta Protozool.* 30: 13–18.

Häder, D.P., Rhiel, E., and Wehrmeyer, W. 1987. Phototaxis in the marine flagellate *Cryptomonas maculata. Journal of Photochemistry and Photobiology,* B: Biology, 1(1): 115–22.

Häder, D.P. and Vogel, K. 1991. Simultaneous tracking of flagellates in real time by image analysis. *J. Math. Biol.* 30: 63–72.

Häder, D.P., Watanabe, M., and Furuya, M. 1986. Inhibition of motility in the cyanobacterium, *Phormidium uncinatum,* by solar and monochromatic UV irradiation. *Plant Cell Physiol.* 27(5): 887–894.

Häder, D.P., Worrest, R.C., Kumar, H.D., and Smith, R.C. 1995. Effect of increased solar ultraviolet radiation on aquatic ecosystems. *Ambio.* 24(3): 174–180.

Halldal, P. 1957. Calcium and magnesium ions in the phototaxis of motile green algae. *Nature* 179: 215–16.

Halldal, P.1958. Action spectrum of phototaxis and related problems in *Volvocales, Ulva*-gametes and *Dinophyceae. Physiol. Plant.* 11: 118–153.

Halldal, P. 1961. Ultraviolet action spectra of positive and negative phototaxis in *Platymonas subcordiformis. Physiol. Plant.* 14: 133–140.

Halldal, P. 1967. Ultraviolet action spectra in algology. (Review) *Photochemistry and Photobiology* 6(7): 445–60.

Hamburger C. 1905. Zur Kenntnis der *Dunaliella salina* und einer Amöbe aus Salinenwasser von Cagliari. *Arch f Protistenkd* 6: 111–131.

Han, Y.S. and Han, T. 2005. UV-B induction of UV-B protection in *Ulva pertusa* (Chlorophyta). *J. Phycol.* 41(3): 523–530.

Hand, W.E.G. and Davenport, D. 1970. The experimental analysis of phototaxis and photokinesis in flagellates. pp. 253–282. In: *Photobiology of Microorganisms.* P. Halldal (Ed.),Wiley, London.

Hansgirg, A.1866. *Prodromus d. Algenflora von Bohmen.*

Haouazine, Y., Rmiki, N., Riyahi, J., Givernaud, T., Mouradi, A., and Lemoine, Y. 1999. Massive accumulation of β-carotene in green alga *Dunaliella salina* induced by variation of nitrate and salinity. *Mededelingen - Faculteit Landbouwkundige en Toegepaste Biologische Wetenschappen* (Universiteit Gent) 64(5b): 435–438.

Happel, J. and Brenner, H. 1973. *Low Reynolds Number Hydrodynamics.* Rev. 2$^{nd}$ ed. Leyden: Noordhoff Int., Gröningen.

Harz, H., Nonnengässer, C., and Hegemann P. 1992. The photoreceptor current of the green alga *Chlamydomonas. Phil. Trans. R. Soc. Lond.* B. 338: 39–52.

Hatanaka, Y. 2002. Application of halo-tolerant microalgae Dunaliella as a food resource. *Foods & Food Ingredients Journal of Japan* 203: 9–16.

Haupt, W. 1959. *Die Phototaxis der Algen. Handbuch der Pflanzenphysiologie.* W. Ruhland (Ed.). Springer, Berlin, 17: 318–370.

Haupt, W. 1966. Phototaxis in Plants. *Int. Rev. Cytol.* 19: 267–299.

Haupt W. and Feinleib M.E. (Eds.). 1979. *Physiology of Movements.* Springer, Berlin.

Haupt, W. 1983. Photoreception and photomovement. *Phil. Trans. Soc.* London. B. 303: 467–478.

Haupt, W. 1986. Photomovement. pp. 1-14. In: *Photomorphogenesis in Plants.* R.E. Kendrick and G.H.M. Kronenberg (Eds.). M. Nijhoff, Dr. W. Junk Publ., Dordrecht.

Haupt, W. 2001. Photomovement: past and future. In: *Photomovement.* D.P. Häder and M. Lebert (Eds.). Elsevier Sci.B.V., Amsterdam

Haupt, W. and Häder, D. P. 1994. Photomovement. pp. 707–732. In: *Photomorphogenesis in Plants.* 2$^{nd}$ Ed., R.E. Kendrick, G.H.M. Kronenberg, and G. Colombetti (Eds.), Kluwer Acad. Publ., Netherlands

Haupt, W. and Seitz, K. 1984. Physiology of movement. pp.172–182. In: *Progress in Botany,* Springer-Verlag, Berlin, Heidelberg.

Hegemann, P. 1997. Vision in microalgae. *Planta* 203: 265–274.

Hegemann, P. and Deininger, W. 1999. Algal eyes and their rhodopsin photoreceptors. In: *Comprehensive Series in Photoscience.* Elsevier Press.

Hegemann, P. and Deininger, W. 2001. Algal eyes and their rhodopsin photoreceptors. pp. 229–243. In: *Photomovement.* Häder, D.P. and Lebert, M. (Eds.). Elsevier Sci.

Hegemann P. and Fischer M. 2001. Algal Eyes. pp. 1–6. In: *Encyclopedia of Life Science.* Nature Publ. Group.

Hegemann P., Fuhrmann M., and Kateriya S. 2001. Algal sensory photoreceptors. *J. Phycol.* 37: 668–676.

Hegemann, P., Gaertner, W., and Uhl, R. 1991. All-trans retinal constitutes the functional chromophore in *Chlamydomonas* rhodopsin. *Biophysical Journal* 60(6): 1477–89.

Hegemann P. and Harz H. 1998. How microalgae see the light. pp. 95-105. In: *Microbial Response to Light and Time.* Univ. Press, Cambridge.

Hegemann, P. and Harz, H. 1993. Photoreception in *Chlamydomonas.* In: *Signal Transduction.* Kurjan, J. and Taylor, B.L. (Eds.), pp. 279–307.

Hegemann, P., Hegemann, U., and Foster, K.W. 1987. Retinal membrane proteins in *Chlamydomonas.* Retinal Proteins, Proc. Int. Conf., Meeting Date 1986, pp. 467–468.

Hegemann, P., Hegemann, U., and Foster, K.W. 1988. Reversible bleaching of *Chlamydomonas reinhardtii* rhodopsin *in vivo. Photochemistry and Photobiology* 48(1): 123–128.

Hegemann P., Neumeier K., Hegemann U., and Kuelnle E. 1990. The role of calcium in *Chlamydomonas* movement responses as analysed by calcium channel inhibitors. *Photochem. Photobiol.* 52: 575–578.

Heidari, R., Riahi, H., and Saadatmand, S. 2000. Effects of salt and irradiance stress on photosynthetic pigments and proteins in *Dunaliella salina* Teodoresco. *J. Sci.,* Islamic Republic of Iran, 11(2): 73–77.

Hejazi, M.A., Andrysiewicz, E., Tramper, J., and Wijffels, R.H. 2003. Effect of mixing rate on β-carotene production and extraction by *Dunaliella salina* in two-phase bioreactors. *Biotechnology and Bioengineering* 84(5): 591–596.

Hejazi, M.A. and Wijffels, R.H. 2004. Milking of microalgae. *Trends in biotechnology* 22(4): 189–94.

Hertz, W., Heck, B., and Koop, H.U. 1981. The flagellar root system in the gamete of *Acetabularia mediterranea. Protoplasma* 109: 257–269.

Hill, N A. and Plumpton, L.A. 2000. Control strategies for the polarotactic orientation of the microorganism *Euglena gracilis. Journal of theoretical biology* 203(4): 357–65.

Hirsch, R., De Guia, M., Falkner, G., and Gimmler, H. Julius von Sachs 1993. Flexible coupling of phosphate uptake in *Dunaliella acidophila* at extremely low pH values. *J. Exp.Bot.* 44(265): 1321–30.

Hirschberg, R. and Hutchinson, W. 1980. Effect of chlorpromazine on phototactic behavior in *Chlamydomonas. Can. J. Microbiol.* 26(2): 265–7.

Hoek Van den C., Mann D.G., Jahns H.M. 1995. *Algae. An Introduction to phycology.* Cambridge, Univ. Press.

Holland, E.M., Braun, F.J., Nonnengaesser, C., Harz, H., and Hegemann, P. 1996. The nature of rhodopsin-triggered photocurrents in *Chlamydomonas.* I. Kinetics and influence of divalent ions. *Biophysical Journal* 70(2): 924–31.

Holland, E.M., Harz, H., Uhl, R., and Hegemann, P. 1997. Control of phobic behavioral responses by rhodopsin-induced photocurrents in *Chlamydomonas. Biophysical J.* 73: 1395–1401.

Honda, S. 1997. Applied technology of foods of microalgae "*Dunaliella*" and its market. *Food Style* 21, 1(2): 28–33.

Hong, S.P., Kim, M.H., and Hwang, J.K. 1998. Biological functions and production technology of carotenoids. *Han'guk Sikp'um Yongyang Kwahak Hoechi* 27(6): 1297–1306.

Horigushi, T., Kawai, H., Kubota, M., Takahashi, T., and Watanabe, M. 1999. Phototactic responses of four marine dinoflagellates with different types of eyespot and chloroplast. *Phycol. Res.* 47: 101–107.

Horiguchi, J.I., Ohba, I., Tada, K., Kobayashi, M., Kanno, T., and Kishimoto, M. 2003. Effective cell harvesting of the halotolerant microalga *Dunaliella tertiolecta* with pH control. *Journal of bioscience and bioengineering* 95(4): 412–5.

Horike, N., Nishikawa, A., Tanaka, H., and Nakamura, S. 2002. Usefulness of Flagellar Regeneration in *Dunaliella* sp. as an Endpoint for the Bioassay of Seawater Pollution. *J. of Japan Soc. On Water. Env.* 23 (8): 485–490.

Hoshaw, R.W. and Maluf, L.Y. 1981. Ultrastructure of the green flagellate *Dunaliella tertiolecta* (*Chlorophyceae, Volvocales*) with comparative notes on three other species. *Phycologia* 20(2): 199–206.

Huang B., Watterson D.M., Lee V.D., and Schiller M.J. 1988. Purification and characterization of a basal body-associated $Ca^{2+}$-binding protein. *J. Cell Biol.* 107(1): 121–131.

Huber, M.E. and Lewin, R.A. 1987. Ethanol-induced flagellar autotomy in *Dunaliella tertiolecta* Butcher (*Chlorophyta: Volvocales*). *Phycologia* 26(1): 138-141.

Huovinen, P., Gomez, I., and Lovengreen, C. 2006. A five-year study of solar ultraviolet radiation in southern Chile ($39^0$ S): potential impact on physiology of coastal marine algae? *Photochem. and Photobiol.* 82: 515–522.

Hutchinson, W. and Hirschberg, R. 1985. Transport of calcium by cells and flagella of *Chlamydomonas. Curr. Microbiol.* 12(1): 27–30.

Hyams, J.S. and Borisy, G.G. 1978. Isolated flagellar apparatus of *Chlamydomonas*: characterisation of forward swimming and alteration of waveform and reversal of motion by calcium ions *in vitro. J. Cell. Sci.* 33: 235–253.

Iseki, M., Matsunaga, S., Murakami, A., and Watanabe, M. 2006. Photoactivated adenylyl cyclase (PAC), the photoreceptor flavoprotein with intrinsic effector function mediating euglenoid photomovements. *Comprehensive Series in Photochemical & Photobiological Sciences* 6: 271–286.

Iseki, M. and Watanabe, M. 2004 Molecular mechanism of photoreaction: photosensor in relation to photodynamics of *Euglena gracilis. Iden* 58(6): 36–41.

Isogai, N., Kamiya, R., and Yoshimura, K. 2000. Dominance between the two flagella during phototactic turning in *Chlamydomonas*. *Zool. Science*. 17: 1261–1266.

Izard, C. and Testa, P. 1968. Effects of cigarette smoke and certain of its components on the motility and the multiplication of *Dunaliella bioculata*. Section 1, 6: 121–56. In: *Annales de la Direction des Etudes et de l'Equipement, Service d'Exploitation Industrielle des Tabacs et des Allumettes*.

Izard, C., Lacharpagne, J., Testa, P., and Testa, A. 1967. Effects of cigarette smoke vapors on *Dunaliella bioculata* strain A1 motility. *Comptes Rendus des Seances de l'Academie des Sciences, Serie D: Sciences Naturelles* 264(1): 47–9.

Izard, C., Testa, P., and Lacharpagne, J. 1967*b*. Effects of cigaret smoke condensates on *Dunaliella bioculata* motility. *Comptes Rendus des Seances de l'Academie des Sciences, Serie D: Sciences Naturelles* 264(20): 2379–81.

Jagger, J. 1983. Effects of near-UV radiation on bacteria. P.1. In: *Photochemical and Photobiological Reviews*. Smith K.C. (Ed.). Plenum Press, New York.

Jahn, T.L. and Bovee, B.C. 1968. Locomotive and motile response in *Euglena*. Vol. 1: 45–107. In: *The Biology of Euglena*. D.E. Buetow (Ed.). Acad. Press, N.Y., London.

James, T.W., Crescitelli, F., Loew, E.R., and McFarland, W.N. 1992. The eyespot of *Euglena gracilis*: a microspectrophotometric study. *Vision Research* 32: 1583–1591.

Jeanne, N. 1979. Effects of lindane on division, cell cycles, and biosynthesis in two unicellular algae. *Can. J.Bot.* 57(13): 1464–72.

Jennings, H.S. 1906. *Behaviour of the lower organisms*. Columbia Univ. Press, New York.

Jimenez, C., Figueroa, F.L., Aguilera, J., Lebert, M., and Häder, D.P. 1996. Phototaxis and gravitaxis in *Dunaliella bardawil*: Influence of UV radiation. *Acta Protozoologica* 35: 287–295.

Jimenez, C. and Niell, F.X. 1991. Influence of temperature and salinity on carbon and nitrogen content in *Dunaliella viridis* Teodoresco under nitrogen sufficiency. *Bioresource Technology* 38(2-3): 91–4.

Jin, E. and Melis, A. 2003. Microalgal biotechnology: Carotenoid production by the green algae *Dunaliella salina*. *Biotechnology and Bioprocess Engineering* 8(6): 331–337.

Joly, N. 1840. Histoire d'un petit crustace (*Artemia salina*) auquel on a faussement attribute la coloration en rouge des marais salants mediterraneens, suivie de recherches sur la cause reelle de cette coloration. *Ann Sc Nat Zool,* Ser 2, 13: 225.

Josef, K., Saranak, J., and Foster, K.W. 2005. An electro-optic monitor of the behavior of *Chlamydomonas reinhardtii* cilia. *Cell motility and the cytoskeleton* 61(2): 83–96.

Josef, K., Saranak, J., and Foster, K.W. 2006. Linear systems analysis of the ciliary steering behavior associated with negative phototaxis in *Chlamydomonas reinhardtii. Cell motility and the cytoskeleton* 63(12): 758–77.

Joyon, L. and Fott, B. 1964. Quelques particularités infrastructurales du plaste des *Carteria* (*Volvocales*). *J. Microsc.* (Paris). 3:159–166.

Kalinichenko, N.F., Starobinets, Z.G., and Biryukova, S.V. et al. 1986. Antibacterial activity of SAS. pp. 99–103. In: *Microbiological epidemiology and clynics of infectious diseases*. Kharkov (In Russian)

Kalugina-Gutnik, O.A. and Belyaev, B.M. 1987. Development of intense aquaculture of algae-macrophytes in Black Sea. *Visn. AN URSR*, 1: 48–54. (In Ukrainian).

Kamiya, R. 2002. Functional diversity of axonemal dyneins as studied in *Chlamydomonas* mutants *Int. Review of Cytology* 219: 115–155.

Kamiya, R. and Hasegawa, E. 1987. Intrinsic difference in beat frequency between the two flagella of *Chlamydomonas reinhardtii. Exp. cell research* 173(1): 299-304.

Kamiya, R. and Witman, G B 1984. Submicromolar levels of calcium control the balance of beating between the two flagella in demembranated models of *Chlamydomonas. The J.Cell Biol.* 98(1): 97–107.

Kang, Y., Xie, W., Gao, Y., Lin, S., Ouyang, L., and Liu, G. 2006. Effect of different concentrations of NaCl and light intensities on β-carotene content in *Dunaliella. Zhiwu Shenglixue Tongxun* 42(2): 315–318.

Karsten, U., Sawall, T., Hanelt, D., Bischof, K., Figueroa, F.L., Flores-Moya, A., and Wiencke, C. 1998. An inventory of UV-absorbing mycosporine-like amino acids in macroalgae from polar to warm-temperate regions. *Botanica Marina* 41(5): 443–453.

Kateriya, S., Nagel, G., Bamberg, E., and Hegemann, P. 2004. "Vision" in single-celled algae. *News in Physiological Sciences* 19: 133–137.

Katz, A., Bental, M., Begani, H., and Avron, M. 1991. *In vivo* pH regulation by a sodium/hydrogen ion antiporter in the halotolerant alga *Dunaliella salina. Plant Physiology* 96(1): 110–15.

Katz, A., Jimez, C., and Pick, U. 1965. Isolation and characterization of aprotein associated with carotene globuls in the alga *Dunaliella bardawil. Plant Physiol.* 108(4):1657–1664.

Katz, A., Pick, U., and Avron, M. Modulation of sodium/hydrogen ion antiporter activity by extreme pH and salt in the halotolerant alga *Dunaliella salina. Plant Physiology* 100(3): 1224–9.

Kawai, H. 1988. A flavin-like autofluorescent substance in the posterior flagellum of golden and brown algae. *J. Phycol.* 24: 171–184.

Kawai, H. 1989. Flagellar autofluorescence in forty-four chlorophyll *c*-containing algae. *Phycologia.* 28: 222–227

Kawai, H. 1992. Green flagellar autofluorescence in brown algal swarmers and their phototactic responses. *Bot. Mag. Tokyo.* 105: 171–184.

Kawai, H. and Kreimer, G. 1992. Sensory mechanisms – phototaxis and light perception in algae. pp. 124–146. In: *The Flagellates.* B.S.C. Leadbeater and J.C. Green (Eds.), Taylor and Francis, London.

Kawai, H., Kubota, M., Kondo, T., and Watanabe, M. 1991. Action spectra for phototaxis in zoospores of the brown alga *Pseudochorda gracilis.* Protoplasma 161: 17–22.

Kawai, H., Nakamura, S., Mimuro, S., Furuya, M., and Watanabe, M. 1996. Microspectrofluorometry of the autofluorescent flagellum in phototactic brown algal zooids. *Protoplasma* 191: 172–177.

Kay, R.A. 1991. Microalgae as food and supplement. *Critical Reviews in Food Science and Nutrition* 30(6): 555–73.

Kessler, J.O., Hill, N.A., and Häder, D.P. 1992. Orientation of swimming flagellates by simultaneously acting external factors. *J. Phycol.* 28: 816–822.

Kivic, P.A. and Vesk, M. 1972. Structure and function in the euglenoid eyespot apparatus: the fine structure, and response to environmental changes. *Plants* 105:1–14.

Kivic, P.A. and Vesk, M. 1974. Structure of the eyespot apparatus in bleached strains of *Euglena gracilis. Cytologie* 10(10): 88–101.

Kivic, P.A. and Walne, P.L. 1983. Algal photosensory apparatus probably represent multiple parallel evolutions. *BioSystems* 16: 31–38.

Ko, J.H. and Lee S.H. 1996. Nucleotide sequence of a cDNA (Accession No. U53812) encoding a contractin-like protein from *Dunaliella salina. Plant Physiology* 112: 445.

Koblenz, B., Schoppmeier, J., Grunow, A., and Lechtreck, K.F. 2003. Centrin deficiency in *Chlamydomonas* causes defects in basal body replication, segregation and maturation. *J. Cell Sci.* 116: 2635–2646.

Kombrink, E. and Wöber, G. 1980. Preparation of intact chloroplasts by chemically induced lysis of the green alga *Dunaliella marina*. *Planta* 149: 123–129.

Kondo, T., Kubota, M., Aono, M., and Watanabe, M. 1988. A computerized video system to automatically analyse movements of individual cells and its application to the study of circadian rhythms in phototaxis and motility in *Chlamydomonas*. *Protoplasma* 1: 185–192.

Konev, S.V. and Volotovsky, I.L. 1979. *Photobiology*. Minsk. Belorus Univ. Publ. (In Russian).

Korol'kov, S.N. and Rychkova, M.P. 1996. Influence of some metabolic process modulators on phototaxis of *Chlamydomonas reinhardtii*. *Zhurnal Evolyutsionnoi Biokhimii i Fiziologii* 32(2): 225–228.

Korshikov, O.A. 1938. *Volvocineae* /Determinant of freshwater algae of URSR. Kiev, AN URSR Publ. (In Russian).

Kostyuk, P.G., Grodzinsky, D.M., Zima, V.L., Magura I.S., Sidorik, E.P., and Shuba, M.F. 1988. *Biophysics*. Kiev, Vysshaya shkola. (In Russian).

Krasnovsky, A.A. 1988. Mechanisms of formation and role of singlet oxygen in photobiological reactions. pp.23–41. In: *Molecular mechanisms of biological action of optical radiation*. Moscow, Nauka. (In Russian).

Kraynukova, A.N. 1988. Biotesting in protection of waters from pollution. In: *Methods of biotesting waters*, Chernogolovka. (In Russian).

Kreimer, G. 1994. Cell biology of phototaxis in flagellates algae. *Int. Rev. Cytol.* 148: 229–310.

Kreimer, G. 1995. Flagellar shock responses in green algae. *Botanica Acta* 108(3): 169–71.

Kreimer, G. 2001. Light perception and signal modulation during photoorientation of flagellate green algae. pp.193–227. In: *Comprehensive Series in Photosciences*. V.1 (Photomovement).

Kreimer, G., Overlaender, C., Sineshchekov, O.A., Stolzis, H., Nultsch, W., and Melkonian, M. 1992. Functional analysis of the eyespot in *Chlamydomonas reinhardtii* mutant ey 627, mt-. *Planta* 188(4): 513–21.

Krishnaswamy, C.S. 1997. Biflagellated unicellular algae as detectors of pesticide contamination in water. *Environmental Monitoring and Assessment* 44(1-3): 487–502.

Kritsky, M.S. 1982. Photoregulation of metabolism and ontogenesis in heterotrophic microorganisms. *Usp. Microbiologii* 17: 41–61. (In Russian)

Kroger, P. and Hegemann P. 1994. Photophobic responses and phototaxis in *Chlamydomonas* are triggered by a single rhodopsin photoreceptor. *FEBS Letters* 341: 5–9.

Krol, M., Maxwell, D.P., and Huner, N.P.A. 1997. Exposure of *Dunaliella salina* to low temperature mimics the high light-induced accumulation of carotenoids and the carotenoid binding protein (Cbr). *Plant and Cell Physiology* 38(2): 213–216.

Kuchitsu, K., Katsuhara, M., and Miyachi, Sh. 1989. Rapid cytoplasmic alkalization and dynamics of intracellular compartmentation of inorganic phosphate during adaptation against salt stress in a halotolerant unicellular green alga *Dunaliella tertiolecta*: $^{31}$P-nuclear magnetic resonance study. *Plant and Cell Physiology* 30(3): 407–414.

Kusakin, O.G. and Drozdov, F.L. 1998. *Phylema of organic world*. Part 2. Procariota and lower eucariota, S. Petersbourg, Nauka. (In Russian).

Kuznicki, O.L. and Mikolajczyk, E. 1982. Motility and behaviour: Contributed paper session in memory of professor Theodore L. Jahn. pp. 149–157. In: *Progress In Protozoology*, Proc. VI Intern. Congr. Protozoology Spec. Congr. Vol. of Acta Protozoologica, Part I.

Kvitko, K.V., Matveev, V.V., and Chunaev, A.S. 1978. Motile and behavioral forms of activity of *Chlamydomonas* and their changes induced by mutations. In: motion and behaviour of unicellular animals. 2: 130–158. (In Russian).

Labbé, A. 1925. Les cycles biologiques des *Dunaliella. Arch. Anat. Microsc.* 21: 313.

Lakin, G.F. 1973. *Biometry*, Moscow, Vysshaya shkola. (In Russian).

Lamare, M.D., Lesser, M.P., Barker, M.F., Barry, T.M., and Schimanski, K.B. 2004. Variation in sunscreen compounds (mycosporine-like amino acids) for marine species along a gradient of ultraviolet radiation transmission within Doubtful Sound, *New Zealand Journal of Marine and Freshwater Research* 38(5): 775–793.

Landau, L.D. and Lifshits, E.M. 1959. *Electrodynamics of continuous media.* Moscow, AN SSSR Publ. (In Russian).

Lauger, P. 1984. Mechanistic properties of ion channels and pumps. pp. 133–138. In: *Information and Energy Transduction in Biological Membranes.* Proc. of the Int. Cong. on Biol. Membranes, Switserland.

Lawson, M.A., Zacks, D. N., Derguini, F., Nakanishi, K., and Spudich, J.L. 1991. Retinal analog restoration of photophobic responses in a blind *Chlamydomonas reinhardtii* mutant. Evidence for an archaebacterial like chromophore in a eukaryotic rhodopsin. *Biophysical Journal* 60(6): 1490–8.

Lebert, M. 2001. Phototaxis of *Euglena gracilis* – flavins and pterins. pp. 299–341. In: *Photomovement.* D.P. Häder and M. Lebert (Eds.), Elsevier Sci., Amsterdam: B.V.

Lebert, M. and Häder, D.P. 1996. How *Euglena* tells up from down. *Nature* 379: 590.

Lebert, M. and Häder, D.P. 2000. Photoreception and phototaxis in flagellated algae. *Res. Adv. Photochem. Photobiol.* 1: 201–226.

Lebert, M., Richter, P., and Häder, D.P. 1997. Signal perception and transduction of gravitaxis in the flgallate *Euglena gracilis.* J. Plant Physiol. 150: 685–690.

Leedale, G.F. 1974. How many are kingdoms of organisms? *Taxon* 23: 261–270.

Lembi, C.A. and Lang W.J. 1965. Electron microscopy of *Carteria* and *Chlamydomonas. Amer. J. Bot.* 52: 464–477.

Lenci, F. 1975. Photochemistry of flavins. pp. 164–183. In: *Biophysics of Photoreceptors and Photobehaviour of Microorganisms.* Proc. of the Intern. School of the CNR of Italy. Badia Fiesolana (Firenze), 1–5 September.

Lenci, F. 1982. Photomovement of microorganisms. pp. 421–435. In: *Trends in Photobiology.* C. Helene (Ed.). Plenum Press, N.Y., London.

Lenci, F., Angelini, N., and Sgarbossa, A. 1996. Molecular basis of photoreception. NATO ASI Series, Series A: Life Sciences 287(Light as an Energy Source and Information Carrier in Plant Physiology): 147–157.

Lenci, F. and Colombetti, G. 1978. Photobehaviour of microorganisms: a biophysical approach. *Ann. Rev. Bioeng.* 7: 341–361.

Lenci, F., Colombetti, G., and Häder, D.P. 1983. Role of flavin quenchers and inhibitors in the sensory transduction of the negative phototaxis in the flagellate, *Euglena gracilis. Current Microbiology* 9(5): 285–9.

Lenci, F., Häder, D.P., and Colombetti, G. 1984. Photosensory responses of freely motile microorganisms. pp. 199–229. In: *Membrane and Sensory Transduction.* F. Lenci and G. Colombetti (Eds.). Plenum Press, N.Y., London.

Lenci, F., Sgarbossa, A., and Angelini, N. 1997. Molecular basis in photomotile microorganisms. pp. 25–37. In: Proc. of the Intern. School of Biophysics *Biophysics of Photoreception. Molecular and Phototransductive Events*, Casamicciola, Napoli, Italy, 10-16 October, 1994.

Lenci, F. and Ghetti, F. 1989. Photoreceptor pigments for photomovement of microorganisms: some spectroscopic and related studies. *J. Photochem. Photobiol.*, B: Biology 3(1): 1–16.

Lenova, K.J., Borisova, O.V., Lukinov, D.I., and Vasser, S.P. 1989. Study of dynamics of cell distribution bu size for characteristics of the state of population of microalgae. *Dop. AN USSR.* 5: 67–69. (In Ukrainian).

Leon, R., Martin, M., Vigara, J., Vilchez, C., and Vega, J.M. 2003. Microalgae mediated photoproduction of beta-carotene in aqueous-organic two phase systems. *Biomolecular engineering* 20(4-6): 177–82.

Leonardo da Vinci (1452–1519) Leonardo's Notebooks. 5 January 2010. At: <http://www.unmuseum.org/leosketch.htm>.

Leontiev, D.V. and Akulov, A.Y. 2002. Revolution in megataxonomy: preconditions and results. *J. Gen. Biology.* 63: 158-176. (In Russian).

Lestgaft, P.F. (1837–1930) Lesgaft P.F. (1837-1909), anatomist, teacher. 5 January 2010. At: <http://www.encspb.ru/en/persarticle.php?kod=2803926042>.

Lerche, V. 1937. Untersuchungen über Entwicklung und Fortpflanzung in der Gattung *Dunaliella. Arch. Protistenk.* 88: S. 236–268.

Levain, N. and Marano-Le Baron, F. 1973. Effects of an organochlorine insecticide lindane, on the growth and structure of two unicellular algae, *Euglena gracilis* Z and *Dunaliella bioculata. Comptes Rendus des Seances de l'Academie des Sciences,* Serie D: Sciences Naturelles 276(1): 37–40.

Levandowsky M. and Hutner S.H. (Eds.) *Biochemistry and Physiology of Protozoa.* Acad. Press, New York. 4: 6–64.

Levy, J.L., Stauber, J.L., and Jolley, D.F. 2007. Sensitivity of marine microalgae to copper: The effect of biotic factors on copper adsorption and toxicity. *Science of the Total Environment* 387(1–3): 141–154.

Linder, J.C. and Stachelin, L.A. 1979. A novel model for fluid secretion by the trypanosomatid contractile vacuole apparatus. *J. Cell Biol.* 83: 371–382.

Litvin, F.F., Sineshchekov, O.A., and Sineshchekov, V.A. 1978. Photoreceptor electric potential in the phototaxis of the alga *Haematococcus pluvialis. Nature* (London) 271: 476–478.

Loeb, S. and Maxwell, S.S. 1910. Further proof of the identity of heliotropism in animals and plants. *Univ. Calif. Pub. Physiol.* 3: 3–195.

Loeblich, L.A. 1969. Aplanospores of *Dunaliella salina (Chlorophyta). J. Protozool.* 16 (Suppl.).

Loeblich, L.A. 1982. Photosynthesis and pigment influenced by light intensity and salinity in the halophilic *Dunaliella salina (Chlorophyta). J. Mar. Biol. Ass. U.K.* 62: 493.

López-Archilla, A.I. and Amils, R. 1999. A comparative ecological study of two acidic rivers in southwestern Spain. *Microb. Ecol.* 38: 146–156.

López-Archilla, A.I., Marin, I., and Amils, R. 2001. Microbial community composition and ecology of an acidic aquatic environment: The Tinto River, Spain. *Microb. Ecol.* 41: 20–35.

Lukas, W.I., Keifer, D.W., and Pesacreta, T.C. 1986. Influence of culture medium pH on charosome development and chloride transport in *Chara corallina. Protoplasma* 130 (1): 5–11.

Luntz, A. 1931*a*. Untersuchungen über die Phototaxis. I. Die absoluten Schwellenwerte und relative Wirksamkeit von Specktralfarben bei grünen und farblosen Einzelligen. *Zeit. Vergl. Physiol.* 14: 68–92.

Luntz, A. 1931*b*. Untersuchungen über die Phototaxis. II. Lichtintensität und Schwimmgeschwindigkeit bei *Eudorina elegans. Zeit. Vergl. Physiol.* 15: 652–678.

Lustigman, B., Lee, L.H., and Weiss-Magasic, C. 1995. Effects of cobalt and pH on the growth of *Chlamydomonas reinhardtii. Bull. Environ. Contam. Toxicol.* 55: 65–72.

Lustigman, B; McCormick, J.M., Dale, G., and McLaughlin, J.J. 1987. Effect of increasing copper and salinity on glycerol production by Dunaliella salina. *Bulletin of environmental contamination and toxicology* 38(2): 59–62.

Lustigman, B., Korky, J., Zabady, A., and McCormick, J.M. 1985. Absorption of copper(2+) by long-term cultures of *Dunaliella salina, D. tertiolecta,* and *D. viridis. Bulletin of Environmental Contamination and Toxicology* 35(3): 362–7.

Lynch, D.V. 1984. Effects of temperature on the membranes of *Dunaliella salina.* Avail. Univ. Microfilms Int., Order No. DA8414410, (1983), 160 pp, From: Diss. Abstr. Int. B 45(3): 740.

Lynch, D.V., Norman, H.A., and Thompson, G.A., Jr. 1984. Changes in membrane lipid molecular species during acclimation to low temperature by *Dunaliella salina. Developments in Plant Biology*, 9 (Struct., Funct. Metab. Plant Lipids), pp. 567–70.

Ma, S.Y., Huang, Y.C., Yang, X.H., and Wu, W.H. 1999. Effects of high $K^+$ and alkaline pH on ultrastructure of *Dunaliella salina* chloroplasts. *Zhiwu Xuebao* 41(12): 1342–1344.

Mainx, F. and Wolf, H. 1929. Reaktionsintensität und Stimmungsänderung in ihrer Bedeutung für eine Theorie der Phototaxis. *Arch. Protistenk.* 68: 105–176.

Malis-Arad, S., Freidlander, M., Ben-Aris, B., and Richmond, A.E. 1980. Alcalinity-induced aggregation in *Chlorella vulgaris.* I. Changes in cell volume and cell-wall structure. *Plant and Cell Physiology* 21(1): 27–35.

Manton I. and Parke M. 1965. Observations on the fine structure of two species of *Platymonas* with special reference to flagellar scales and the mode of origin of the theca. *J. Mar. Biol. Ass. U.K.* 45: 743–754.

Marangoni, R., Lorenzini, E. and Colombetti, G. 1996. Photosensory transduction in flagellated algae. pp. 263–274. In: *Light and Energy Source and Information Carrier in Plant Physiology.* Plenum Press, New York.

Marano, F. 1992. Flagellar Apparatus, Cell Motility and Phototaxis. pp.17–44. In: *Dunaliella: Physiologyt, Biochemistry, and Biotechnology.* A. Ben-Amotz and M. Avron, (Eds.) CRC Press.

Marano, F., Krishnaswamy, S., Betrencourt, C., Schoevaert, D., Provost, J.M., and Volochine, B. 1988. Control of ciliary beat by calcium: the effects of lindane, a potent insecticide. *Biology of the Cell* 63(2): 143–50.

Marbach, J. and Mayer, A.M. 1970. Direction of phototaxis in *Chlamydomonas reinhardii* and its relation to cell metabolism. *Phycologia* 9(3–4): 255–60.

Marbach, J. and Mayer, A.M. 1971. Effect of electric field on the phototactic response of *Chlamydomonas reinhardtii. Isr. J. Bot.* 20: 96–100.

Mardia, K.V. 1972. *Statistic of directional data.* Acad. Press, New York, London.

Margaritis, A., Kennedy, K., and Zajic, J.E. et al. 1979. Biosurfactant production by *Nocardia erythropolis*. *Develop. Ind. Microbiol.* Vol. 20. Proc. 35th Gen. Meet. Soc. Ind. Microbiol. (Houston, Tex., 1978), Arlington, 5a: 424–428.

Markert B.A.,. Breure A.M, and Zechmeister H.G. (Eds.). 2003. *Bioindicators and Biomonitors. Principles, Concepts and Applications.* Elsevier.

Markovits, A., Gianelli, M.P., Conejeros, R., and Erazo, S. 1993. Strain selection for β-carotene production by *Dunaliella*. *World Journal of Microbiology & Biotechnology* 9(5): 534–7.

Martin, R.L., Wood, C., Baehr, W., and Applebury, M. 1986. Visual pigment homologies refealed by DNA hybridization. *Science* 232: 1266–1269.

Martynenko, A.I., Posudin, Y.I., Massjuk, N.P., and Lilitskaya, G.G. 1996. Effect of external factors on photomovement of algae of genus *Dunaliella* Teod. *Algologia* 6(2): 150–156. (In Russian).

Masashi, T. and Teruo, S. 1982. Artificial control of cytoplasmic pH and its bearing on cytoplasmic streaming, electrogenesis and excitability of *Characeae* cells. *Bot. Mag. Tokyo* 95 (1038): 147–154.

Massjuk, N.P. 1961*a*. Investigation of *Dunaliella salina* in nature and in conditions of laboratory culture. All-Union Meeting on culture of unicellular algae, March 6–11, Leningrad. (In Russian).

Massjuk, N.P. 1961*b*. Carotene-containing alga *Dunaliella salina* Teod. in salt reservoirs of Crimea region. *Ukr.Bot.J.* 18(1): 100–109 (In Ukrainian).

Massjuk, N.P. 1965*a*. The ways of enrichment of natural stocks of *Dunaliella* as source of production of carotene in industrial scales. Problems of hydrobiology. I Congress of Hydrobiological Society. Moscow, February 1–6 (In Russian).

Massjuk, N.P. 1965*b*. Effect of ions of Na, Mg, Cl and $SO_4$ on growth, reproduction and carotene synthesis of alga *Dunaliella salina* Teod. *Ukr.Bot.J.* 22(5): 3–11 (In Ukrainian).

Massjuk, N.P. 1965*c*. Carbonates and bicarbonates as stimulators of growth and accumulation of carotene in culture *Dunaliella salina* Teod. *Ukr.Bot.J.* 22(6): 18–22 (In Ukrainian).

Massjuk, N.P. 1966. Mass culture of carotene-containing alga *Dunaliella salina* Teod. *Ukr. Bot. J.* 23(2): 12–19 (In Ukrainian).

Massjuk, N.P. 1967. Estimation of suitability of rapa of Saki reservoirs for cultivation of carotene-containing algae. *Ukr.Bot.J.* 24(4): 37–43 (In Ukrainian).

Massjuk, N.P. 1973. *Morphology, Systematic, Ecology, Geographical Distribution of the Genus Dunaliella Teod and Perspectives of its Practical Utilization.* Kiev, Naukova Dumka. (In Russian).

Massjuk, N.P. 1993. *Evolutionary aspects of morphology of eucariotic algae.* Kiev, Naukova Dumka. (In Russian).

Massjuk, N.P. 2006. *Chlorodendrophyceae* class. nov. (*Chlorophyta, Viridiplantae*) in flora of Ukraine.I. Capacity, phylogenitic links, systematic position. *Ukr.Bot.J.* 63(5): 662–675. (In Ukrainian).

Massjuk, N.P. and Abdula, S.G. 1969. First experience in cultivation of carotene-containing algae in semi-industrial conditions. *Ukr.Bot.J.* 26(3): 21–27. (In Ukrainian)

Massjuk, N.P. and Abdula, S.G. 1971. About results and perspectives of artificial cultivation of carotene-containing algae in semi-industrial conditions. *Materials of I Conf. on Study of SporePplants of Ukraine.* Kiev, Naukova Dumka. (In Russian)

Massjuk, N.P., Abdula, S.G., and Radchenko, M.I. 1970. Effect of some extraneous organisms on culture of *Dunaliella salina* Teod. in semi-industrial conditions. *Ukr. Bot. J.* 27(4): 456–461. (In Russian).

Massjuk, N.P. and Kostikov, I.Y. 2002. *Algae in the System of the Organic World*. Kiev. Academperiodika. (In Russian).

Massjuk, N.P.and Posudin Y.I. 1991*a*. *Logical Basis of Classification of the Behaviour of Motile Microorganisms*. Kiev. Ukr.Agr.Acad. Publ. (In Russian).

Massjuk, N.P.and Posudin, Y.I. 1991*b*. *Photoreceptor Systems of Monad Algae*. Kiev. Ukr.Agr.Acad. Publ. (In Russian).

Massjuk, N.P. and Posudin, Y.I. 2002. Rigorous fundamentals of classification of light-induced behaviour from freely motile microorganisms. *Acta Botanica Malacitana* 27: 147–158.

Massjuk, N.P. and Posudin, Y.I. 2007. Effect of pH on photomovement parameters of *Dunaliella salina* Teod. (*Chlorophyta*) Algologia 17(1): 14–20 (In Russian).

Massjuk, N.P., Posudin, Y.I., Radchenko, M.I., and Sheiko, O.M. 1988. Logical basis of classification of the behaviour of motile microorganisms. *Ukr.Bot.J.* 45(5): 1–7 (In Ukrainian).

Massjuk, N.P., Posudin, Y.I., Radchenko, M.I., and Sheiko, A.N. 1991. The parametrical principle of the classification of photomovement of organisms. *J. Photochem. Photobiol.* 10: 269–271.

Massjuk N., Posudin Yu., Lilitskaya G. 2007. Photomovement of the cells of *Dunaliella* Teod. (Dunaliellales, Chlorophyceae, Viridiplantae). Akademperiodika, Kiev (In Russian).5 January 2010. At: http://window.edu.ru/window_catalog/files/r56606/posudin.pdf

Massjuk, N.P. and Tereshuk, O.A. 1983. Collection of algal cultures of M. Kholodny Institute of Botany AN USSR. pp.104–114. In: *Cultivation of Collection Strains of Algae*. Leningrad. (In Russian).

Massjuk, N.P.and Yurchenko, V.V. 1962. Effect of concentration of hydrogen ions on alga *Dunaliella salina* Teod. *Ukr.Bot.J.* 19(4): 91–95 (In Ukrainian).

Mast, S.O. 1911. *Light and Behavior of Organisms*. John Wiley and Sons, New York, Chapman and Hall, London.

Mast, S.O. 1927. The structure and function of the eye-spot in unicellular and colonial organisms. *Arch. Protistenk.* 60(2): 197–220.

Matsuda, A., Yoshimura, K., Sineshchekov, O.A., Hirono, M., and Kamiya, R. 1998. Isolation and characterization of novel *Chlamydomonas* mutants that display phototaxis but not photophobic response. *Cell Motility and the Cytoskeleton* 41: 353–362.

Matsunaga, S., Takahashi, T., Watanabe, M., Sugai, M., and Hori, T. 1999. Control by ammonium ion of the change from step-up to step-down photophobically responding cells in the flagellate alga *Euglena gracilis*. *Plant and Cell Physiology* 40(2): 213–221.

Mattox, K.R. and Stewart, K.D. 1984. Classification of the green algae: A concept based on comparative cytology. In: *Systematics of the Green Algae*. D.E.G. Irvine and D.M. John (Eds.) Acad. Press, London, pp. 29–72.

Maurette, M.T., Oliveros, E., Infelta, P.P., Ramsteiner, K., and Braun, A.M.1983. Singlet oxygen and superoxide: experimental differentiation and analysis. *Helv. Chim. Acta.* 66: 722–733.

McFadden, G.J., Schulze, D., Surek, B., Salisbury, J.L., and Melkonian, M. 1987. Basal body reorientation mediated by a $Ca^{2+}$-modulated contractile protein. *J. Cell Biol.* 105: 903-912.

McLachlan, J. and Parke, M. 1967. *Platymonas impellucida* sp. nov. from Puerto Rico. *J. Mar. Biol. Ass. U.K.* 47: 723–733.

Melkonian, M. 1978. Structure and significance of cruciate flagellar root systems in green algae: comparative investigations in species of *Chlorosarcinopsis* (*Chlorosarcinales*). *Plant Syst. Evol.* 130: 265–292.

Melkonian, M. 1980. Ulrtastructural aspects of basal body associated fibrous structures in green algae: a critical review. *BioSystems* 12: 85–103.

Melkonian, M. 1981. Fate of eyespot lipid globules after zoospore settlement in the green alga *Pleurastrum terrestre* Fritsch et John. *Br. Phycol. J.* 16: 247–255.

Melkonian, M. 1983. Evolution of green algae in relation to endosymbiosis. pp. 1003–1007. In: *Endocytobiology* II. H.N.A. Schenk and W. Schwemmler (Eds.). De Gruyter, Berlin, New York.

Melkonian, M. 1984. Flagellar apparatus ultrastructure in relation to green algae classification. pp. 73–120.In: *Systematics of the Green Algae.* D.E.G. Irvine and D.M. John (Eds.). Acad. Press, London.

Melkonian, M. 1989*a*. The functional analysis of the flagellar apparatus in green algae. *Symposia of the Society for Experimental Biology* 35: 589–606.

Melkonian, M. 1989*b*. Centrin-mediated motility: a novel cell motility mechanism in eucaryotic cells. *Botanica Acta* 102: P. 3–4.

Melkonian, M. 1990. Phylum *Chlorophyta*. Class *Chlorophyceae*. In: *Handbuch of Protoctista*. Margulis L. et al. (Eds.). Jones and Bartlett Publ., Boston, pp. 608-616.

Melkonian, M. and Preisig, H.R. 1984. Ultrastructure of the flagellar apparatus in the green flagellate *Spermatozopsis similis*. *Pl. Syst. Evol.* 146: 145–162.

Melkonian, M. and Robenek, H. 1979. The eyespot of the flagellate *Tetraselmis cordiformis* Stein (*Chlorophyceae*): structural specialization of the outer chloroplast membrane and its possible significance in phototaxis of green alga. *Protoplasma* 100: 183–197.

Melkonian, M. and Robenek, H. 1980. Eyespot membranes of *Chlamydomonas reinhardtii*: A freeze-fracture study. *J. Ultrastr. Res.* 72: 90–102.

Melkonian, M. and Robenek, H. 1984. The eyespot apparatus of flagellated green algae: a critical review. *Progress in Phycol. Res.* 3: 195–268.

Melkonian, M., Schulze, D., McFadden, G.J., and Robenek, H. 1988. A polyclonal antibody (anticentrin) distinguishes between two types of fibrous flagellar roots in green algae. *Protoplasma* 144: 56–61.

Melnikov, S.V., Alyoshkin, V.R., and Roshin, P.M. 1972. *Planning Experiment in Investigation of Agricuyltural Processes*. Leningrad, Kolos. (In Russian).

Mendez-Alvarez, S., Leisinger, U., and Eggen, R. I. L. 1999. Adaptive responses in *Chlamydomonas reinhardtii*. *International Microbiology* 2(1): 15–22.

Menzel, D.W., Anderson, J., and Randtke, A. 1970. Marine phytoplankton vary in their response to chlorinated hydrocarbons. *Science* 167(926): 1724–6.

Merten, P., Lechtreck, K.F., and Melkonian, M. 1995. Nucleus basal body connector of Dunaliella: threshold concentration of calcium necessary for in vitro contraction. *Botanica Acta* 108(1): 2–6.

Messerli, M.A., Amaral-Zettler, L.A., Zettler, E., Jung, S.K., Smith, P.J.S., and Sogin, M.L. 2005. Life at acidic pH imposes an increased energetic cost for a eukaryotic acidophile. *J. Exp. Biol.* 208: 2569–2579.

Metzner, P. 1929. Bewegungstudien an Peridineen. *Z. Bot.* 22: 287–299.

Meyer, R. and Hildebrand, E. 1988. Phototaxis of *Euglena gracilis* at low external calcium concentration. *J. Photochem. Photobiol.* B: Biology. 2: 443–453.

Mikolajczyk, E. 1986. $Na^+/K^+$ transport and photosensitivity of the colorless flagellate *Peranema trichophorum* (*Euglenida*). *Photochem. Photobiol.* 43: 455–459.

Mikolajczyk, E. 1984*a*. Photophobic responses in *Euglenina*. I. Effects of excitation wavelength and external medium on the step-up response of light- and dark-grown *Euglena gracilis*. *Acta Protozool.* 23: 1–10.

Mikolajczyk, E. 1986. $Na^+-K^+$ transport and photosnsitivity of the colorless flagellate *Peranema trichophorum* (*Euglenida*). *Photochem. Photobiol.* 43: 455–459.

Mikolajczyk, E. 1984*b*. Photophobic responses in *Euglenina*. II. Sensitivity to light of the colorless flagellate *Astasia longa* in low and high viscosity medium. *Acta Protozool.* 23: 85–92.

Mikolajczyk, E. and Diehn, B. 1975. The effect of potassium iodide on photophobic responses in *Euglena*: evidence for two photoreceptor pigments. *Photochem.Photobiol.* 22(6): 269–71.

Mikolajczyk, E. and Diehn, B. 1978. Morphological alterations in *Euglena gracilis* induced by treatment with CTAB (cetyltrimethylammonium bromide) and Triton X-100: correlations with effects on photophobic behavioural responses. *J. Photozool.* 25: 461–470.

Mikolajczyk, E. and Diehn, B. 1979. Mechanosensory responses and mechanoreception in *Euglena gracilis*. *Acta Protozoologica* 18: 591-602.

Mikolajczyk, E. and Walne, P.L 1990. Photomotile responses and ultrastructure of the euglenoid flagellate *Astasia fritschii*. *J. Photochem.Photobiol.*, B: Biology 6(3): 275-82.

Mil'ko, E.S. 1963. Effect of illuminance and temperature on synthesis of pigments in *Dunaliella salina*. *Microbiologiya* 32(4): 307. (In Russian).

Mitchell, D.R. 2000. *Chlamydomonas* flagella. *J. Phycol.* 36: 261–273.

Miyoshi, Y. 1979. Photomovements. *Koseibutsugaku* 2: 141–51.

Mizoguchi, T. 2006. *Dunaliella salina* (Dunal) Teodoresco powders as hepatopathy inhibitors. Jpn. Kokai Tokkyo Koho JKXXAF JP 2006213613 A 20060817.

Moestrup, Ø. and Thomsen, H.A. 1974. An ultrastructural study on the flagellate *Pyramimonas orientalis* with particular emphasis on Golgy apparatus activity and the flagellar apparatus. *Protoplasma* 81: 247–269.

Mohammad, R.H. and Mansour, S. 2003. Acquaintance with biotechnological abilities of *Dunaliella* algae. Majmoa-i Maghalat-i Sevomin Hemayesh Maliy Biotechnology Jomhoriy-i Islame-i Iran, Mashhad, Islamic Republic of Iran, Sept. 9-11, 2003 4: 30–33.

Mohn, F.H. and Contreras, O.C. 1990. Harvesting of the alga *Dunaliella*: some considerations concerning its cultivation and impact on the production costs of β-carotene. Antofagasta, Chile: Forschungszentrum Jüllich, 55 p.

Morel-Laurens, N. 1987. Calcium control of phototactic orientation in *Chlamydomonas reinhardtii*: Sign and strength of response. Photochem. Photobiol. 45: 119–128.

Morens, H.M.L. and Feinleih, M. 1983. Photomovement in an "eyeless" mutant of *Chlamydomonas*. *Photochem. Photobiol.* 37: 189–194.

Mornin, L. and Francis, D. 1967. The fine structure of *Nematodium armatum*, a naked Dinoflagellate. *J. Microsc.* 6: 759–772.

Mosolov, A. and Belkin, A. (1980). "Secret of Antony van Leeuwenhoek?" *Nauka i Zhizn* (Science and Life) 9: 80–2

Mosteller, F. and Tukey, J.W. 1978. Data Analysis and Regression. Massachusets: Addison-Wisley Publ. Comp. Inc., Reading.

Moulton, T.P., Borowitzka, L.J., and Vincent, D.J. 1987. The mass culture of *Dunaliella salina* for β-carotene: from pilot plant to production plant. *Hydrobiologi* 151/152: 99–105.

Murthy, K.N.C., Rajesha, J., Swamy, M.M., and Ravishankar, G.A. 2005. Comparative Evaluation of Hepatoprotective Activity of Carotenoids of Microalgae. *Journal of Medicinal Food* 8(4): 523–528.

Murthy, K.N.C., Vanitha, A., Rajesha, J., Swamy, M.M., Sowmya, P.R., Ravishankar, and G.A. 2005. *In vivo* antioxidant activity of carotenoids from *Dunaliella salina* – a green microalga. *Life Sciences* 76(12): 1381–1390.

Mustacich, R.V. and Ware B.R. 1976. A study of protoplasmic streaming in *Nitella* by laser Doppler spectroscopy. *Biophysical journal* 16(5): 373–88.

Myroniuk, V.I. and Kurinna, S.M. 2000. Influence of azide and salicylhydroxamate on the respiration of *Chlamydomonas snowiae* Printz and *Dunaliella salina* Teod. *Ukrains'kii Botanichnii Zhurnal* 57(3): 302-305.

Myronyuk, V.I., Massjuk, N.P., and Akopyants, N.S. 1980. Effect of pH and some inhibitors on activity of catalase oligo- and hyperhalobic algae. *Ukr.Bot.J.* 39(3): 60–62. (In Ukrainian).

Nagel, W.A. 1901. Phototaxis, Photokinesis und Unterschiedsempfindlichkeit. *Bot. Ztg.* 59: 287–299.

Nagel, G., Ollig, D., Fuhrmann, M., Kateriya, S., Musti, A. M., Bamberg, E., and Hegemann, P. 2002. Channelrhodopsin-1: a light-gated proton channel in green algae. *Science* 296(5577): 2395–2398.

Nagel, G., Szellas, T., Kateriya, S., Adeishvili, N., Hegemann, P., and Bamberg, E. 2005. Channelrhodopsins: directly light-gated cation channels. *Biochemical Society Transactions* 33(4): 863–866.

Nakajima, T. and Takahashi, M. 1991. A photo-bioreactor using algal phototaxis for solids-liquid separation. *Water Research* 25(10): 1243–7.

Nakamura, K., Bray, D.F., Costerton, J.W. and Wagenaar, E.B. 1973. The eyespot of *Chlamydomonas eugametos*: a freezeetch study. *Can. J. Bot.* 51: 817–819.

Nakamura, S., Ogihara, H., Jinbo, K., Tateishi, M., Takahashi, T., Yoshimura, K., Kubota, M, Watanabe, M. and Nakamura, S. 2001. *Chlamydomonas reinhardtii* Dangeard (Chlamydomonadales, Chlorophyceae) mutant with multiple eyespots. *Phycol.Res.* 49:    115–122.

Nakanishi, K., Derguini, F., Rao, V. J., Zarrilli, G., Okabe, M., Lien, T., Johnson, R., Foster, K.W., and Saranak, J. 1989. Theory of rhodopsin activation:  probable charge redistribution of excited state chromophore. *Pure and Applied Chemistry* 61(3): 361–4.

Nakanishi, K. 1985. Bioorganic studies with rhodopsin. *Pure and Applied Chemistry* 57(5): 769–76.

Neumann, R. and Hertel, R. 1994. Purification and characterization of a riboflavin-binding protein from flagella of *Euglena gracilis*. *Photochemistry and Photobiology* 60(1): 76–83.

Neuscheler, W. 1967. Bewegung und Orientierung bei *Micrasterias denticulata* Bréb. im Licht. I. Zur Bewegungs- und Orientierungsweise. *Z. Pflanzenphysiol.* 57: 46–59.

Nickel, M., Leininger, S., Proll, G., and Brummer, F. 2001. Comparative studies on two potential methods for the biotechnological production of sponge biomass. *Journal of Biotechnology* 92(2): 169–178.

Nichols, K.M. and Rikmensoel, R. 1978. Control of flagellar motion in *Chlamydomonas* and *Euglena* by mechanical microinjection on $Mg^{2+}$ and $Ca^{2+}$ and by electric current injection. *J. Cell Sci.* 29: 233–247.

Nikookar, K., Moradshahi, A., and Hosseini, L. 2005. Physiological responses of *Dunaliella salina* and *Dunaliella tertiolecta* to copper toxicity. *Biomolecular Engineering* 22(4): 141–146.

Noaman, N.H., Shaalan, S.H., Khaleafa, A.M., and Abd El Aziz, W.M. 2002. Physiological responses of *Synechococcus leopoliensis* and *Dunaliella salina* to pyrethroids and organophosphorus insecticides. *Egyptian Journal of Biotechnology* 11: 124–137.

Nobel, P. 1973. *Physiology of plant cell (physical-chemical approach).* Moscow. Mir. (In Russian).

Norman, H.A. and Thompson, G.A., Jr.1985. Effects of low-temperature stress on the metabolism of phosphatidylglycerol molecular species in *Dunaliella salina. Archives of biochemistry and biophysics* 242(1): 168–75.

Norris, R.E. and Pearson, B.R. 1975. Fine structure of *Pyramimonas parkeae,* sp. nov. (*Chlorophyta, Prasinophyceae*). *Arch. Protistenk.* 117: 192–203.

Nosova, L. P., Goronkova, O.I., Mikhailova, L.I., Spektorova, L.V. 1979. Intensive culturing of *Dunaliella tertiolecta.* Kordyum, V.A. (Ed.) *Rol Nizshikh Org. Krugovorote Veshchestv Zamknutykh Ekol. Sist.,* Mater. Vses. Soveshch., 10th, pp. 85–9.

Novikova, I.P., Parshykova, T.V., Vlasenko, V.V., and Zubenko, I.B. 2007. Effect of $K_2Cr_2O_7$ on the photosynthetic activity and mobility of *Euglena gracilis* Klebs cells. *International Journal on Algae* 9(3): 224–235.

Ntefidou, M., Iseki, M., Richter, P., Streb, C., Lebert, M., Watanabe, M., and Haeder, D.P. 2003*a*. RNA interference of genes involved in photomovement in *Astasia longa* and *Euglena gracilis* mutants. *Recent Research Developments in Biochemistry* 4(Pt. 2): 925–930.

Ntefidou, M., Iseki, M., Watanabe, M., Lebert, M., and Häder, D.P. 2003*b*. Photoactivated adenylyl cyclase controls phototaxis in the flagellate Euglena gracilis. *Plant Physiology* 133(4): 1517–1521.

Nultsch, W. 1962. Phototactische Actionsspectren von Cyanophyceen. *Ber. Dtsch. Bot. Ges.* 75: 43–453.

Nultsch, W. 1970. Photomotion of microorganisms and its interaction with photosynthesis. In: *Photobiology of microorganisms.* P. Halldal (Ed.), Wiley, London, New York, pp. 213–251.

Nultsch, W. 1971. Phototactic and photokinetic action spectra of the diatom *Nitzschia communis. Photochem Photobiol.* 14: 705–712.

Nultsch, W. 1973. Relation between photomotion and photosynthesis. pp. 245–273. In: *Primary Molecular Events in Photobiology.* A. Checcucci and R.A. Weale (Eds.). NATO Advanced Study Institute, Badia Fiesolana, 4-16 September, 1972. Elsevier Sci. Publ. Comp., Amsterdam.

Nultsch, W. 1975. Phototaxis and photokinesis. pp. 29–90. In: *Primitive Sensory and Communication Systems: The Taxes and Tropisms of Microorganisms and Cells.* M.J. Carlile (Ed.). Acad. Press, N.Y., San. Francisco.

Nultsch, W. 1977. Effect of external factors on phototaxis of *Chlamydomonas reinhardtii.* II. Carbon dioxide, oxygen and pH. *Archives of Microbiology* 112(2): 179–85.

Nultsch, W. 1979. Effect of external factors on phototaxis of *Chlamydomonas reinhardtii.* III. Cations. *Arch. Microbiol.* 123: 93–99.

Nultsch, W. 1980. Photomotile responses in gliding organisms and bacteria. pp. 69–88. In: *Photoreception and Sensory Transduction in Aneural Organisms.* F. Lenci and G. Colombetti (Eds.). Plenum Press, New York.

Nultsch, W. 1983. The photocontrol of movement of *Chlamydomonas*. pp. 521–539. In: *The Biology of Photoreception*. D.J. Cosens and D. Vince-Price (Eds.), Soc. Experiment. Biology, Symp. XXXVI.

Nultsch, W. 1985. Photosensing in cyanobacteria. pp. 147–164. In: *Sensory Perception and Transduction in Aneural Organisms*. G. Colombetti, F. Lenci, and P.S. Song, (Eds.). Plenum Press, New York, London.

Nultsch, W. 1991. Survey of photomotile responses in microorganisms. pp. 1–5. In: *Biophysics of Photoreceptors and Photomovement in Microrganisms*. F. Lenci et al. (Eds.). Plenum Press, New York

Nultsch, W. and Häder, D.P. 1979. Photomovement of motile microorganisms. *Photochem. Photobiol.* 29: 423–437.

Nultsch, W. and Häder, D.P. 1988. Photomovement in motile microorganisms II. *Photochem. Photobiol.* 47: 837–869.

Nultsch, W., Pfau, J., and Dolle, R. 1986. Effects of calcium channel blockers on phototaxis and motility of *Chlamydomonas reinhardtii*. *Arch. Microbiol.* 144: 393–397.

Nultsch, W. and Rueffer, U. 1994. The orientation of free-swimming organisms towards light using the example of the flagellate Chlamydomonas reinhardtii. *Naturwissenschaften* 81(4): 164–74.

Nultsch, W., Schuchart,H., and Koenig, F. 1983. Effects of sodium azide on phototaxis of the blue-green alga *Anabaena variabilis* and consequences to the two-photoreceptor systems-hypothesis. *Arch. Microbiol.* 134: 33–37.

Nultsch, W. and Throm, G. 1975. Effect of external factors on phototaxis of *Chlamydomonas reinhardtii*. *Arch. Microbiol.* 103: 175–179.

Nultsch, W., Throm, G., and Rimscha, I. 1971. Phototactische Untersuchungen an *Chlamydomonas reinhardtii* Dangeard in homokontinuierlicher Kultur. *Arch. Microbiol.* 80: 351–369.

Ohta, H., Shirakowa, H., Uchida, K., Yoshida, M., Matuo, Y., and Enami, I. Cloning and sequencing of the gene encoding the plasma membrane $H^+$-ATPase from an acidophilic red alga, *Cyanidium caldarium*. *Biochim. Biophys. Acta* 1319: 9–13.

O'Kelly, C.G. and Floyd, G.L. 1983-1984. Flagellar apparatus absolute orientations and the phylogeny of the green algae. *Bio Systems* 16(3-4): 227–51.

Okita, N., Isogai, N., Hirono, M., Kamiya, R., and Yoshimura, K. 2004. Phototactic activity in *Chlamydomonas* 'non-phototactic' mutants deficient in $Ca^{2+}$-dependent control of flagellar dominance or in inner-arm dynein. *Journal of Cell Science* 118(3): 529–537.

Omodeo, P. 1975. Morphology of the phototactic apparatus in eucaryotic flagellated cells. pp. 24-50. In: *Biophysics of Photoreceptors and Photobehaviour of Microorganisms*. Proc. of the Intern. School of the CNR of Italy, Badia Fiesolana (Firenze), 1–5 September, 1975.

Oren, A. 2005. A hundred years of Dunaliella research: 1905-2005. *Saline Systems* 1: 1–17.

Orme, S. and Kegley, S. 2004. PAN pesticide database. pp. 119–146. In: *Pesticide Action Network*. P.S. Song (Ed.), CA Plenum Publ. Corp., San Francisco.

Orset, S. and Young, A.J. 1999. Low-temperature-induced synthesis of β-carotene in the microalga *Dunaliella salina* (Chlorophyta). *J. Phycol.* 35(3): 520–527.

Pace, F., Ferrara, R., and Del Corratore, G. 1977. Effects of sub-lethal doses of copper sulphate and lead nitrate on growth and pigment composition of *Dunaliella salina* Teod. *Bull. Environm. Contam. Toxicology* 17: 679–685.

Parke, M. and Manton, I. 1965. Preliminary observations on the fine structure of *Prasinocladus marinus. J. Mar. Biol. Ass. U.K.* 45: 525–536.

Parke, M. and Manton, I. 1967. The specific identity of the algal symbiont in *Convoluta roscoffensis*. *J. Mar. Biol. Ass. U.K.* 47: 445–464.

Parshikova, T.V. and Pakhomova, M.N. 1988. Sensibility of different species of algae to the action of surface-active substances. *Hydrobiological J.* Kiev, 1988. Deposit in VINITI 26.04.88, № 3215–B88. (In Russian).

Parshikova, T.V. 1988. Effect of surface-active substances of various chemical nature on viability of natural populations of algae. *Hydrobiological J.* Kiev, 1988. Deposit in VINITI 26.04.88, № 3216–B88. (In Russian)

Parshikova, T.V. 2004. *Surface-Active Substances as Regulating Factors of the Development of Algae*. Kiev, Phytosociocentre.

Parshikova, T.V., Lipnitska, G.P., Liltskaya, G.G., and Posudin, Y.I. 1990. Effect of SAS on photomovement and chlorophyll fluorescence of two species of *Dunaliella* Teod. *Ukr.Bot.J.* 47(5): 60–63. (In Russian).

Parshikova, T.V. 2003. Surfactant role in the regulation of microscopic algae development. Kiev. Nats. Univ. Kiev, Ukraine. *Gidrobiologicheskii Zhurnal* 39(1): 64–70.

Parshikova, T.V., Vlasenko, V.V., Shchegoleva, T.Y., and Musienko, M.M. 2004. Approbation of express-methods for nondestructive control over the functioning of algae as bioindicators under environment contamination. *Dopovidi Natsional'noi Akademii Nauk Ukraini* 11: 172–178.

Pazour, G.J., Sineshchekov, O.A., and Witman, G.B. 1995. Mutational analysis of the phototransduction pathway of Chlamydomonas reinhardtii. *Journal of Cell Biology* 131(2): 427–40.

Pavlov, I.P. (1849–1936) Ivan Pavlov. 4 January 2010. At: <http://en.wikipedia.org/wiki/Ivan_Pavlov>.

Perfiliev, B.V. 1915. About the movement of blue-green algae *Synechococcus*. *J. Microbiology*. 2: 283–293. (In Russian).

Peskov V.A. 1979. Biopolymers of blue-green algae and some ways of their use. Dissertation Candidate Biol. Sci. Kiev. (In Russian).

Pfau, J., Nultsch, W., and Ruffer, U.A. 1983. A fully automated and computerized system for simultaneous measurement of motility and phototaxis in *Chlamydomonas*. *Arch. Microbiol.* 135: 259–264.

Pfeffer, W. 1904. *Pflanzenphysiologie*. Vol. II. 2$^{nd}$ ed., Engelmann, Leipzig.

Piccini, E. and Omodeo, P. 1975. Photoreceptors and phototactic programs in protista. *Boll. Zool.* 42: 57–79.

Pick, U. 1992. ATPases and ion transport in *Dunaliella*. pp. 63-93. In: *Dunaliella: Physiology, Biochemistry and Biotechnology*. M. Avron and A. Ben-Amotz (Eds.). Boca Raton etc. CRC Press.

Pick, U. 1999. *Dunaliella acidophila* – a most extreme acidophilic alga. pp. 465–476. In: *Enigmatic Microorganisms in Extreme Environments*. J. Seckbach (Ed.). Dordrecht Kluwer Acad. Press, The Netherlands.

Pick, U., Bannet, G., Fisher, M., Katz, A., Weiss, M., and Zchut, S. Molecular mechanisms of adaptation to extreme conditions in the alga *Dunaliella*. At: "http://www.weizmann.ac.il/Biological Chemistry/Pick/uripick.html"

Pick, U., Katz, A., Levin, E., Paz, K., Ventrela, R., and Weiss, M. Molecular basis of salinit-tolerance in the halotolerant alga *Dunaliella*. 5 January 2010. At: http://www.weizmann.ac.il/Biology/open_day_2000/images/upic.pdf

Pileri, D. 1988. Application of laser light scattering to intracellular streaming. Eastman Kodak Co., Rochester, NY, USA. Proceedings of the International Conference on Lasers, Volume Date 1987, pp. 791–801

Pisal, D.S. and Lele, S.S. 2005. Carotenoid production from microalga, *Dunaliella salina*. *Indian Journal of Biotechnology* 4(4): 476–483.

Poff, K.L. 1985. Temperature sensing in microorganisms. p. 299. In: *Sensory Perception and Transduction in Aneural Organisms*. G. Colombetti, F. Lenci, and P.S. Song (Eds.). Plenum Press, New York.

Poff, K.L. and Hong, C.B. 1982. Photomovement and photosensory transduction in microorganisms. *Photochem. Photobiol.* 36: 749–752.

Ponomarenko, S., Zhminko, P., and Zhminko, O. 2001. The use of Dopler's laser spectroscopy for toxicity's screening and monitoring of chemical substances. Prague 2000, International Symposium & Exhibition on Environmental Contamination in Central & Eastern Europe, Proceedings, 5th, Prague, Czech Republic, Sept. 12–14, Meeting Date 2000, pp.1616–1620.

Posudin, Y.I. 1982. Photobehaviour of *Euglena gracilis*. *Usp.sovr.biologii* 93(2): 230–235. (In Russian).

Posudin, Y.I. 1985. *Laser microfluorometry of biological objects*. Kiev, Vyssha shkola. (In Russian).

Posudin, Y.I. 1989. *Laser photobiology*. Kiev, Vyssha shkola (In Russian).

Posudin, Y.I. 1992*a*. Optical methods of investigation of photobiological reactions of higher and lower plants. Dissertation Dr. Biol. Sci. St. Petersbourg. (In Russian).

Posudin, Y.I. 1992*b*. Method of biotesting the presence of chemical compounds in aquatic medium. Invention Sertificate 173992288, № 4737382. (In Russian).

Posudin, Y.I. 1995. *Biophysicist Sergei Tschachotin*. Kiev, Nat.Agr.Univ.Publ. (In Russian).

Posudin, Y.I. 1998. Measurement of gravitaxis of algae as method of biomonitoring of aquatic media. *Nauk.visnyk NAU*. 3: 15–21. (In Russian).

Posudin Yu.I. 2007–3. Experimental investigation of photomovement of algae. Naukovi dopovidi NAU (In Ukrainian).http://nd.nauu.kiev.ua/2007-3/07pyioaf.pdf

Posudin, Y.I. 2009. Principle of microphotometry of flagella beating . Patent of Ukraine № 45376, Registration N U200905299 (In Ukrainian).

Posudin, Y.I. *Biophysisist Sergei Tschachotin* (electronic version). A.Y. Budantsev (Ed.) 5 January 2010. At: "http://www.edu.ru/db/portal/e-library/00000051/Чахотин.pdf" (In Russian).

Posudin Y.I. and Massjuk, N.P. 1993. Photomovement of *Dunaliella* Teod. *Biol. Nauki*. 2: 28–32. (In Russian).

Posudin, Y.I. and Massjuk, N.P. 1996. Diffractional mechanism of photoreception of unicellular green flagellates. *Algologia* 6(4): 368–376. (In Russian).

Posudin, Y.I. and Massjuk, N.P. 1997. Diffraction mechanism of photoreception in green unicellular flagellated algae. *Hydrobiol. J.* 33(6): 124–131.

Posudin, Y.I. and Suprun, A.D. 1992. To the problem of photoregulation mechanism of flagellates. *Biol. Nauki*. 11–12: 27–32. (In Russian).

Posudin Yu.I., Didyk A.V. 2007-2(7). Experimental confirmation of ptrotective functions of carotene during photosensitization *Dunaliella salina*. Naukivi dopovidi of NAU. At: http://www.nbuv.gov.ua/e-Journals/nd/2007-2/07pyiods.pdf

Posudin, Y.I., Kononchuk, V.R., Massjuk, N.P., and Lilitskaya, G.G. 1991. To studying the mechanisms of photoreception of *Dunalilla salina* Teod. *Algologia* 1(3): 24–34. (In Russian)

Posudin, Y.I., Massjuk, N.P., and Lilitskaya, G.G. 1993. Effect of calcium ions on photomovement of two species of genus *Dunaliella* Teod. *Algologia* 3(3): 16–23. (In Russian)

Posudin, Y.I., Massjuk, N.P., and Lilitskaya, G.G. 2004. Effect of ultraviolet radiation on photomovement of two species *Dunaliella* Teod. *Algologia* 14(2): 113–126. (In Russian).

Posudin, Y.I., Massjuk, N.P., and Lilitskaya, G.G. 1996a. Videomicrography of algae photomovement and vectorial method of biomonitoring. Proceedings of SPIE-The International Society for Optical Engineering 2628 (Optical and Imaging Techniques for Biomonitoring), pp. 382–388.

Posudin, Y.I., Massjuk, N.P., and Lilitskaya, G.G. 1996b. Vector method of biotesting aquatic media. *Algologia* 1(1): 15–25.

Posudin, Y.I., Massjuk, N.P., and Lilitskaya, G.G. 1996c. Photomovement parameters as test-functions during biomonitoring. *Polish J. Env. Sci.* 5(3): 51–57.

Posudin, Y.I., Massjuk, N.P., Lilitskaya, G.G., and Golubkova, M.G. 1992. Action of ionizing radiation on photomovement of algae. *Radiobiology* 32(2): 292–298. (In Russian).

Posudin, Y.I., Massjuk, N.P., Lilitskaya, G.G., and Radchenko, M.I. 1990a. Phototopotaxis of two species *Dunaliella* Teod. in ultraviolet part of spectrum. *Biofisica* 35(6): 968–971. (In Russian).

Posudin, Y.I., Massjuk, N.P., Lilitskaya, G.G., and Radchenko, M.I. 1991. Phototopotaxis of two species *Dunaliella* Teod. *Ukr.Bot.J.* 48(4): 48–53. (In Russian).

Posudin, Y.I., Massjuk, N.P., Lilitskaya, G.G., and Radchenko, M.I. 1992b. Photomovement of two species of *Dunalialla* Teod. (*Chlorophyta*). *Algologia* 2(2): 37–47.

Posudin, Y.I., Massjuk, N.P., Lilitskaya, G.G., and Shevchenko, A.I. 1995. Effect of sodium azide on photomovement of two species *Dunaliella* Teod. *Plant Physiology.* 42(3): 432–434. (In Russian).

Posudin, Y.I., Massjuk, N.P., Radchenko, M.I., and Lilitskaya, G.G.1988. Photokinetic reactions of two species *Dunaliella* Teod. *Microbiologia* 57(6): 1001–1006. (In Russian).

Posudin Yu., Hanelt D., and Wiencke C. 2004a. Effect of ultraviolet and photosynthetically active radiation on green alga *Ulva lactuca*// // Naukovyi visnyk NAU. 72: 13–31.

Posudin Yu., Murakami A., Kamiya M., and Kawai H. 2004b. Effect of light of different intensity on chlorophyll fluorescence of *Ulva pertusa* Kjellman (*Chlorophyta*).*Int. J. on Algae.* 6(3):235–250.

Posudin, Y.I. and Repetskii, N.S. 1988. The destructive reactions of lower organisms to the action of light and sensitizers. *Biologicheskie Nauki* (Moscow) 1: 33–6.

Prasanna, R., Sood, A., Suresh, A., Nayak, S., and Kaushik, B.D. 2007. Potentials and applications of algal pigments in biology and industry. *Acta Botanica Hungarica* 49(1–2): 131–156.

Proceedings of the Third Asia-Pacific Conference on Algal Biotechnology, held 7–10 May 1997, Phuket, Thailand. *J. Appl. Phycol.* 9(5) Borowitzka, M.A. (Ed.) Neth., 90 pp.

Proschöld, T., Marin, B., Schlösser, U.G., and Melkonian, M. 2001. Molecular phylogeny and taxonomic revision of *Chlamydomonas* Ehrenberg and *Chloromonas* Gobi, and description of *Oogamochlamys* gen. nov. and *Lobochlamys* gen. nov. *Protist* 152: 265–300.

Pun, K.C., Cheung, R.Y.H., and Wong, M.H. 1995. Characterization of sewage sludge and toxicity evaluation with microalgae. *Marine Pollution Bulletin* 31(4–12): 394–401.

Quarmby, L.M. 1996. $Ca^{2+}$ influx activated by low pH in *Chlamydomonas*. *Journal of General Physiology* 108(4): 351–361.

Racey, T.J., Hallett, F.R., and Nickel, B. 1981. A quasi-elastic light scattering and cinematographic investigation of motile *Chlamydomonas reinhardtii*. *Biophys. J.* 35: 557–571.

Raja, R., Hemaiswarya, S., and Rengasamy, R. 2007. Exploitation of *Dunaliella* for beta-carotene production. *Applied microbiology and biotechnology* 74(3): 517–23.

Ramazanov, Z. M., Klyachko-Gurvich, G. L., Ksenofontova, A. L., and Semenenko, V. E. 1988. Influence of suboptimal temperature on the content of $\beta$-carotene and lipids in the halophilic alga *Dunaliella salina*. *Fiziologiya Rastenii* (Moscow) 35(5): 864–9.

Rao, P.S.N., Chauhan, V.D., and Rao, K.S. 1982. Effects of sodium chloride, pH and carbon source on the growth of brine alga, *Dunaliella* sp. *Indian Journal of Marine Sciences* 11(3): 262–3.

Rashkova, S. and Vlakhov, S. 1988. Marine biotechnology. *Priroda* (Sofia) 37(2): 45–9.

Raven, J.A. 1990. Sensing pH? *Plant, Cell and Environment* 13(7): 721–9.

Remis, D., Brunovska, A., Ziegler, W., and Gimmler, H. 1994. Photoinduced pH changes in suspensions of the acid-resistant green alga *Dunaliella acidophila*. *Biologia* (Bratislava, Slovakia) 49(4): 645–50.

Remis, D., Simonis, W., and Gimmler, H. 1992. Measurments of the transmembrane electrical potential of *Dunaliella acidophila* by microelectrodes *Arch. Microbiol.* 158: 350–355.

Rhiel, E., Häder, D.P., and Wehrmeyer, W. 1988. Diaphototaxis and gravitaxis in a freshwater *Cryptomonas*. *Plant Cell Physiol.* 29: 755–760.

Richter, P.R., Hader, D.P., Goncalves, R.J., Marcoval, M A., Villafane, V.E., and Helbling, E.W. 2007. Vertical migration and motility responses in three marine phytoplankton species exposed to solar radiation. *Photochemistry and photobiology* 83(4): 810–7.

Richter, P., Helbling, W., Streb, C., and Häder, D.P. 2007. PAR and UV effects on vertical migration and photosynthesis in Euglena gracilis. *Photochemistry and Photobiology* 83(4): 818–823.

Richter, P., Lebert, M., Tahedl, H., and Häder, D.P. 2001. Calcium is involved in the gravitactic orientation in colorless flagellates. *J. Plant. Phys.* 158: 689–697.

Ridge, K.D. 2002. Algal rhodopsins: phototaxis receptors found at last. *Current biology: CB* 12(17): R588–90.

Riisgaard, H.U., Noergaard, N.K., and Soegaard-Jensen, B.1980. Further studies on volume regulation and effects of copper in relation to pH and EDTA in the naked marine flagellate *Dunaliella marina*. *Marine Biology* (Berlin, Germany) 56(4): 267–76.

Ringo, D.L. 1967. The arrangement of subunits in flagellar fibers. *J. Ultrastruct. Res.* 17: 266–277.

Ristori, T., Ascoli, C., Banchetti, R., Parrini, P., and Petracci, D. 1981. Localization of photoreceptor and active membrane in the green alga *Haematococcus pluvialis*. In: *Proc. of the Sixth Int. Congress on Protozoology*, Warsaw, p. 314.

Riyahi, J., Haouazine, Y., Akallal, R., Givernaud, T., Lemoine, Y., and Mouradi, A. 2006. Valorization attempt of a Moroccan salt pond: $\beta$-carotene production by the halotolerant green alga *Dunaliella salina*. *Archiv fuer Hydrobiologie*, Supplement 163: 51–62.

Rodriguez, M.C., Barsanti, L., Passarelli, V., Evangelista, V., Conforti, V., and Gualtieri, P. 2007. Effects of chromium on photosynthetic and photoreceptive apparatus of the alga *Chlamydomonas reinhardtii*. *Environmental Research* 105(2): 234–239.

Rokitsky, P.F. 1973. *Biological statistics*. Minsk.Vysh. Shkola. (In Russian).

Rosenbaum, J.L. and Child, F.M. 1967. Flagellar regeneration in protozoan flagellates. *J. Cell Biol.* 147: 179–183.

Rothert, W. 1901. Beobachtungen und Betrachtungen über taktische Reizerscheinungen. *Flora* (Jena). 88: 371–421.

Rosenberg, D.M. and V.H. Resh, (Eds.), 1993. *Freshwater Biomonitoring and Benthic Macroinvertebrates*. Chapman and Hall, New York.

Ruediger, W. and Lopez-Figueroa, F. 1992. Photoreceptors in algae. *Photochemistry and Photobiology* 55(6): 949–54.

Rüffer, U. and Nultsch, W. 1985. High-speed cinematographic analysis of the movement of *Chlamydomonas*. *Cell Motility* 5: 251–263.

Rüffer, U. and Nultsch, W. 1990. Flagellar responses of *Chlamydomonas* cells held on micropipettes: change in flagellar beat frequency. *Cell Mot. Cytoskel*. 15: 162–167.

Ruffer, U and Nultsch, W 1998. Flagellar coordination in Chlamydomonas cells held on micropipettes. *Cell motility and the cytoskeleton* 41(4): 297–307.

Ruinen, J. 1938. Notizen über Salzflagellaten. II. Über die Verbreitung der Salzflagellaten. *Arch. Protistenk*. 90: 210–258.

Sacan, M.T. and Balcioglu, I.A. 2001. Bioaccumulation of aluminium in *Dunaliella tertiolecta* in natural seawater: aluminium-metal (Cu, Pb, Se) interactions and influence of pH. *Bulletin of Environmental Contamination and Toxicology* 66(2): 214–221.

Sacan, M.T., Balcioglu, I.A., and Ercan, C. 2000. Laboratory bioaccumulation of copper, lead and selenium in the marine alga *Dunaliella tertiolecta*: metal pair situation. *Toxicological and Environmental Chemistry* 76(1-2): 17–27.

Sacan, M.T., Oztay, F., and Bolkent, S. 2007. Exposure of *Dunaliella tertiolecta* to lead and aluminum: toxicity and effects on ultrastructure. *Biological Trace Element Research* 120(1–3): 264–272.

Saeko, T. and Ritsu, K. 1997. Beat Frequency Difference Between the Two Flagella of *Chlamydomonas* Depends on the Attachment Site of Outer Dynein Arms on the Outer-Doublet Microtubules. *Cell Motility and the Cytoskeleton* 36: 68–75.

Salisbury, J.L. 1988. The lost neuromotor apparatus of *Chlamydomonas*: rediscovered. *J. Protozool*. 35(4): 574–577.

Samson, G., Morisette, J.C., and Popovich, R. 1988. Copper quenching of the variable fluorescence in *Dunaliella tertiolecta*. New evidence for a copper inhibition effect on PSII photochemistry. *Photochem. Photobiol*. 48: 329–332.

Samson, G. and Popovic, R. 1988. Use of algal fluorescence for determination of phytotoxicity of heavy metals and pesticides as environmental pollutants. *Ecotoxicology and environmental safety* 16(3): 272–8.

Sanders, M.A., Boron, A.T., and Salisbury, J.L. 1991. Flagellar excitation in *Chlamydomonas reinhardtii* II: Centrin-mediated microtubule severing. Abstr. 31st Annu. Meet. Amer. Soc. Cell Biol., Boston, Mass., 8–12 Dec., 1991.

Sanders, D., Hansen, U.P., and Slayman, C.L. 1981. Role of the plasma membrane proton pump in pH regulation in non-animal cells. *Proc. Natl. Acad. Sci. U.S.A*. 78: 5903–5907.

Saraiva, M.C. 1972. Influence of gamma radiation (cobalt 60) on cultures of a chlorophyte, *Dunaliella bioculata*. Marine Biology 15(1): 74–80.

Sarma, Y.S.R.K. and Chowdhury, T.K. 1985. Influence of pre- and post-treatment with thiourea, caffeine and cysteine on UV-induced damage in four desmid taxa. *Phykos* 24(1–2): 98–107.

Schmidt, J. 1978. Studies on the role of calcium in processes of photoreception in *Chlamydomonas reinhardtii* and *Limulus polyphemus*. Avail. Univ. Microfilms Int., Order No. 7901395. 116 pp.

Schmidt, J.A. and Eckert, R. 1976. Calcium couples flagellar reversal to photostimulation in *Chlamydomonas reinhardtii. Nature* 262: 1824-1827.

Schmidt, W., Galland, P., Senger, H., and Furuya, M. 1990. Microspectrophotometry of *Euglena gracilis*. Pterin- and flavin-like fluorescence in the paraflagellar body. *Planta* 182(3): 375–81.

Schmidt, M., Gener, G., Luff, M., Heiland, I., Wagner, V., Kaminski, M., Geimer, S., Eitzinger, N., Reienweber, T., Voytsekh, O., Fiedler, M., Mittag, M., and Kreimer, G. 2006. Proteomic analysis of the eyespot of Chlamydomonas reinhardtii provides novel insights into its components and tactic movements. *Plant Cell* 18(8): 1908–1930.

Schnabel, R., Huet, J., and Thomm, M. et al. 1983. Phylogeny of the archaebacteria and eukaryotes: homology of the DNA-dependent RNA polymerases. pp. 895–912. In: *Endocytobiology*. H.E.A. Schenk and W. Schwemmler (Eds.). Berlin, New York: Du Gruyter.

Schoppmeier, J. and Lechtreck, K.F. 2002. Localization of p210-related proteins in green flagellates and analysis of flagellar assembly in the green alga Dunaliella bioculata with monoclonal anti-p210. *Protoplasma* 220(1–2): 29–38.

Sechenov, I.M. (1829–1905) Ivan Sechenov. 28 December 2009. At: <http://en.wikipedia.org/wiki/Ivan_Sechenov>.

Sedova T.V. 1977. *Fundamentals of cytology of algae*. Leningrad, Nauka. (In Russian).

Sekler, I., Glaser, H.U., and Pick, U. 1991. Characterization of a plasma membrane $H^+$-ATPase from the extremely acidophilic alga *Dunaliella acidophila. J. Membr. Biol.* 121: 51–57.

Sekler, I., Weiss, M., and Pick, U. 1994. Activation of the *Dunaliella acidophila* plasma membrane $H^+$-ATPase by tripsin cleavage of a fragment that contains a phosphorylation sity. *Plant Physiol.* 105: 1125–1132.

Semenenko, V.E. and Abdullayev, A.A. 1980. Parametrical control of biosynthesis of β-carotene in the cells of *Dunaliella salina. Plant Physiology* 27: 22. (In Russian).

*Sensory Perception and Transduction in Aneural Organisms* 1985. G. Colombetti, F. Lenci, and P.S. Song (Eds.), Plenum Press, New York, 330 p.

Serritti, A., Ferrara, R., Barghigiani, C., Petrosino, A., Del Carratore, G., and Norti, M.1981. A preliminary study on the distribution of ionic cadmium in batch cultures of *Dunaliella salina* by differential pulse anodic stripping voltammetry. *Thalassia Jugoslavica* 17(1): 55–59.

Sgarbossa, A., Checcucci, G., and Lenci, F. 2002. Photoreception and photomovements of microorganisms. *Photochemical & Photobiological Sciences* 1(7): 459–467.

Shen, L., Jiang, J., Lin, Q., and Lai, F. 2006. Test of toxicity of nitryl-aromatic hydrocarbon to *Dunaliella salina. Haiyang Kexue* 30(5): 15–17.

Shitanda, I. and Tatsuma, T. 2006. Electrochemical System for the Simultaneous Monitoring of Algal Motility and Phototaxis. *Analytical Chemistry* 78(1): 349–353.

Shitanda, I., Takada, K., Sakai, Y., and Tatsuma, T. 2006. Algal biosensing systems for the evaluation of chemical toxicity. *Seisan Kenkyu* 58(2): 156–160.

Shiu, C.T. and Lee, T.M. 2005. Ultraviolet-B-induced oxidative stress and responses of the ascorbate-glutathione cycle in a marine macroalga *Ulva fasciata. J. Exp. Bot.* 56(421): 2851–2865.

Shoevaert, D., Krishnaswamy, S., Couturier, M., and Marano, F. 1988. Ciliary beat and cell motility of *Dunaliella*: computer analysis of high speed micro-cinematography. *Biol.Cell.* 62: 229.

Shubert L.E. (ed.). 1984. *Algae as Ecological Indicators*. Academic Press, New York.

Siebert, F. 2003. *Retinal proteins.* Duerr, H. and Bouas-Laurent, H. (Eds.) *Photochromism* (Revised Edition) pp. 756–792.

Simonet, D.E., Knausenberger, W.I., Townsend, L.H., Jr., and Turner, E.C., Jr. 1978. A biomonitoring procedure utilizing negative phototaxis of first instar *Aedes aegypti* larvae. *Archives of Environmental Contamination and Toxicology* 7(3): 339–47.

Sineshchekov, O.A. 1991*a*.Electrophysiology in flagellated algae. pp. 191–202. In: *Biophysics of Photoreceptors and Photomovement in Microorganisms.* F. Lenci, F. Ghetti, G. Colombetti, D.P. Häder, and Song P.S. (Eds.), Plenum Press, New York.

Sineshchekov, O.A. 1991*b*. Photoreception in unicellular flagellates: bioelectric phenomena in phototaxis. pp. 523–532. In: *Light in Biology and Medicine.* Vol. II. Proc. Of the III Congress of the Eiropean Society for Photobiology. Budapest. H.R.H. Duglas (Ed.). New York: Plenum Press.

Sineshchekov, O.A., Geiβ, D., Sineshchekov, V.A., Galland, P., and Senger, H. 1994. Fluorometric characterization of pigments associated with isolated flagella of *Euglena gracilis:* evidence for energy migration. *J. Photochem. Photobiol.* 23: 225–237.

Sineshchekov, O.A. and Govorunova, E.G. 1999. Rhodopsin-mediated photosensing in green flagellated algae. In: *Trends in Plant Science* 4: 58–63.

Sineshchekov, O.A. and Govorunova, E.G. 2001*a*. Electrical events in photomovement of green flagellated algae. pp. 245–280. In: *Photomovement.* D.P. Häder and M. Lebert (Eds.). Elsevier Science B.V., Amsterdam.

Sineshchekov, O.A. and Govorunova, E.G. 2001*b*. Rhodopsin receptors of phototaxis in green flagellate algae. *Biochemistry* (Moscow, Russian Federation) (Translation of Biokhimiya (Moscow, Russian Federation) 66(11): 1300–1310.

Sineshchekov, O.A. and Govorunova, E.G. 2001*c*. Electrical events in photomovement of green flagellated algae. Biol. Dep., Moscow State Univ., Moscow, Russia. *Comprehensive Series in Photosciences* (Photomovement) 1: 245–280.

Sineshchekov, O.A., Lebert, M., and Häder, D.P. 2000. Effects of light on gravitaxis and velocity in *Chlamydomonas reinhardtii. J. Plant. Physiol.* 157: 247–254.

Sineshchekov, O.A. and Litvin F.F. 1982. Photoregulation of movement of microorganisms. *Usp. Microbiol.* 17: 62–87. (In Russian).

Sineshchekov, O.A. and Litvin F.F. 1987. Fluorescence of rhodopsin and its connection with the primary processes of light energy transformation. *Biofisika* 32: 540–555.

Sineshchekov, O.A. and Litvin F.F. 1988. Mechanisms of phototaxis of microorganisms. Pp.212–227. In: *Molecular mechanisms of biological action of optical radiation.* Moscow, Nauka. (In Russian).

Sineshchekov, O.A., Govorunova, E.G., and Litvin F.F. 1989. Role of photosynthetic apparatus and stigma in the formation of spectral sensitivity of phototaxis in flagellated green algae. *Biofizika* 34(2): 255–8.

Sineshchekov, O.A., Sineshchekov, V.A, and Litvin F.F. 1978. Photoinduced bioelectric reactions during phototaxis of unicellular flagellate. *DAN SSSR.* 239: 471–474. (In Russian)

Sineshchekov, O.A., Govorunova, E.G., Der, A., Keszthelyi, L., and Nultsch, W. 1994. Photoinduced electric currents in carotenoid-deficient Chlamydomonas mutants reconstituted with retinal and its analogs. *Biophysical Journal* 66(6): 2073–84.

Sineshchekov, O.A., Jung, K.H., and Spudich, J.L. 2002. Two rhodopsins mediate phototaxis to low- and high-intensity light in *Chlamydomonas reinhardtii. Proceedings of the National Academy of Sciences of the United States of America* 99(13): 8689–8694.

Sineshchekov, O.A. and Spudich, J.L. 2005. Sensory rhodopsin signaling in green flagellate algae. pp. 25–42. In: *Handbook of Photosensory Receptors*. Briggs, Winslow R. and Spudich, John L. (Eds.).

Sineshchekov, O.A., Sudnitsin, V.V., Govorunova, E.G., and Litvin, F.F. 2001. Rhythmic activity in the green flagellated alga Haematococcus pluvialis and its role in regulation of cell motility. *Biologicheskie Membrany* 18(2): 83–91.

Sirenko, L.A. and Kositskaya, V.N. 1988. *Biologically active substances of algae and water quality*. Kiev, Naukova Dumka. (In Russian).

Song, P.S. 1983. Protozoans and related photoreceptors: Molecular aspects. *Ann. Rev. Biophys. Bioeng.* 12: 35–68.

Song, P.S. 1985. Primary molecular events in aneural cell photoreceptors. In: *Sensory Perception and Transduction in Aneural Organisms*. G. Colombetti and Pill-Soon Song (Eds.). Plenum Publ. Corp., New York. 89: 47.

Sperling, P.G., Walne, P.L., Schwarz, O.J., and Triplet, L.L. 1973. Studies on characterization of pigments from isolated eyespots of euglenoid flagellates. *J. Phycol. Suppl.* 9: 20.

Spikes, J.D. 1977. Photosensitization. pp. 87-112. In: *The Science of Photobiology*. Smith K.C. (Ed.). Plenum Press, New York.

Spudich, J.L., Zacks, D.N., and Bogomolni, R.A. 1995. Microbial sensory rhodopsins: photochemistry and function. *Israel Journal of Chemistry* 35(3-4): 495–513.

Stallwitz, E. and Häder, D.P. 1993. Motility and phototactic orientation of the flagellate *Euglena gracilis* impaired by heavy metal ions. *J. Photochem. Photobiol.* B: Biol. 18: 67–74.

Starobagatov, Y.I. 1986. To the problem of a number of kingdoms of eucariotic organisms. *Trudy Zool.Inst.* 144: 4–25. (In Russian)

Stavis, R.L. 1974. The effect of azide on phototaxis in *Chlamydomonas reinhardtii*. *Proc. Nat. Acad. Sci. USA.* 71(5): 1824–1827.

Stavis, R.L. 1975. Phototaxis in Chlamydomonas: a sensory receptor system. Avail. Xerox Univ. Microfilms, Ann Arbor, Mich., Order No. 75–27, 884, 212 pp. From: Diss. Abstr. Int. B 36(6): 2583.

Stavis, R.L. and Hirschberg, R. 1973. Phototaxis in *Chlamydomonas reinhardtii*. *J. Cell Biol.* 59(2): 367–377.

Stavskaya, S.S. 1981. *Biological destruction of anionic SAS*. Kiev. Naukova Dumka (In Russian).

Stom, D.I., Balayan, A.E., Kobzhitskaya, N.Z., and Kozhova, O.M. 1984. Loss of motility in *Dunaliella salina* cells as a criterion of toxic action. *Gidrobiologicheskii Zhurnal* 20(5): 46–9.

Stom, D.I. and Roth, R. 1981. Some effects of polyphenols on aquatic plants: I. Toxicity of phenols in aquatic plants. Res. Inst. Biol., State Univ., Irkutsk, USSR.*Bulletin of Environmental Contamination and Toxicology* 27(3): 332–7.

Strasburger, E. 1878. Wirkung des Lichtes und der Wärme auf Schwärmsporen *Jena Z. Naturw* 12: 551.

Suzuki, T., Yamasaki, K., Fujita, S., Oda, K., Iseki, M., Yoshida, K., Watanabe, M., Daiyasu, H., Toh, H., Asamizu, E., Tabata, S., Miura, K., Fukuzawa, H., Nakamura, S., and Takahashi, T. 2003. Archaeal-type rhodopsins in *Chlamydomonas*: model structure and intracellular localization. *Biochemical and Biophysical Research Communications* 301(3): 711–717.

Tafreshi, A.H. and Shariati, M. 2006. Pilot culture of three strains of *Dunaliella salina* for β-carotene production in open ponds in the central region of Iran. *World Journal of Microbiology & Biotechnology* 22(9): 1003–1006.

Tahedl, H. and Häder, D.P. 2001. Automated biomonitoring using real time movement analysis of *Euglena gracilis*. *Ecotoxicol. and Environ. Safety* 48: 161–169.

Takahashi, T., Kubota, M., Watanabe, M., Yoshihara, K., Derguini, F., and Nakanishi, K. 1992. Diversion of the sign of phototaxis in a *Chlamydomonas reinhardtii* mutant incorporated with retinal and its analogs. *FEBS Letters* 314(3): 275–9.

Takahashi, T., Yoshihara, K., Watanabe, M., Kubota, M., Johnson, R., Derguini, F., and Nakanishi, K. 1991. Photoisomerization of retinal at 13-ene is important for phototaxis of *Chlamydomonas reinhardtii*: simultaneous measurements of phototactic and photophobic responses. *Biochemical and Biophysical Research Communications* 178(3): 1273–9.

Takahashi, T. and Watanabe, M. 1993. Photosynthesis modulates the sign of phototaxis of wild-type *Chlamydomonas reinhardtii*. Effects of red background illumination and 3-(3',4'-dichlorophenyl)-1,1-dimethylurea. *FEBS Letters* 336(3): 516–20.

Takimura, O., Fuse, H., and Yamaoka, Y. 1989. Effect of various environmental factors on arsenic accumulation in *Dunaliella* sp. *Nippon Kagaku Kaishi* 4: 740–3.

Tang, Q., Guo, X., Chen, X., Xu, G., and Wang, C. 2006. Preparation of β-carotene microemulsion from *Dunaliella salina* and its stability. *Zhongguo Haiyang Yaowu* 25(6): 38–41.

Taranova, L.A. 1987. Biological destruction of cationic surface-active substances. *Visn. AN USSR*. 8: 40–50. (In Russian).

Teodorescu, E.C. 1905. Organization et développement du *Dunaliella*, nouveau genere de *Volvocacée Polyblepharidée. Beih. Bot. Centralbl.* 18: 215–232.

Téodoresco, E.C. 1906. Observations morphologiques et biologiques sur le genre Dunaliella. *Revue Générale de Botanique* 18: 353–371, 409–427 & figs. 1–75.

Thakur, A. and Kumar, H.D. 1998*a*. Glycerol production by *Dunaliella salina* in response to various combinations of organic carbon compounds. *Cytobios* 93(373): 129–134.

Thakur, A. and Kumar, A. 1998*b*. Effect of pH, temperature, salinity and nitrogen concentration on the growth and carotenoid content of Dunaliella salina. *Acta Botanica Hungarica* 41(1-4): 293–297.

Thakur, A. and Kumar, H.D. 1999. Effect of different nitrogen sources on growth and glycerol production by *Dunaliella salina. Cytobios* 97(385): 79–86.

Thakur, A., Kumar, H.D., and Cowsik, S.M. 2000. Effect of pH and inorganic carbon concentration on growth, glycerol production, photosynthesis and dark respiration of *Dunaliella salina. Cytobios* 102(400): 69–74.

Thurston, A. 1999. Giovanni Borelli and the Study of Human Movement: An Historical Review. *Aust. N. Z. J. Surg.* Vol. 69.

Tollin, G. 1969. Energy transduction in algal phototaxis. *Current Topics in Bioenergetics*. 3: 417–446.

Tollin, G. 1973. Phototaxis in *Euglena gracilis*. II. Biochemical aspects. pp. 91–105. In: *Behavior of Microorganisms*. A. Perez-Miravete (Ed.). Plenum Press, New York.

Topachevsky, A.V. and Massjuk, N.P. 1984. *Freshwater algae of Ukrainian SSR*. Kiev. Vyssha shkola. (In Russian).

Topics of current research. Extremely acidic mining lakes (pH<3). 5 January 2010. At: "http://www.bio.uni-potsdam.de/oeksys/fsvte.htm".

Treviranus, L.C. (1779–1864) Ludolph Christian Treviranus. 23 December 2009. At: <http://en.wikipedia.org/wiki/Ludolph_Christian_Treviranus>.

Treviranus L.C. 1817. Beobachtungen über die Bewegung der grünen Materie im Pflanzenreich. *Vermischte Schriften Anat. und Physiol. Inhalsts.* 2: 71.

Tsuji, N., Hirayanagi, N., Okada, M., Miyasaka, H., Hirata, K., Zenk, M.H., and Miyamoto, K. 2002. Enhancement of tolerance to heavy metals and oxidative stress in *Dunaliella tertiolecta* by zinc-induced phytochelatin synthesis. *Biochemical and Biophysical Research Communications* 293(1): 653–659.

Tsukida, K., Saiki, K., Takii, and T., Koyama, Y. 1982. Separation and determination of *cis/trans*-β-carotene by high preformance liquid chromatography. J. Chromat. 245: 359.

Turker, S.M. and Balcioglu, I.A. 2001. Bioaccumulation of aluminium in *Dunaliella tertiolecta* in natural seawater: aluminium-metal (Cu, Pb, Se) interactions and influence of pH. *Bulletin of environmental contamination and toxicology* 66(2): 214–21.

Tzvylev, O.P. and Tkachenko, V.N. 1981. *Application of unicellular algae for biological analysis.* Moscow. Legkaya I pishevaya promyshlennost. (In Russian).

Ukhtomsky, A.A. (1875–1942) Ukhtomsky A.A., (1875–1942), physiologist . 5 January 2010.<http://www.encspb.ru/en/persarticle.php?kod=2803926576>.

Vakhmistrov, D.B. and Bogorov, LV. 1987. Secretion of surface-active substances by plant cells. *Dokl. Akademii nauk.* 297(5): 1273–1275. (In Russian).

Vanitha, A., Murthy, K.N.C., Kumar, V., Sakthivelu, G., Veigas, J.M., Saibaba, P., and Ravishankar, G.A. 2007. Effect of the carotenoid-producing alga, *Dunaliella bardawil*, on CCl4-induced toxicity in rats. *International Journal of Toxicology* 26(2): 159–167.

Vendt, V.P. 1963. New perspective source of carotene production (alga Dunaliella salina). In: *Vitamin resources and their application* 6: 156. (In Russian).

Ventosa, A. and Nieto, J.J. 1995. Biotechnological applications and potentialities of halophilic microorganisms. *World Journal of Microbiology & Biotechnology* 11(1): 85–94.

Veselova, T.V., Veselovskii, V.A., Vlasenko, V.V., Matskivskii, V.I., Pen'kov, F.M., and Chernavskii, D.S. 1990. Variability of functional parameters as a test of [algal] cell stress following intoxication. *Fiziologiya Rastenii* (Moscow) 37(4): 733–8.

Villar, R., Laguna, M.R., Calleja, J.M., and Cadavid, I. 1992. Effects of *Phaeodactylum tricornutum* and *Dunaliella tertiolecta* extracts on the central nervous system. *Planta medica* 58(5): 405–9.

Vinegla, B., Segovia, M., and Figueroa, F.L. 2006. Effect of Artificial UV Radiation on Carbon and Nitrogen Metabolism in the Macroalgae *Fucus spiralis* L. and *Ulva olivascens* Dangeard. *Hydrobiologia* 560: 31–42.

Vladimirova, M.G. 1978. Ultrastrutural organization of the cell of *Dunaliella salina* Teod. and its functional changes depending on the intensity of light and temperature. *Plant Physiology* 25 (3): 571–583. (In Russian).

Vlasenko, V.V. 2004. The study of photoresponses of unicellular motile microalgae by Doppler spectrometry. *Dopovidi Natsional'noi Akademii Nauk Ukraini* 4: 159–164.

Voskresensky, K.A. 1960. *Artemia – valuable forage in industrial fish-breeding.* Moscow, Moscow State Univ. Publ. (In Russian)

Walne, P.L., Passarelli, V., Barsanti, L., and Gualtieri, P. 1998. Rhodopsin: a photopigment for phototaxis in *Euglena gracilis. Crit. Rev. Plant. Sci.* 17: 569–574.

Walne, P.L. and Arnott, H.J. 1967. The comparative ultrastructure and possible function of the eyespot *Euglena granulata* and *Chlamydomonas eugametos. Planta* 77: 325–353.

Walsh, G.E. 1983. Cell death and inhibition of population growth of marine unicellular algae by pesticides. *Aquatic Toxicology* 3(3): 209–14.

Wang, S., Wang, Y., Zhang, Z., Hong, H., and Zheng, W. 2001. Effects of $Cu^{2+}$, $Zn^{2+}$ and $Cd^{2+}$ on the phototaxic response of *Artemia salina* and two shrimp (*Penaeus penicillatus* and *Penaeus monodon*) nauplius. *Xiamen Daxue Xuebao, Ziran Kexueban* 40(6): 1333–1336.

Watanabe, M. and Erata, M. 2001. Yellow-light sensing phototaxis in cryptomonad algae. pp. 345-373. In: *Photomovement.* D.P. Häder and M. Lebert (Eds.), Elsevier Sci. B.V., Amsterdam.

Watanabe, M. and Furuya, M. 1974. Action spectrum of phototaxis in a cryptomonad alga, *Cryptomonas* sp. *Plant Cell Physiol.* 15: 413-420.

Watanabe, M., Furuya, M., Miyoshi, Y., Inoue, Y., Iwahashi, I., and Matsumoto, K. 1982*a*. Design and performance of the Okazaki Large Spectrograph for photobiological research. *Photochem. Photobiool.* 36: 491.

Watanabe, M. and Furuya, M. 1982*b*. Phototactic behavior of individual cells of *Cryptomonas* sp. *Plant Cell Physiol.* 35: 559–563.

Watson, M.W. 1975. Flagellar apparatus, eyespot and behavior of *Microthamnion kuetzingianum* (Cholophyceae) zoospores. *J. Phycol.* 11: 439–448.

Wayne, R., Kadota, A., Watanabe, M., and Furuya, M. 1991. Photomovement in *Dunaliella salina*: fluence rate-response curves and action spectra. *Planta* 184: 515–524.

Wegmann, K. 1979. Biochemische Anpassung von *Dunaliella* an wechselude salinität und Temperatur. *Ber. Dtsch. Bot. Ges.* 92(1): 43–54.

Wegmann, K. 1968. Des Weg des Kohlenstoffs bei der Photosynthese und Dunkelfixierung in *Dunaliella* spec. *Tubingen* 45 S.

Wegmann, K. and Metzner, K. 1971. Synchronization of *Dunaliella* cultures. *Arch. Microbiol.* 78: 360–367.

Wegmann, K. 1971. Osmotic regulation of photosynthetic glycerol production in *Dunaliella.* *Biochimica et Biophysica Acta, Bioenergetics* 234(3): 317–23.

Wegmann, K., Ben-Amotz, A., and Avron, M. 1980. Effect of temperature on glycerol retention in the halotolerant algae *Dunaliella* and *Asteromonas.* *Plant Physiology* 66(6): 1196–7.

Weiss, M. and Pick, U. 1990. Transient sodium flux following hyperosmotic shock in the halotolerant alga *Dunaliella salina:* a response to intracellular pH changes. *Journal of Plant Physiology* 136(4): 429–38.

Weiss, M. and Pick, U. 1996. Primary structure and effect of pH on the expression of the plasma membrane H+-ATPase from *Dunaliella acidophila* and *Dunaliella salina. Plant Physiology* 112(4): 1693–1702.

Williams, R.M. and Braslavsky, S.E. 2001. Triggering of photomovement - molecular basis. *Comprehensive Series in Photosciences* 1(Photomovement): 15-50.

Witman, G.B. 1994. *Chlamydomonas* phototaxis. *Trends Cell Biol.* 3: 403–408.

Witman, G.B. 1990. Introduction to cilia and flagella. *Ciliary and Flagellar Membranes.* R.A. Bloodgood (Ed.). Plenum Press, New York.

Wolken, J.J. 1967. *Euglena: an experimental organism for biochemical and biophysical studies.* Appleton-Century-Crofts, New York.

Wolken, J.J. 1971. *Invertebrate Photoreceptors.* Acad. Press, N.Y., London.

Wolken, J.J. 1975. *Photoprocesses, Photoreceptors, and Evolution.* Acad. Press, New York, San Francisco, London, pp. 118–136.

Wolken, J.J. 1977. *Euglena:* the photoreceptor system for phototaxis. *J. Protozool.* 24: 518–522.

Wolken, J.J. and Shin, E. 1958. Photomotion in *Euglena gracilis* I. Photokinesis. II. Phototaxis. *J. Protozool.* 5: 39–46.

Wong, C.K.C., Cheung, R.Y.H., and Wong, M.H. 1999. Toxicological assessment of coastal sediments in Hong Kong using a flagellate, *Dunaliella tertiolecta. Environmental Pollution* 105(2): 175–183.

Wu, Y., Xiong, Y., Lin, C., and Yuan, L. 2006. The joint-biotoxicity effect of the different forms of phosphorus on heavy metal (Cu, Zn, Cd). *Huanjing Kexue Xuebao* 26(12): 2045–2051.

Xenopoulos, M.A. and Frost, P.C. 2003. UV radiation, phosphorus, and their combined effects on the taxonomic composition of phytoplankton in a boreal lake. *Journal of Phycology* 39(2): 291–302.

Xie, R., Tang, X., and Li, Y. 1999. Study on joint toxicity of heavy metal and organophosphorus pesticide to marine microalgae *Haiyang Huanjing Kexue.* 18(2): 16-20.

Xu, F., Zhang, P., Yu, D., and Li, Y. 2006. Effect of enhanced UV-B radiation to growth of *Ulva pertusa* Kjellman and *Platymonas helgolandica* Kylin var. tsingtaoensis. *Qingdao Daxue Xuebao, Gongcheng Jishuban* 21(2): 49–53.

Xu, J. and Gao, K. 2007. Effects of solar UVR on the effective quantum yield of *Ulva lactuca. Haiyang Xuebao (Zhongwenban)* 29(1): 127–132.

Yamano K., Saito H., Ogasawara Y., Fujii Sh., Yamada H., Shirahama H., and Kawai H. 1996. The autofluorescent substance in the posterior flagellum of swarmer of the brown alga *Scytosiphon lomentaria. Z. Naturforsch.* 51: 155–159.

Yamaoka, Y. 1994. Manufacture of beta-carotene by culturing algae (*Dunaliella*). (Kogyo Gijutsuin, Japan). *Jpn. Kokai Tokkyo Koho,* 4 pp.

Yamaoka, Y., Takimura, O., Fuse, H., and Kamimura, K. 1992. Effect of magnetism on growth of *Dunaliella salina. Res. Photosynth.,* Proc. Int. Congr. Photosynth., 9th 3: 87–90.

Yamaoka, Y., Takimura, O., Fuse, H., and Kamimura, K. 1992. Recovery of gallium from seawater by *Dunaliella* sp. *Nippon Kagaku Kaishi* 5: 506–8.

Yamaoka, Y., Takimura, O., Fuse, H., and Kamimura, K. 1994. β-Carotene production by *Dunaliella salina* in fed-batch and semi-continuous cultures under nutrient supplement. *Seibutsu Kogaku Kaishi* 72(2): 111–14.

Yamaoka, Y., Takimura, O., Fuse, H., Kamimura, K., Manabe, E., Takano, H., and Hirano, M. 1992. Effects of various environmental factors on β-carotene production by *Dunaliella salina. Hakko Kogaku Kaishi* 70(1): 25–8.

Yamaoka, Y., Takimura, O., Fuse, H., Kamimura, K., Murakami, K., Ueda, S., and Saiki, M. 1996. β-Carotene production by *Dunaliella salina* using glass powder containing nutrients. *Seibutsu Kogaku Kaishi* 74(4): 269–272.

Yamaoka, Y., Takimura, O., Fuse, H., Murakami, K., Kamimura, K., Ueda, S., and Saeki, M. 1997. β-Carotene production by *Dunaliella salina* using boron-glass powder containing iron. *Seibutsu Kogaku Kaishi* 75(3): 181–184.

Yammamoto, K.M., Satake M., Shinogawa H., and Fujiwara Y. 1983. Amelioration of the ultraviolet sensitivity of an *Escherichia coli* recA mutant in the dark by photoreactivating enzyme. *Mol. Gen. Genet.* 190: 511–515.

Yang, Q. 1988. The effect of temperature on the fatty acid composition of marine phytoplankton. *Haiyang Yu Huzhao* 19(5): 439–46.

Yarden, O., Freund, M., and Rubin, B. *Dunaliella salina*: a convenient test organism for detection of pesticide residues in water and soil. *Fresenius Environmental Bulletin* 2(1): 31-6.

Yoshikawa, S. 2005. Archaeal type *Chlamydomonas* opsins and photo signal transduction in *Chlamydomonas*. *Sorui* 53(3): 247–249.

Yoshimura, K. 1994. Chromophore orientation in the photoreceptor of *Chlamydomonas* as probed by simulation with polarized light *Photochemistry and Photobiology* 60(6): 594–597.

Yoshimura, K. and Kamiya, R. 2001. The sensitivity of *Chlamydomonas* photoreceptor is optimized for the frequency of cell body rotation. *Plant and Cell Physiology* 42(6): 665–672.

Yoshimura, K., Matsudo, Y., and Kamiya, R. 2003. Gravitaxis in *Chlamydomonas reinhardtii* studied with novel mutants. *Plant Cell Physiol.* 44(10): 1112–1118.

Yu, Z., Zhang, J., Shi, F., and Wu, C. 1999. New method for evaluating toxicity of heavy metals on marine macroalgae. *Haiyang Yu Huzhao* 30(2): 199–205.

Yurina, E.V. 1966. Experience of cultivation of halobiontic algae *Asteromonas gracilis* Artari and *Dunaliella salina* Teod. *Vestnic of MGU.* 4(6): 76–83. (In Russian).

Yurkova, G.N. 1965. Effect of thermal factor on *Dunaliella salina* Teod. *Ukr.Bot.J.* 22(6): 51–57. (In Russian).

Zacks, D.N., Derguini, F., Nakanishi, K., and Spudich, J.L. 1993. Comparative study of phototactic and photophobic receptor chromophore properties in *Chlamydomonas reinhardtii*. *Biophysical Journal* 65(1): 508–18.

Zelnichenko, A.T., Kovalchuk, V.S., and Posudin, Y.I. 1988. Effect of electromagnetic fields on movement of microorganisms. *Biofisika* 33(5): 841–844. (In Russian).

Zhao, X., Wang, Z., Wu, Y., Zhang, W., and Zou, J. 1992. Studies on laser mutagenesis of β-carotene production in *Dunaliella salina*. *Zhongguo Jiguang* 19(6): 463–6.

Zhang, P., Tang, X., Cai, H., Yu, J., and Xiao, H. 2005. Effects of the enhanced UV-B radiation on the growth of interaction competition between marine macro-algae and microalgae population. *Shengtai Xuebao* 25(12): 3335–3342.

Zhang, X., Meng, Z., Shi, Y., and Wang, P. 2006. Effect of light, temperature and nutrition on growth and pigment accumulation of three strains of *Dunaliella salina*. *Zhongguo Haiyang Daxue Xuebao, Ziran Kexueban* 36(5): 754–762.

Zhang, Z., Chen, X., Zhang, L., Jiang, G., and Wang, C. 2006. Antiaging effects of beta-carotene from *Dunaliella salina* on fruit flies and rats *Zhongguo Yaolixue Tongbao* 22(11): 1324–1328.

Zhou, Q. 1995. Culture of algae in the mother liquor of salt-making and its recovery technology. *Haihuyan Yu Huagong* 24(6): 11–6.

Zhu, Y. and Wang, D. 2001. Toxicity and bioaccumulation of germanium in two microalgae *Spirulina platensis* and *Dunaliella salina*. *Haiyang Kexue* 25(10): 5–7.

Zimmerman, M.A. 1981. Beginner's guide to spectral analysis. Part 2. *Byte* 3: 166–198.

Zlochevskaya, I.V., Absalyamov, S.Y., and Galimova L.M. et al. 1981. To the mechanism of phungicidal action of quartery ammonia compounds. *Nauch. doklady vyshei shkoly* 3: 82–87. (In Russian).

Zmitrovich, I.V. 2003. Revision of phylogenic tree of eukariota: variant of eugenosoid ancestor. *Algology* 13(3): 227–268. (In Russian).

# Index of Latin Names

# Subject Index

# Author's Index

# Chemicals – CAS nomenclature

A23187 =        4-benzoxazolecarboxylic acid

acetal =        1,1-diethoxy-ethane acetamide

acetazine =     1-[10-[3-(dimethylamino)propyl]-10H-phenothiazin-2-yl]- ethanone

alachlor =      2-chloro-N-(2,6-diethylphenyl)-N-(methoxymethyl)-acetamide

anion surface-active substance (ASAS) or sodium salt of dodecyl sulphoacid (NaSDS)

arylon =        pesticide from Collection of Nat. Agr.Univ., Kiev, Ukraine

basta =         2-amino-4-(hydroxymethylphosphinyl)-butanoic acid

catamine or cationic surfactant (alkyldimethylbensylammonium chloride)

cinnarizine =   1-(diphenylmethyl)-4-(3-phenyl-2-propen-1-yl)-piperazine

db-cAMP =       3',5'-cyclic AMP dibutyrate

DCMU =           1,1-dimethyl urea

diltiazem =     3-(acetyloxy)-5-[2-(dimethylamino)ethyl]-2,3-dihydro-2-(4-
                methoxyphenyl)- 1,5-benzothiazepin-4(5H)-one hydrochloride

diuron =        N'-(3,4-dichlorophenyl)-N,N-dimethyl-urea

DPC =           diphenylcarbazide

dual =          2-chloro-N-(2-ethyl-6-methylphenyl)-N-(2-methoxy-1-methylethyl)-piperazine

eradicane =     2,2-dichloro-N,N-di-2-propen-1-yl-acetamide

flunarizine =   1-[bis(4-fluorophenyl)methyl]-4-[(2E)-3-phenyl-2-propenyl]- piperazine

harmoni =       pesticide from Collection of Nat. Agr.Univ., Kiev, Ukraine

IBMX =          isobutylmethylxanthine

ionophore A23187 = 4-benzoxazolecarboxylic acid

isoptin =	benzeneacetonitrile = α [3-[[2-(3,4-dimethoxyphenyl) ethyl] methylamino] propyl]-3,4dimethoxy-α-(1-methylethyl) hydrochloride

lindane =	1,2,3,4,5,6-hexachloro-cyclohexane

natural surface-active substances of polysaccharide origin (PSAS) [Peskov, 1979]

nimodipine =	1,4-dihydro-2,6-dimethyl-4-(3-nitrophenyl)-, 3-(2- methoxyethyl) 5-(1-methylethyl) ester 3,5-pyridinedicarboxylic acid

non-ionogenic surface-active substance (NSAS) or hydropol (from Collection of Institute of Colloidal Chemistry and Chemistry of Water of Nat. Acad. Sci. of Ukraine)

ouabain =	3-[(6-deoxy-α-L-mannopyranosyl)oxy]-1,5,11,14,19- pentahydroxy-card-20(22)-enolide

sodium azide = $NaN_3$

tecto =	2-(4-thiazolyl)-1H-benzimidazole

verapamil =	α- [3-[[2-(3,4-dimethoxyphenyl)ethyl]methylamino] propyl]-3,4- dimethoxy-α-(-methylethyl)-benzeneacetonitrile

# About the Authors

**Professor Yuriy Posudin**, Doctor of Biological Sciences, National University of Life and Environmental Sciences of Ukraine, Kiev, Ukraine. He graduated from the Radiophysical Faculty of Kiev State University in 1969. Dr. Posudin's principal scientific interests are the investigation of photobiological reactions of algae and plants, and the non-destructive quality evaluation of agricultural products. He is an author of 18 textbooks, and over 120 scientific papers and has 20 patents for inventions.

**Professor Nadiya Massjuk**, Doctor of Biological Sciences, Leading Research Fellow, M.G. Kholodny Institute of Botany of National Academy of Sciences of Ukraine, Kiev, Ukraine. Dr. Massjuk has studied the biology of algae for many years (biodiversity, evolution, phylogeny, its place in the world of living organisms, and applied algology). She has been interested in the biology of photomovement in relation to the study of the diversity, phylogeny of phytoflagellates, classification, biotechnology of growth, and synthesis of carotene. Dr. Massjuk has authored over 260 scientific papers.

**Dr. Galyna Lilitskaya,** Leading Engineer, M.G. Kholodny Institute of Botany of the National Academy of Sciences of Ukraine, Kiev, Ukraine. Dr. Lilitskaya graduated from the Biological Faculty of Kiev State University in 1986. She has authored over 40 scientific papers on algal photomovement, biodiversity, and flora.